Basic Training in Mathematics

A Fitness Program for Science Students

Basic Training in Mathematics

A Fitness Program for Science Students

R. SHANKAR

Yale University
New Haven, Connecticut

PLENUM PRESS • NEW YORK AND LONDON

Library of Congress Cataloging-in-Publication Data

On file

ISBN 0-306-45035-6 (Hardbound)
ISBN 0-306-45036-4 (Paperback)

© 1995 Plenum Press, New York
A Division of Plenum Publishing Corporation
233 Spring Street, New York, N. Y. 10013

10 9 8 7 6 5 4 3 2

For

UMA

PREFACE

This book is based on a course I designed a few years ago and have been teaching at Yale ever since. It is a required course for physics majors, and students wishing to skip it have to convince the Director of Undergraduate Studies of their familiarity with its contents. Although it is naturally slanted toward physics, I can see a large part of it serving the needs of anyone in the physical sciences since, for the most part, only very basic physics ideas from Newtonian mechanics are employed. The *raison d'être* for this book and the course are identical and as follows.

While teaching many of the core undergraduate courses, I frequently had to digress to clear up some elementary mathematical topic which bothered some part of the class. For instance, I recall the time I was trying to establish how ubiquitous the harmonic oscillator was by showing that the Taylor series of any potential energy function at a stationary point was given to leading order by a quadratic function of the coordinate. At this point some students wanted to know what a Taylor series was. A digression to discuss Taylor series followed. At the next stage, when I tried to show that if the potential involved many coordinates, the quadratic approximation to it could be decoupled into independent oscillators by a change of coordinates, I was forced to use some form of matrix notation and elementary matrix ideas, and that bothered some other set of students. Once again we digressed. Now, I was not averse to the idea that in teaching physics, one would also have to teach some new mathematics. For example, the course on electricity and magnetism is a wonderful context in which to learn about Legendre polynomials. On the other hand, it is not the place to learn for the first time what a complex exponential like $e^{im\phi}$ means. Likewise, in teaching special relativity one does not want to introduce sinh and cosh, one wants to use them and to admire how naturally they serve our purpose. To explain what these functions are at this point is like explaining a pun. In other words, some of the mathematical digressions were simply not desirable and quite frustrating for the teacher and student alike.

Now, this problem was, of course, alleviated as the students progressed through the system, since they were taking first-rate courses in the mathematics department in the meantime and could soon tell you a surprising thing or two about the edge-of-the-wedge theorem. But one wished the students would have a grasp of the basics of each essential topic at some rudimentary level from the outset, so that instructors could get on with their job with the least amount of digressions. From the student's point of view, this allowed more time to think about the subject proper and more freedom to take advanced courses.

When this issue was raised before the faculty, my sentiments were shared by many. It was therefore decided that I would design and teach a course that would deal with topics in differential calculus of one or more variables (including trigonometric, hyperbolic, logarithmic, and exponential functions), integral calculus of one and many variables, power series, complex numbers and function of a complex variable, vector calculus, matrices, linear algebra, and finally the elements of differential equations.

In contrast to the mathematical methods course students usually take in the senior year, this one would deal with each topic in its simplest form. For example, matrices would be two-by-two, unless a bigger one was absolutely necessary (say, to explain degeneracy). On the other hand, the treatment of this simple case would be thorough and not superficial. The course would last one semester and be self-contained. It was meant for students usually in the sophomore year, though it has been taken by freshmen, upper-class students, and students from other departments.

This book is that course.

Each department has to decide if it wants to devote a course in the sophomore year to this topic. My own view (based on our experience at Yale) is that such a preventive approach, which costs one course for just one semester, is worth hours of curing later on. Hour for hour, I can think of no other course that will yield a higher payoff for the beginning undergraduate embarked on a career in the physical sciences, since mathematics is the chosen language of nature, which pervades all quantitative knowledge. The difference between strength or weakness in mathematics will subsequently translate into the difference between success and failure in the sciences.

As is my practice, I directly address the student, anticipating the usual questions, imagining he or she is in front of me. Thus the book is ideal for self-study. For this reason, even a department that does not have, as yet, a course at this level, can direct students to this book before or during their sophomore year. They can turn to it whenever they run into trouble with the mathematical methods employed in various courses.

Acknowledgments

I am pleased to thank all the students who took Physics 301a for their input and Ilya Gruzberg, Sentil Todadri, and George Veronis for comments on the manuscript.

As always, it has been a pleasure to work with the publishing team at Plenum. My special thanks to Senior Editor Amelia McNamara, Editor Ken Howell, and Senior Production Editor Joseph Hertzlinger.

I thank Meera and AJ Shankar for their help with the index.

But my greatest debt is to my wife Uma. Over the years my children and I have been able to flourish, thanks to her nurturing efforts, rendered at great cost to herself. This book is yet another example of what she has made possible through her tireless contributions as the family muse. It is dedicated to her and will hopefully serve as one tangible record of her countless efforts.

R. Shankar

Yale University
New Haven, Connecticut

NOTE TO THE INSTRUCTOR

If you should feel, as I myself do, that it is not possible to cover all the material in the book in one semester, here are some recommendations.

- To begin with, you can skip any topic in fine print. I have tried to ensure that this does not do violence to continuity. The fine print is for students who need to be challenged, or for a student who, long after the course, begins to wonder about some subtlety; or runs into some of this material in a later course, and returns to the book for clarification. At that stage, the student will have the time and inclination to read the fine print.

- The only chapter which one can skip without any serious impact on the subsequent ones, is that on vector calculus. It will be a pity if this route is taken; but it is better to leave out a topic entirely rather than rush through everything. More moderate solutions like stopping after some sections, are also possible.

- Nothing teaches the student as much as problem solving. I have given a lot of problems and wish I could have give more. When I say more problems, I do not mean more which are isomorphic to the ones given, except for a change of parameters, but genuinely new ones. As for problems that are isomorphic, you can generate any number (say for a test) and have them checked by a program like *Mathematica*.

- While this course is for physical scientists, it is naturally slanted towards physics. On the other hand, most of the physics ideas are from elementary Newtonian mechanics and must be familiar to anyone who has taken a calculus course. You may still have to customize some of the examples to your specialty.

I welcome your feedback.

NOTE TO THE STUDENT

In American parlance the expression "basic training" refers to the instruction given to recruits in the armed forces. Its purpose is to ensure that the trainees emerge with the fitness that will be expected of them when they embark on their main mission.

In this sense the course provides basic training to one like yourself, wishing to embark on a program of study in the physical sciences. It has been my experience that incoming students have a wide spectrum of preparation and most have areas that need to be strengthened. If this is not done at the outset, it is found that the results are painful for the instructor and student alike. Conversely, if you cover the basic material in this book you can look forward to a smooth entry into any course in the physical sciences. Of course, you will learn more mathematics while pursuing your major and through courses tailored to your specialization, as well as in courses offered by the mathematics department. This course is not a substitute for any of that.

But this course is *unlike* a boot camp in that you will not be asked to do things without question; no instructor will bark at you to "hit that desk and give me fifty derivatives of e^x." You are encouraged to question everything, and as far as possible everything you do will be given a logical explanation and motivation.

The course *will be* like a boot camp in that you will be expected to work hard and struggle often, and will emerge proud of your mathematical fitness.

I have done my best to simplify this subject as much as possible (but no further), as will your instructor. But finally it is up to you to wrestle with the ideas and struggle for total mastery of the subject. Others cannot do the struggling for you, any more than they can teach you to swim if you won't enter the water. Here is the most important rule: do as many problems as you can! Read the material before you start on the problems, instead of starting on the problems and jumping back to the text to pick up whatever you need to solve them. This leads to patchy understanding and partial knowledge. Start with the easy problems and work your way up. This may seem to slow you down at first, but you will come out ahead. Look at other books if you need to do more problems. One I particularly admire is *Mathematical Methods in the Physical Sciences*, by M. Boas, published by Wiley and Sons, 1983. It is more advanced than this one, but is very clearly written and has lots of problems.

Be honest with yourself and confront your weaknesses before others do, as they invariably will. Stay on top of the course from day one: in mathematics, more than anything else, your early weaknesses will return to haunt you later in the course. Likewise, any weakness in mathematical preparation will trouble you during the rest of you career. Conversely, the mental muscles you develop here will stand you in good stead.

CONTENTS

DIFFERENTIAL CALCULUS OF ONE VARIABLE

1.1. Introduction

Students taking the course on mathematical methods generally protested vigorously when told that we were going to start with a review of calculus, on the grounds that they knew it all. Now, that proved to be the case for some, while for many it was somewhat different: either they once knew it, or thought they once knew it, or actually knew someone who did, and so on. Since everything hinges on calculus rather heavily, we will play it safe and review it in the first three chapters. However, to keep the interest level up, the review will be brief and only subtleties related to differential and integral calculus will be discussed at any length. The main purpose of the review is to go through results you probably know and ask where they come from and how they are interrelated and also to let you know where you really stand. If you find any portion where you seem to be weak, you must find a book on calculus, work out many more problems than are assigned here, and remedy the defect at once.

1.2. Differential Calculus

Let us assume you know what a function $f(x)$ is: a machine that takes in a value of x and spits out a value f which depends on x. For example $f(x) = x^2 + 5$ is a function. You put in $x = 3$ and it gives back the value $f = 3^2 + 5 = 14$. We refer to f as the *dependent variable* and x as the *independent variable*. Note that in some other problem x can be the location of a particle at time t in which case $x(t)$ is the dependent variable and t is the independent variable.

We will assume the function is continuous: this means that you can draw the graph of f without taking the pen off the paper. More formally:

Definition 1.1. *A function $f(x)$ is continuous at $x = a$ if for any $\epsilon > 0$, however small, we can find a δ such that $|f(x) - f(a)| < \varepsilon$ for $|x - a| < \delta$.*

For example the function

$$f(x) = |x| \qquad x \neq 0 \tag{1.2.1}$$
$$f(0) = 66 \qquad x = 0 \tag{1.2.2}$$

is not continuous at the origin even though $|f(x) - 0|$ can be made as small as we want as we approach the origin, i.e., $f(x)$ has a nice limit as we approach the origin, but the limiting value is not the value of the function at the origin. On the other hand if we choose $f(0) = 0$, the function becomes continuous.

In other words, as we approach the point in question, not only must the values encountered approach a limit, the limit must equal the value ascribed to that point by the function, if the function is to be declared continuous.

The *derivative* of the function, denoted by $f'(x)$, $f^{(1)}$, Df or $\frac{df}{dx}$, is defined by the limit

$$\frac{df}{dx} = \lim_{\Delta x \to 0} \frac{f(x + \Delta x) - f(x)}{\Delta x}$$
$$= \lim_{\Delta x \to 0} \frac{\Delta f}{\Delta x}. \tag{1.2.3}$$

Thus the derivative measures the rate of change of the dependent variable with respect to the independent variable. For example in the case of $x(t)$, which is the position of a particle at time t, dx/dt is the instantaneous velocity.

Let us now compute a derivative taking as an example, $f(x) = x^2$. We have

$$f(x + \Delta x) = x^2 + 2x\Delta x + (\Delta x)^2$$
$$\Delta f = 2x\Delta x + (\Delta x)^2$$
$$\frac{\Delta f}{\Delta x} = 2x + \Delta x. \tag{1.2.4}$$

If we now take the limit $\Delta x \to 0$, we get

$$\frac{d(x^2)}{dx} = 2x.$$

Clearly the function has to be continuous before we can carry out the derivative operation. However, continuity may not be enough. Consider for example $f(x) = |x|$ at the origin. If we choose a positive Δx, we get one value for the derivative $(+1)$, while if we choose a negative Δx, we get a different slope (-1). This fact is also clear if one draws a graph of $|x|$ and notices that there is no unique slope at the origin.

Once you know how to differentiate a function, i.e., take its derivative, you can take the derivative of the derivative by appealing repeatedly to the above definition of the derivative. For example, the derivative of the derivative of x^2, also called its second derivative, is 2. The second derivative of $|x|$ is zero everywhere, except at the origin, where it is ill defined.

The second derivative is denoted by

$$\frac{d^2 f}{dx^2} = f''(x) = D^2 f = f^{(2)}(x).$$

The extension of this notation to higher derivatives is obvious.

Let us note that if f and g are two functions

$$D(af(x) + bg(x)) = aDf + bDg, \tag{1.2.5}$$

where a and b are constants. One says that taking the derivative is a *linear* operation. One refers to $L = af + bg$ as a *linear combination*, where the term *linear* signifies that f and g appear linearly in L, as compared to, say, quadratically. *Eqn. (1.2.5) tells us that the derivative of a linear combination of two functions is the corresponding linear combination of their derivatives.* To prove the above, one simply goes back to the definition Eqn. (1.2.3). One changes x by Δx and sees what happens to L. One finds that ΔL is a linear combination of Δf and Δg, with coefficients a and b. Dividing by Δx and using the definition of Df and Dg, the result follows.

One can also deduce from the definition that

$$D[fg] = gDf + fDg \tag{1.2.6}$$

as well as the *chain rule*:

$$Df(u(x)) = \frac{df}{du}\frac{du}{dx}. \tag{1.2.7}$$

Problem 1.2.1. *Demonstrate these two results from first principles.*

For example if $u(x) = x^2 + 1$ and $f(u) = u^2$, then

$$\frac{df}{dx} = (2u)(2x) = 2(x^2 + 1)(2x).$$

You can check the correctness of this by brute force: express u in terms of x *first* so that f is explicitly a function of just x, and then take its derivative.

Problem 1.2.2. *Show from first principles that $D(1/x) = -1/x^2$.*

Similarly, one can deduce from the definition of the derivative that

$$D(f/g) = \frac{gDf - fDg}{g^2}. \tag{1.2.8}$$

Problem 1.2.3. *Prove the above by applying the rule for differentiating the product to the case where f and $1/g$ are multiplied. (In taking the derivative of $1/g$ you must use the chain rule.)*

Another useful result is

$$\frac{df}{dx}\frac{dx}{df} = 1 \tag{1.2.9}$$

to be understood as follows. First we view f as a function of x and calculate $\frac{df}{dx}$, the derivative at some point. Next we invert the relationship and write x as a function of f and compute $\frac{dx}{df}$ *at the same point.* The above result connects these two derivatives. (For example if $f(x) = x^2$, the inverse function is $x = \sqrt{f}$.) The truth of this result is obvious geometrically. Suppose we plot f versus x. Let us then take two nearby points separated by $(\Delta x, \Delta f)$, with both points lying on the graph. Now, the increments Δf and Δx satisfy

$$\frac{\Delta f}{\Delta x} \cdot \frac{\Delta x}{\Delta f} = 1$$

and they will continue to do so as we send them both to zero. But in the limit they turn into the corresponding derivatives since Δf is the change in f due to a change Δx in x (that is to say, the amount by which we must move in the vertical direction to return to the graph if we move away from it horizontally by Δx) and vice versa.

After these generalities, let us consider the derivatives of some special cases which are frequently used. First consider $f(x) = x^n$ for n a positive integer. From the binomial theorem, we have

$$\begin{aligned}
\Delta f &= f(x + \Delta x) - f(x) \\
&= (x + \Delta x)^n - x^n \\
&= \sum_{r=0}^{n} \frac{n!}{r!(n-r)!} x^{n-r}(\Delta x)^r - x^n \\
&= nx^{n-1}\Delta x + \mathcal{O}(\Delta x)^2
\end{aligned} \tag{1.2.10}$$

where $\mathcal{O}(\Delta x)^2$ stands for terms of order $(\Delta x)^2$ or higher. If we now divide both sides by Δx and take the limit $\Delta x \to 0$, we obtain

$$Dx^n = nx^{n-1}. \tag{1.2.11}$$

It is useful to see what one would do if one did not know the binomial theorem. *To find the derivative, all we need is the change in f to first order in Δx, since upon dividing by Δx, and taking the limit $\Delta x \to 0$, all higher order terms will vanish.* With this in mind, consider

$$(x + \Delta x)^n = \underbrace{(x + \Delta x)(x + \Delta x)\cdots(x + \Delta x)}_{n \text{ times}}. \tag{1.2.12}$$

The leading term comes from taking the x from each bracket. There is clearly just one contribution to this term. To order Δx, we must take an x from every bracket except one, where we will pick the Δx. There are clearly n such terms, corresponding to the n brackets. Thus

$$\Delta x^n = n x^{n-1} \Delta x$$

to this order. The result now follows.

Armed with this result, we can now calculate the derivative of any polynomial

$$P_n(x) = \sum_{m=0}^{n} a_m x^m \qquad (1.2.13)$$

by using the linearity of the differentiation process.

What about the derivative of $x^{\frac{1}{2}}$? If we blindly use Eqn. (1.2.11), we get

$$D x^{\frac{1}{2}} = \frac{1}{2 x^{\frac{1}{2}}} \qquad (1.2.14)$$

However, we must think a bit since the binomial theorem, as we learned it in school, is established (say by induction) for positive integer powers only. (Since x^n is defined as the product of n factors of x, this makes sense only for positive integer n.) We will return to the question of raising x to any power and then show that Eqn. (1.2.11) holds for all cases, even irrational n. For the present let us just note that Eqn. (1.2.14) follows from Eqn. (1.2.9). Let $f(x) = x^2$. Then $\frac{df}{dx} = 2x$. Inverting, we begin with $x = f^{\frac{1}{2}}$. According to equation (1.2.9)

$$\frac{dx}{df} = \frac{df^{\frac{1}{2}}}{df} = \frac{1}{df/dx} = \frac{1}{2x} = \frac{1}{2f^{\frac{1}{2}}}$$

which is just Eqn. (1.2.14). Notice that in checking that the two derivatives are indeed inverses of each other, we evaluate them at the same point in the (x, f) plane, i.e., we replace x by $f^{\frac{1}{2}}$ which is the value assigned to that x by the functional relation $f = x^2$.

1.3.   Exponential and Log Functions

We now turn to the broader question of what x^p means, where the power p is not necessarily a positive integer. (Until we understand this, we cannot address the question of what the derivative of x^p is.) What does it mean to multiply x by itself p times in this case? One proceeds to give a meaning to this as follows.[1]

The key step is to demand that the fundamental principle, which is true for positive integer powers:

$$x^m x^n = \underbrace{x \cdots x}_{m \text{ times}} \bullet \underbrace{x \cdots x}_{n \text{ times}}$$

$$= x^{m+n}, \tag{1.3.1}$$

i.e., that exponents add upon multiplication, be true even when they are no longer positive integers. With this we can now define $x^{\frac{1}{p}}$, where p is an integer, as that number which satisfies

$$\underbrace{x^{\frac{1}{p}} \cdots x^{\frac{1}{p}}}_{p \text{ times}} = x. \tag{1.3.2}$$

Thus we define $x^{\frac{1}{p}}$ as that number which when multiplied by itself p times yields x. (There may well be more than one choice that works, as you know from the square root case. We then stick to any one branch, say the positive branch in the case of the square root.) Note that this definition does not tell us a systematic way to actually find $x^{\frac{1}{p}}$. At present, all we see is a trial and error search. Later we shall find a more systematic way to find $x^{\frac{1}{p}}$. For the present let us note that the above definition gives us enough information to find the derivative of $x^{\frac{1}{p}}$. Let $y = x^{\frac{1}{p}}$. We want $\frac{dy}{dx}$. Let us find $\frac{dx}{dy}$ and invert it:

$$\frac{dx}{dy} = \frac{d(y^p)}{dy}$$

$$= py^{p-1} \text{ (valid since } p \text{ is an integer)}$$

$$\frac{dy}{dx} = \frac{1}{py^{p-1}}$$

$$= \frac{1}{p} x^{(\frac{1}{p}-1)}. \tag{1.3.3}$$

Thus we find that Eqn. (1.2.11) is valid for the exponent $\frac{1}{p}$. Once we know $x^{\frac{1}{p}}$, we can define $x^{\frac{q}{p}}$ for integer q as the q-th power of $x^{\frac{1}{p}}$. We can find its derivative by using the chain rule, Eqn. (1.2.7) and verify that Eqn. (1.2.11) holds for any rational power $\frac{p}{q}$.

---

[1] This example is very instructive since it tells us how a familiar concept is to be generalized. The basic idea is to list the properties of the familiar case and ask if a more general set of entities can be found satisfying these conditions. For example in mathematical physics one sometimes needs to define integrals in p dimensions, where p is not integer. Clearly the notion of the integral as the area or volume bounded by some curve or surface has to be abandoned. Instead some other features of integration have to be chosen for the generalization.

We can also use Eqn. (1.3.1) to define negative powers. What does x^{-m} mean, for integer m? We demand that it satisfy

$$x^n \bullet x^{-m} = x^{(n-m)}.$$

It is clear that if we set

$$x^{-m} = \frac{1}{\underbrace{x \cdots x}_{m \text{ times}}}$$

the desired result obtains. Thus negative powers are the inverses of positive powers. It also follows that

$$
\begin{aligned}
x^0 &= x^{(m-m)} \\
&= x^m \bullet \frac{1}{x^m} \\
&= 1 \qquad\qquad\qquad\qquad (1.3.4)
\end{aligned}
$$

Thus we have managed to give a meaning to any rational power of x. We are not done yet for two reasons. First, we still want to finish off the irrational powers, powers not of the form $\frac{p}{q}$. Second, we do not want a trial and error definition of powers, we want something more direct, that is to say, a scheme by which given the base (x in our case) and a power, there is a direct algorithm for computing it.

To this end let us now ask what a^x means for any x. Note that x is now the exponent, not the base. This is because we want to vary the exponent continuously over all real values, and wish to denote this variable by x. To define a^x, we compute its derivative with respect to x. You may wonder how we can compute the derivative of something before having defined it! Watch!

$$
\begin{aligned}
\Delta a^x &= a^{(x+\Delta x)} - a^x & (1.3.5) \\
&= a^x(a^{\Delta x} - 1) & (1.3.6) \\
&= a^x(1 + \ln(a)\Delta x + \cdots - 1) & (1.3.7) \\
\frac{da^x}{dx} &= a^x \ln(a). & (1.3.8)
\end{aligned}
$$

The above steps need some explanation. In Eqn. (1.3.5) we are just implementing the definition of Δy for the case $y = a^x$. In the next equation we are using the law for adding exponents. Eqn. (1.3.7) is the most subtle. There we are trying to write an expression for $a^{\Delta x}$. It is clear that it is very close to 1. This is because $a^0 = 1$. The deviation from 1 has a term linear in Δx, with a coefficient that depends on a, and we call it the function $\ln(a)$, pronounced "Ellen of a". Higher order terms in Δx will not matter for the derivative. Let us get a feel for $\ln(a)$, which is also called the *natural logarithm of a*. Compare for example $2^{.001}$ to $3^{.001}$. The first quantity, when raised to the 1000-th power gives

2, while the second gives 3. Approximating $2^{.001}$ and $3^{.001}$ as above, in terms of $\ln 2$ and $\ln 3$, we find $(1 + .001 \ln 2)^{1000} = 2$ while $(1 + .001 \ln 3)^{1000} = 3$. Clearly $\ln(3) > \ln(2)$. It is also clear that $\ln(a)$ grows monotonically with a. In addition $\ln(1) = 0$ because 1 raised to any power will never leave the value 1. On the other hand, if $a < 1$, raising it to a positive power will lower its value, thus $\ln a$ will be negative. There is clearly some $a > 1$ for which $\ln a = 1$. Let us call this number e. So by definition

$$\frac{de^x}{dx} = e^x. \tag{1.3.9}$$

We do not know the value of e yet, but let us proceed. By taking higher derivatives, we find the amazing result

$$D^n e^x = e^x. \tag{1.3.10}$$

Now I will reveal our strategy. We are trying to find out what a^x means for all a. We first take the case $a = e$. What we do know about $f(x) = e^x$ is the following:

1. At $x = 0$, the function equals 1, since anything to power 0 equals 1.

2. At the same point, all derivatives equal 1.

It turns out that this is all we need to find the function everywhere. The trick is to use what is called a Taylor series, and it goes as follows. Let $f(x)$ be some function which we are trying to reconstruct based on available information at the origin. Let us say the function is given by

$$f(x) = 6 + 2x + 3x^2 + 5x^3 \tag{1.3.11}$$

but we do not know that. Say all we have is some partial information at the origin. To begin with, say we only know $f(0)$, the exact value at the origin. Then the best approximation we can construct is

$$f_0(x) = f(0) = 6 \tag{1.3.12}$$

where the subscript 0 on $f_0(x)$ tells us it is the approximation based on zero knowledge of its derivatives. Our guess does not follow the real $f(x)$ for too long. In general it will not, unless f happens to be a constant.0

Suppose now that we are also given $f^{(1)}(0)$, the first derivative at the origin, whose value is clearly 2 in our example. This tells us how the function changes near the origin. We now come up with the following linear approximation:

$$f_1(x) = f(0) + xf^{(1)}(0) = 6 + 2x \tag{1.3.13}$$

where the subscript on $f_1(x)$ tells us the approximation is based on knowledge of one derivative at the origin. We see that f_1 agrees with f *at* the origin and also grows at the same rate, i.e., has the same first derivative. It therefore approximates

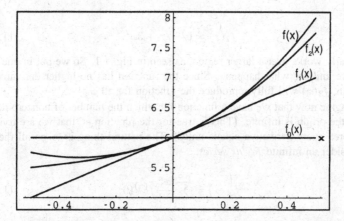

Figure 1.1. Various approximations to the real $f(x)$, based on more and more derivatives at the origin. (The number of known derivatives is given by the subscript.) In the present case, three derivatives are all we need. In general, an infinite number could be required to fully reconstruct the function.

the function f for a small region near the origin and then it too starts differing from it, as shown in Fig. 1.1:

In general this will happen when we approximate a function by its linearized version, unless the function happens to be linear, i.e., the function has a fixed rate of change. But in our case, and in general, the rate of change itself will have a rate of change, given by the second derivative, which in turn can have a rate of change and so on.

Suppose now we are also given $f^{(2)}(0) = 6$. How we do construct the better approximation that incorporates this? The answer is that the approximation, called $f_2(x)$ in our notation, is

$$f_2(x) = f(0) + x f^{(1)}(0) + \frac{x^2}{2} f^{(2)}(0) \tag{1.3.14}$$

Let us check. Set $x = 0$ on both sides of the top equation, and see that $f_2(0) = f(0)$. Next, take the derivative of both sides and then set $x = 0$. The first term on the right gives nothing since it is a constant, while the last one vanishes since a single power of x remains upon differentiating and that vanishes upon setting $x = 0$. Only the middle term survives and gives $f^{(1)}(0)$, the correct first derivative at the origin. Finally consider the second derivative at the origin. Only the last term on the right survives and contributes a value equal to the second derivative of the actual function f at the origin. Thus we have cooked up a function that matches our target function f in three respects at the origin: it has the same initial value, slope, and rate of change of slope.

If we put in the actual derivatives in the above formula, we will of course obtain

$$f_2(x) = 6 + 2x + 3x^2. \tag{1.3.15}$$

This clearly works over a larger region, as seen in Fig. 1.1. So we put in one more derivative and see what happens. Since the function has no higher derivatives at the origin, $f_3(x)$ will fully reproduce the function for all x.

Imagine now that we have a function for which the number of nonzero derivatives at the origin is infinite. (This is true for the function e^x that we are trying to build here: every derivative equals unity.) The natural thing is to go all the way and consider an infinite *Taylor series*:

$$f_\infty(x) = \sum_0^\infty \frac{x^n}{n!} f^{(n)}(0) \tag{1.3.16}$$

What can we say about this sum? What relation does it bear to the function f? First, we must realize that an infinite sum of terms can be qualitatively different from a finite sum. For example, the sum may be infinite, even though the individual terms are finite and progressively smaller. Chapter 4 on infinite series will tell us how to handle this question. For the present we will simply appeal to a result from that chapter, called the *ratio test*, which tells us that an infinite sum

$$S = \sum_{n=0}^\infty a_n x^n \tag{1.3.17}$$

converges (and defines a function of x) as long as

$$\lim_{n \to \infty} \left| \frac{a_{n+1} x^{n+1}}{a_n x^n} \right| < 1 \quad \text{which means} \tag{1.3.18}$$

$$|x| < R \tag{1.3.19}$$

$$R = \lim_{n \to \infty} \left| \frac{a_n}{a_{n+1}} \right| \tag{1.3.20}$$

where R is called the *interval of convergence*. The ratio test merely ensures that each term is strictly smaller in size than the previous term as $n \to \infty$.

With all this in mind, we take the following stance. *We will take the function e^x to be defined by its infinite Taylor series as long as the sum converges.* Since we have no other definition of this function, there is no need to worry if this is "really" the function.

The series for e^x is, from Eqns. (1.3.10-1.3.16):

$$e^x = \sum_0^\infty \frac{x^n}{n!} \tag{1.3.21}$$

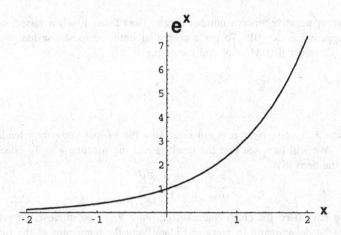

Figure 1.2. Plot of the function e^x. Notice its growth is proportional to the value of the function itself.

(Recall that every derivative of the function is also e^x which equals 1 at $x = 0$.) The ratio test tells us that in this case, since $a_n = 1/n!$,

$$R = \lim_{n \to \infty} \frac{(n+1)!}{n!} \to \infty, \qquad (1.3.22)$$

i.e., that the series converges for all finite x. *Thus we have defined the function for all finite x based on what we knew at $x = 0$.*[2]

We are now ready to find e: simply set $x = 1$ in Eqn. (1.3.21). As we keep adding more terms to the sum we see it converges quickly to a value around 2.7183. We can now raise e to any power. For example, to find $e^{95/112}$ we just set $x = 95/112$ in the sum and compute as many terms as we want to get any desired accuracy. There is no trial and error involved. We may choose x to be any real number, say π or even e!

Figure 1.2 shows a plot of the exponential function for $-2 < x < 2$.

There is a second way to define the exponential function. Consider the following function defined by two integers M and N:

$$e^x_{N,M} = (1 + \frac{x}{N})^M \qquad (1.3.23)$$

If we fix M and let $N \to \infty$, the result is clearly 1. On the other hand if we fix N and let $M \to \infty$, the result will either be 0 or ∞ depending on whether x

---

[2] When we study Taylor series later, we will see that this situation is quite unusual. Take for example, the function $1/(1-x)$. Suppose we only knew its Taylor series about the origin: $1+x+x^2+x^3+\ldots$. The ratio test tells us the series converges only for $|x| < 1$. One then has to worry about how to reconstruct the function beyond that interval. This point will be discussed in Chapter 6. For the present let us thank our luck and go on.

is positive or negative since a number greater (less than) 1, when raised to large powers approaches ∞ (0). To get a nontrivial limit we must consider the case $M \propto N$. Consider first $M = N$ and the object

$$e_N^x \equiv e_{N,N}^x$$
$$= (1 + \frac{x}{N})^N \tag{1.3.24}$$

in the limit $N \to \infty$. Now it is not clear how the various competing tendencies will fare. We will now see that the result is just the function e^x. To check this consider the derivative:

$$\frac{de_N^x}{dx} = \frac{(1 + \frac{x}{N})^N}{(1 + \frac{x}{N})}, \tag{1.3.25}$$

where we have used the chain rule. In the limit $N \to \infty$, there is no problem in setting the denominator to unity and identifying the numerator as the function being differentiated. Since the function equals its derivative, and $f(0) = 1$, it must be the function e^x since these two properties were all we needed to nail it down completely. Let us now trivially generalize to the function e^{ax} which is given by a similar series in ax by choosing $N = M/a$ in Eqn. (1.3.23). Its derivative is clearly ae^{ax}.

The exponential function is encountered very often. If $P(t)$ is the population of a society, and its rate of growth is proportional to the population itself, we say

$$\frac{dP}{dt} = aP(t). \tag{1.3.26}$$

It is clear that $P(t) = e^{at}$. One refers to this growth as exponential growth. On the other hand consider the decay of radioactive atoms. The less there are, the less will be the decay rate:

$$\frac{dP(t)}{dt} = -aP(t), \tag{1.3.27}$$

where a is positive. In this case the function decays exponentially: $P(t) = e^{-at}$.

The second definition of e^x arises in the banking industry as follows. Say a bank offers simple interest of x dollars per annum. This means that if you put in a dollar, a year later you get back $(1 + x)$ dollars. A rival bank can offer the same rate but offer to compound every six months. This means that after six months your investment is worth $(1 + \frac{x}{2})$ which is reinvested at once to give you at year's end $(1 + \frac{x}{2})^2$ dollars. You can see that you get a little more: $x^2/4$ to be exact. If now another bank gets in and offers to compound N times a year and so on, we see that the war has a definite limit: interest is compounded continuously and one dollar becomes at year's end

$$\lim_{N \to \infty} (1 + \frac{x}{N})^N = e^x \text{ dollars.}$$

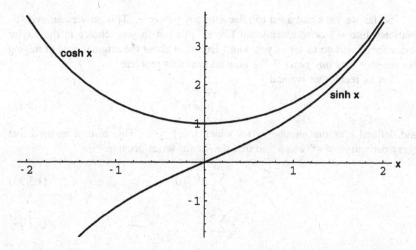

Figure 1.3. Plot of the hyperbolic sinh and cosh functions. Note that they are odd and even respectively and approach $e^x/2$ as $x \to \infty$.

From the function e^x we can generate the following *hyperbolic functions:*

$$\sinh x \;=\; \frac{e^x - e^{-x}}{2} \qquad\qquad (1.3.28)$$

$$\cosh x \;=\; \frac{e^x + e^{-x}}{2} \qquad\qquad (1.3.29)$$

These functions are often called sh x and ch x, where the h stands for "hyperbolic". They are pronounced *sinch* and *cosh* respectively. Figure 1.3 is a graph of these functions.

They obey many identities such as

$$\cosh^2 x - \sinh^2 x \;=\; 1 \qquad\qquad (1.3.30)$$
$$\sinh(x+y) \;=\; \sinh x \;\cosh y + \cosh x \;\sinh y \qquad (1.3.31)$$

(which can be proved, starting from the defining Eqs.(1.3.28-1.3.29)) and numerous others which we cannot discuss in this chapter devoted to calculus. For the present note that these relations look a lot like those obeyed by trigonometric functions. The intimate relation between the two will be taken up later in this book. Readers wishing to bone up on this subject should work through the exercises in this chapter. Note that $\cosh x$ and $\sinh x$ are even and odd, respectively, under $x \to -x$.

Problem 1.3.1. *Verify that* $\sinh x$ *and* $\cosh x$ *are derivatives of each other. Verify Eqns. (1.3.30-1.3.31).*

So far we have managed to raise e to any power x. This power can even be irrational like $\sqrt{2}$ or transcendental like π: just put in your choice in the Taylor series for e^x and go as far as you want. But what about the original goal of raising *any number a to any power*? We now address that problem.

Let us recall that we had

$$\frac{da^x}{dx} = \ln(a)a^x \tag{1.3.32}$$

and defined e as that number a for which $\ln(e) = 1$. This in turn ensured that every derivative of e^x was e^x so that the Taylor series became

$$e^x = \sum_0^\infty f^{(n)}(0)\frac{x^n}{n!} \tag{1.3.33}$$

$$= \sum_0^\infty \frac{x^n}{n!}. \tag{1.3.34}$$

By exactly the same logic we have for general a,

$$Da^x = \ln(a)a^x \tag{1.3.35}$$
$$D^2a^x = (\ln a)^2 a^x \tag{1.3.36}$$

and so on leading to the series

$$a^x = \sum_0^\infty (\ln a)^n \frac{x^n}{n!} \tag{1.3.37}$$

$$= e^{x\,\ln a}. \tag{1.3.38}$$

It appears that we have a formula for a^x, but in terms of the function $\ln a$, the natural logarithm of a. All we know about this function is that $\ln 1 = 0$ and $\ln e = 1$. We will now fully determine this function, solving the problem we set ourselves.

Setting $x = 1$ in Eqn. (1.3.38), we find the relation

$$a = e^{\ln a} \tag{1.3.39}$$

as an identity in a. This equation tells us two things. First $\ln a$ is the power to which e must be raised to give a. Second, it means that the ln function and the exponential function are inverses, just like the square root function and square function are inverses: for any positive x it is true that

$$x = (\sqrt{x})^2. \tag{1.3.40}$$

We now find $\ln a$ by the same trick of writing down its Taylor series. Taking the derivative of both sides of Eqn. (1.3.39) with respect to a, we have

$$1 = e^{\ln a}D\ln a = aD\ln a \tag{1.3.41}$$

which implies

$$D \ln a = \frac{1}{a}. \tag{1.3.42}$$

Since we know how to differentiate negative integer powers, we can deduce that

$$D^n \ln a = \frac{(-1)^{n+1}(n-1)!}{a^n}. \tag{1.3.43}$$

We now know the derivative of the function at any a. However we can't launch a series about the point $a = 0$ (as we did in the case of the exponential function) since all derivatives diverge at this point. However the logic of the Taylor series is unaffected by the point about which we choose to expand the function. We can write in general

$$f(x) = f(a) + f^{(1)}(a)(x-a) + \frac{1}{2!}f^{(2)}(a)(x-a)^2 + \ldots \tag{1.3.44}$$

for any a. In the case of e^x, $a = 0$ was a nice point since all the derivatives equaled 1 there. For the ln function $a = 1$ is a very nice point since

$$D^n \ln a = \frac{(-1)^{n+1}(n-1)!}{a^n} = (-1)^{n+1}(n-1)!. \tag{1.3.45}$$

The Taylor series for $\ln x$ about $x = 1$ is then

$$\ln x = \ln 1 + (x-1) - \frac{(x-1)^2}{2} + \frac{(x-1)^3}{3} + \ldots \tag{1.3.46}$$

where $\ln 1 \doteq 0$. If we apply the test for convergence we find that the series converges for

$$|x - 1| < 1 \text{ or } 0 < x < 2. \tag{1.3.47}$$

It is clear the series cannot go beyond $x = 0$ on the left since $\ln 0 = -\infty$, i.e., $-\infty$ is the power to which e must be raised to give 0.

In terms of $y = x - 1$

$$\ln(1+y) = y - \frac{y^2}{2} + \frac{y^3}{3} + \ldots \tag{1.3.48}$$

In using this formula we must remember that y is a measure of x from the point $x = 1$ and that the series is good for $|y| < 1$. The log tables you used as a child were constructed from this formula. You don't need those tables any more. Say you want $\ln 1.25$. It is given to good accuracy by just the first two terms

$$\ln\left(1 + \frac{1}{4}\right) = \frac{1}{4} - \frac{1}{32} \simeq .2188 \tag{1.3.49}$$

which compares very well with the value of .2231 from the tables. If you add one more term in the series, you get .2240

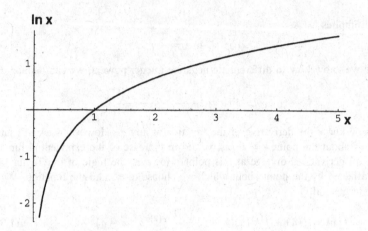

Figure 1.4. Plot of the function $\ln(x)$. Notice that $\ln 1 = 0$, $\ln e = 1$, $\ln x \to -\infty$ as $x \to 0$ and $\ln x \to \infty$ as $x \to \infty$.

How are we to get the ln of a number ≥ 2? There are general tricks which will be discussed in Chapter 6, but we will deal with this by deriving and using a property of the ln that you must know.

From the very definition, (Eqn. (1.3.39), for two numbers a and b,

$$a = e^{\ln a} \tag{1.3.50}$$
$$b = e^{\ln b} \tag{1.3.51}$$
$$ab = e^{\ln a + \ln b} \tag{1.3.52}$$
$$= e^{\ln ab} \quad \text{so that} \tag{1.3.53}$$
$$\ln(ab) = \ln a + \ln b. \tag{1.3.54}$$

Using the property

$$\ln(ab) = \ln a + \ln b \tag{1.3.55}$$

we can obtain the ln of a big number ab starting with the ln of smaller numbers a and b which in turn could be dealt with in the same way until we get to the stage where we need only the ln's of numbers less than 2. For example, knowing $\ln 1.6$ and $\ln 1.8$ we can get $\ln 2.88$ as the sum of the two logarithms. Fig. 1.4 depicts the ln function obtained by this or any other way.

We now know how to calculate a^x as follows:

$$a^x = (e^{\ln a})^x \tag{1.3.56}$$
$$= e^{x \ln a}. \tag{1.3.57}$$

Note that everything above is well defined: for any given a we can find $\ln a$ using the Taylor series, we can then exponentiate the result using the series for the exponential function.

So far we have been considering the possibility of expressing any number a as e raised to some power, namely $\ln a$. One says that e is the *base* for the logarithm. We are however free to choose some other base. For example it is easier to think of 100 as 10^2 rather than as $e^{4.606\cdots}$. To accommodate the possibility of other bases, we introduce the function $\log_b a$, which appears in the identity

$$a = b^{\log_b a} \tag{1.3.58}$$

and call it "log of a to the base b." Thus

$$2 = \log_{10} 100. \tag{1.3.59}$$

Since e had some special properties, and is mathematically the natural choice for base, $\log_e a$ is called the "natural logarithm" and denoted by the symbol we have been using: $\ln a$.[3]

The relation between logarithms with respect to base e and any other base b is easily deduced: for any number y we have two identities:

$$y \equiv e^{\ln y} \tag{1.3.60}$$
$$\equiv b^{\log_b y} \tag{1.3.61}$$
$$= (e^{\ln b})^{\log_b y} \tag{1.3.62}$$
$$= e^{\ln b \cdot \log_b y}. \tag{1.3.63}$$

It follows by comparison of the exponents between the first and last equations that

$$\log_b y = \frac{\ln y}{\ln b} \tag{1.3.64}$$
$$\equiv \frac{\log_e y}{\log_e b}. \tag{1.3.65}$$

In particular $\ln 10 = 2.303..$ serves as the conversion factor between the natural logarithm $\ln$ and $\log_{10}$.

Now that we have given an operational meaning to x^p for *any* p, we can deduce what the derivative of x^p is. We proceed as follows:

$$\frac{dx^p}{dx} = \frac{de^{p \ln x}}{dx} \tag{1.3.66}$$
$$= e^{p \ln x} \frac{dp \ln x}{dx} \tag{1.3.67}$$
$$= x^p \cdot p \frac{1}{x} \tag{1.3.68}$$
$$= p x^{p-1}. \tag{1.3.69}$$

---

[3] Of course, one can argue that 10 is more natural for humans based on our fingers and that, $e = 2.718..$ is not natural, unless you have been playing with firecrackers.

When students in the introductory math course were asked what the derivative of $x^{6.1}$ was, they quickly came up with $6.1x^{5.1}$, but only a few knew the full story recounted above.

If we use the above result we can develop the Taylor series for $(1+x)^p$ about the point 1:

$$(1+x)^p = 1 + px + \frac{(p)(p-1)}{2}x^2 + \frac{p(p-1)(p-2)}{3!}x^3 + \cdots \qquad (1.3.70)$$

This is the generalization of the *binomial theorem* for noninteger p.

Problem 1.3.2. *Derive the series up to four terms as shown above.*

Note two things. First, if p is an integer the series terminates after a finite number of terms and gives the familiar binomial theorem we learned in school for integer powers. Second, if x is very small we can stop after the second term:

$$(1+x)^p \simeq 1 + px \qquad (1.3.71)$$

for any p. This is a very useful result and you should know it all times.

Let us finally ask: if the ln function is the inverse of the e^x function, what are the inverses of $\sinh x$ and $\cosh x$? Let us define $\sinh^{-1} x$ as that number whose sinh is x. From the graph of $\sinh x$ you can see that the answer is unique. We can find the derivative of this function using the inverse function trick:

$$
\begin{aligned}
y &= \sinh^{-1} x \\
x &= \sinh y \\
\frac{dx}{dy} &= \cosh y \\
&= \sqrt{1 + \sinh^2 y} \\
&= \sqrt{1 + x^2} \\
\frac{dy}{dx} &= \frac{1}{\sqrt{1 + x^2}}. \qquad (1.3.72)
\end{aligned}
$$

You can see from the graph that each value of $\cosh x$ has two origins, related by a sign. The inverse function is uniquely defined if we agree, say, to follow the positive branch. (This is analogous to the fact that each number has two square roots.) Unlike the sinh function which always has an inverse which is also unique, the cosh has an inverse only in the interval $1 \le \cosh \le \infty$ and the latter is double valued.

Given sinh and cosh, you can form ratios of these, take their derivatives, their inverses, the derivatives of the inverses, and so on. The fun is endless! We must relegate some of this to the exercises and move on. You must however have on your fingertips the following Taylor series which you can read off from the series for e^x (Eqn. (1.3.21)) (also on your finger tips) and the very definitions of these functions:

$$\sinh x \;=\; \sum_{n=0}^{\infty} \frac{x^{2n+1}}{(2n+1)!} \qquad\qquad (1.3.73)$$

$$=\; x + \frac{x^3}{3!} + \frac{x^5}{5!} + \cdots \qquad\qquad (1.3.74)$$

$$\cosh x \;=\; \sum_{n=0}^{\infty} \frac{x^{2n}}{(2n)!} \qquad\qquad (1.3.75)$$

$$=\; 1 + \frac{x^2}{2!} + \frac{x^4}{4!} + \cdots \qquad\qquad (1.3.76)$$

Problem 1.3.3. *Demonstrate the above. Observe that* sinh *and* cosh *contain only odd and even powers of* x, *which is why they are odd and even, respectively, under* $x \to -x$.

1.4. Trigonometric Functions

Here too we will only deal with some points that involve calculus. Let us begin with the notion of sines and cosines as ratios of the opposite and adjacent sides to the hypotenuse in a right triangle. It is assumed that you are familiar with various identities involving these functions, their ratios (tan, cot, sec), and their addition formulae. Let us recall one that we will need shortly:

$$\sin(A + B) = \sin A \, \cos B + \cos A \, \sin B. \qquad\qquad (1.4.1)$$

You are all no doubt aware that:

$$D \sin x \;=\; \cos x \qquad\qquad (1.4.2)$$
$$D \cos x \;=\; -\sin x. \qquad\qquad (1.4.3)$$

Do you know where this comes from? A first ingredient in the proof is the result

$$\lim_{\theta \to 0} \frac{\sin \theta}{\theta} = 1, \qquad\qquad (1.4.4)$$

valid only if the angle is measured in *radians*. The radian is a way to measure angles just like degrees. It is however a more natural unit of angular measurement, as the following discussion will make clear.

Consider the circle of radius r.

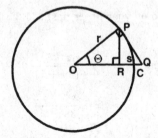

Figure 1.5. Introduction to the radian. Note that if θ is measured in radians, the arc length $s = r\theta$.

By dimensional analysis, its circumference must be given by the formula

$$C(r) = rf \tag{1.4.5}$$

where f is function of any dimensionless variable formed out of r and other dimensionful parameters specifying the circle. Since there exist no such things,

$$C(r) = cr \tag{1.4.6}$$

where c is a constant. By *definition* it is 2π:

$$C(r) = 2\pi r. \tag{1.4.7}$$

By the same logic, the area of the circle is

$$A(r) = \bar{c}r^2 \tag{1.4.8}$$

The constant $\bar{c}$ is now found as follows. Suppose we increase r by Δr. This adds to the area an annulus of circumference $2\pi r$ and thickness Δr, so that the change in the area is

$$\Delta A = 2\pi r \Delta r \tag{1.4.9}$$

which means

$$\frac{dA}{dr} = 2\pi r \tag{1.4.10}$$

which, upon comparing to the derivative of Eqn. (1.4.8), tells us $\bar{c} = \pi$ and

$$A(r) = \pi r^2. \tag{1.4.11}$$

Consider now an arc of the circle which subtends an angle θ at the center as shown in the Fig. 1.5.

The arc length s is a linear function of the angle subtended, θ. That is to say, if you double the angle, you double the arc length. It is also a linear function of the radius: if you blow up the radius by a factor 2, you double the arc length. (The answer also follows from dimensional analysis. Since s has dimensions of length it must be r times a function of a dimensionless variable formed out of r, of which kind there are none.) Thus it is possible to write:

$$s = cr\theta \tag{1.4.12}$$

where c is some constant. Its value depends on the way we measure angle. For example if we measure it in degrees, whereby 360 degrees make a full circle, $c = 2\pi/360$. That is, if you use this c above and set $\theta = 360$, you get $s = 2\pi r$, which is the correct formula for the circumference, from the very definition of π. Let us instead measure the angle in units such that $c = 1$. Call this unit a radian. How many radians is a full circle? It must be 2π since with $c = 1$ this is the value of θ that gives the right circumference. Since $2\pi \simeq 6$, a radian is roughly 60 degrees. (More precisely a radian is 57.2958 degrees.) *It will be assumed hereafter that all angles are being measured in radians so that the arc length numerically equals the product of the radius and the angle subtended.* To prove Eqn. (1.4.4) we turn to Fig. 1.5. It is clear from the figure that

$$POR \leq POC \leq POQ.$$

Using the formulae for the area of the triangles POR and POQ and the segment POC, this becomes

$$\frac{1}{2}r\sin\theta \; r\;\cos\theta \leq \frac{\theta}{2\pi}\pi r^2 \leq \frac{1}{2}r\;r\;\tan\theta.$$

If we divide everything by $\frac{1}{2}r^2 \sin\theta$ we find

$$\cos\theta \leq \frac{\theta}{\sin\theta} \leq \frac{1}{\cos\theta}.$$

If we now let $\theta \to 0$, the ratio in between gets squeezed between two numbers both of which approach 1 (since $\cos 0 = 1$) and the result follows.

A corollary of the above result is that for small angles,

$$\cos\theta \;=\; (1 - \sin^2\theta)^{1/2} \tag{1.4.13}$$

$$=\; 1 - \frac{1}{2}\theta^2 + \cdots \tag{1.4.14}$$

where we have used the generalized binomial theorem $(1 + x)^p \sim 1 + px + \dots$. Let us now find the derivative of $\sin\theta$ from first principles:

$$\Delta\sin\theta \;=\; \sin(\theta + \Delta\theta) - \sin(\theta) \tag{1.4.15}$$

$$=\; \sin(\theta)\cos(\Delta\theta) + \cos(\theta)\sin(\Delta\theta) - \sin(\theta) \tag{1.4.16}$$

$$=\; \cos(\theta)\Delta\theta + \mathcal{O}(\Delta\theta)^2 + \tag{1.4.17}$$

where we have used the approximations for $\sin\theta$ and $\cos\theta$ at a small angle $\Delta\theta$ keeping only terms of linear order since higher order terms do not survive the limit involved in taking the derivative. It now follows that the derivative of $\sin\theta$ is $\cos\theta$. Given this strategy you can work out the derivative of $\cos\theta$, use the rule for derivative of the product or ratio of functions to obtain the derivatives of tan, sec, etc. You are expected to know the definitions of these functions and their derivatives, as well as values of these functions at special values of their arguments such as $\pi/4$, $\pi/6$, etc.

We now have all the information we need to construct the Taylor series for the sine and cosine at the origin:

$$\sin x \;=\; x - \frac{x^3}{3!} + \frac{x^5}{5!} + \cdots \tag{1.4.18}$$

$$=\; \sum_0^\infty \frac{(-1)^n x^{2n+1}}{(2n+1)!} \tag{1.4.19}$$

$$\cos x \;=\; 1 - \frac{x^2}{2!} + \frac{x^4}{4!} + \cdots \tag{1.4.20}$$

$$=\; \sum_0^\infty \frac{(-1)^n x^{2n}}{(2n)!}. \tag{1.4.21}$$

The ratio test gives the same result as in the case of e^x: these series converge for all finite x. As with the logarithm, you can use these series to get a very good approximation to any trigonometric function. Say you want $\sin 30°$ which is exactly .5. The first two terms in the series with $x = \pi/6$(radian) give

$$\sin \pi/6 \simeq \pi/6 - (\pi/6)^3/6 = .499 \tag{1.4.22}$$

Problem 1.4.1. *Derive the above series for the* sin *and* cos, *given* $D \sin x = \cos x$, $D \cos x = -\sin x$, $\cos 0 = 1$, *and* $\sin 0 = 0$. *Show that the series converge for all finite* x.

The above series are remarkable. By knowing all the derivatives at one point, the origin, we know what the functions are going to do a mile away. For example, the series for $\sin x$ will vanish if you set $x = 23445671\pi$ where the sin must vanish. You may wish to try it out on a calculator for just $x = \pi$, using some approximation for π. The series knows that the sine, which starts growing linearly near the origin, is going to turn around, hit zero at π, turn upwards again, hit zero at 2π, and on and on.

Given the trig functions, you can define their inverses in the natural way. For example $\sin^{-1} x$ is the angle whose sin is x. There is some ambiguity in this definition, just as there was in the square root, there being two choices in the latter related by a sign and an infinite number here related by the periodicity and symmetry of the sin function. If however we restrict the angle to lie in the interval $(-\pi/2, \pi/2)$, the inverse sin is unique. (Imagine the sin function in this interval and note that each value of $\sin\theta$ comes from a unique θ.) We can then ask what its derivative is. The trick is to do what we did for the ln earlier:

$$
\begin{aligned}
y &= \sin^{-1} x \\
x &= \sin y \\
\frac{dx}{dy} &= \cos y \\
&= \sqrt{1 - \sin^2 y} \\
&= \sqrt{1 - x^2} \\
\frac{dy}{dx} &= \frac{1}{\sqrt{1 - x^2}}
\end{aligned}
\qquad (1.4.23)
$$

One also writes $\arcsin x$ in place of $\sin^{-1} x$, and likewise for all inverse trigonometric and hyperbolic functions.

1.5. Plotting Functions

You will be frequently called upon to sketch some given functions. For example the solution to some problem may be a complicated function, and it is no use having it if you cannot visualize its key features. In particular, you must be able to locate points where it vanishes, where it blows up, where it has its maxima and minima, its behavior at special points such as 0 and ∞, and so on. This is something that comes with practice and you cannot learn it all here. But here is a modest example. You should draw a sketch as we go along.

Let us look at

$$
f(x) = \frac{x^2 - 5x + 6}{x - 1} e^{-x/5}. \qquad (1.5.1)
$$

Far to the left, as we approach $-\infty$, the numerator in the polynomial can be approximated by the highest power, x^2 and the denominator by x, and the ratio by x. Thus the function behaves as $xe^{-x/5}$ which approaches $-\infty$ as $x \to -\infty$. As for finite x, let us rewrite f as

$$
f(x) = \frac{(x - 2)(x - 3)}{x - 1} e^{-x/5} \qquad (1.5.2)
$$

which tells us that we must focus on three special points: $x = 1$ where the denominator vanishes, $x = 2$ and $x = 3$ where the numerator vanishes. In addition we will focus on $x = 0$. As we move to the right from $-\infty$, we first come to $x = 0$ where the function is still negative and has the value -6 and a slope $1/5$:

$$f(x) = -6 + x/5 + \cdots \qquad (1.5.3)$$

However as we approach $x = 1$, the function must blow up due to the vanishing denominator and the blow up must be towards $-\infty$ since none of the factors in the ratio of polynomials multiplying the exponentials has changed sign. This in turn means f has a local maximum somewhere between 0 and 1. To the right of $x = 1$ the function is large but positive since the denominator has changed sign. The function then decreases until we reach $x = 2$ where it vanishes. Now it turns negative until we get to $x = 3$ where it vanishes and changes sign due to the factor $x - 3$. Thus f has a minimum between $x = 2$ and $x = 3$. To the right of $x = 3$ the function is positive (as are all the factors $x - 1, x - 2, x - 3$) all the way to infinity where it behaves like $xe^{-x/5}$. Now we have to decide who wins: the growing factor x, or the declining factor $e^{-x/5}$. Stated differently, in the ratio $\frac{x}{e^{x/5}}$, which is bigger at large x? The trick to resolving this is to use:

L'Hôpital's rule (given without proof): To determine the ratio of two functions, both of which blow up or both of which vanish at some point (infinity in this example), take the ratio of their derivatives; if these cannot give a definite answer, take the ratio of the next derivatives and so on until a clear limit emerges.

In our example, we are interested in the ratio $x/(e^{x/5})$, where the numerator and denominator blow up as $x \to \infty$.[4] Upon taking one derivative x turns into 1, $e^{x/5}$ becomes $e^{x/5}/5$ and the ratio of the derivatives, $5/e^{x/5}$, clearly vanishes as $x \to \infty$. Thus the function $f(x)$ has a maximum to the right of $x = 3$ where it starts turning downwards to zero.

Note that we did not try to actually locate the maxima and minima too precisely. This can be tedious, but done if we need this information. Usually the caricature painted in Fig. 1.6 is already very useful.

You can now compare your sketch with the plot of the function given in Fig. 1.6.

Problem 1.5.1. *Show that $x^n e^{-x} \to 0$ as $x \to \infty$. Thus the falling exponential can subdue any power. Use L'Hôpital's rule to show that the growth of $\ln x$ is weaker than any positive power x^p, i.e., $\frac{\ln x}{x^p}$ vanishes as $x \to \infty$ for any $p > 0$.)*

---

[4] As an example of a case where both functions go to zero, consider the indeterminate ratio $(1 - \cos^2 x)/x$ as $x \to 0$. Upon taking one derivative, we find the ratio $(2 \sin x \cos x/1)$ which clearly vanishes as $x \to 0$.

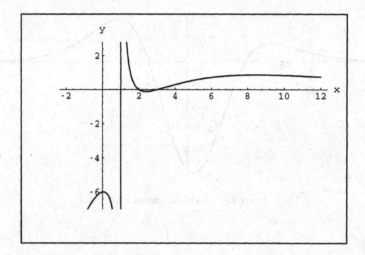

Figure 1.6. Plot of the function $f(x)$.

Problem 1.5.2. *Analyze the function*

$$S(x) = \frac{x^2 + x - 6}{4 + \cosh x}$$

and compare your findings to Fig. 1.7.

1.6. Miscellaneous Problems on Differential Calculus

Besides some of the tricky points we discussed above, you are of course expected to know all the basics of differential calculus as well as the properties of the special functions we encountered. The following set of problems is by no means an exhaustive test of your background. It should however suffice to give you an idea of where you stand. If you find any weak areas while doing them, you should strengthen up those areas by going to a book devoted to calculus.

Problem 1.6.1. *Expand the function* $f(x) = \sin x / (\cosh x + 2)$ *in a Taylor series around the origin going up to* x^3. *Calculate f(.1) from this series and compare to the exact answer obtained by using a calculator.*

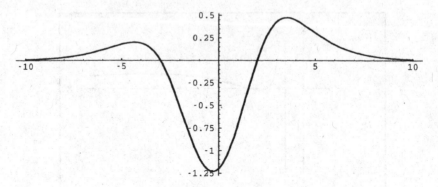

Figure 1.7. Plot of the function $S(x)$.

Problem 1.6.2. *Find the derivatives of the following functions: (i)* $\sin(x^3 + 2)$, *(ii)* $\sin(\cos(2x))$, *(iii)* $\tan^3 x$, *(iv)* $\ln(\cosh x)$, *(v)* $\tan^{-1} x$, *(vi)* $\tanh^{-1} x$, *(vii)* $\cosh^2 x - \sinh^2 x$, *(viii)* $\sin x/(1 + \cos x)$.

Problem 1.6.3. *A bank compounds interest continually at a rate of 6% per annum. What will a hundred dollars be worth after 2 years? Use an approximate evaluation of e^x to order x^2.*

Problem 1.6.4. *According to the Theory of Relativity, if an event occurs at a space–time point (x, t) according to an observer, another moving relative to him at speed v (measured in units in which the velocity of light $c = 1$) will ascribe to it the coordinates*

$$x' = \frac{x - vt}{\sqrt{1 - v^2}} \tag{1.6.1}$$

$$t' = \frac{t - vx}{\sqrt{1 - v^2}}. \tag{1.6.2}$$

Verify that s, the space–time interval is same for both: $s^2 = t^2 - x^2 = t'^2 - x'^2 = s'^2$. Show that if we parametrize the transformation terms of the rapidity θ,

$$x' = x \cosh\theta - t \sinh\theta \tag{1.6.3}$$

$$t' = t \cosh\theta - x \sinh\theta \tag{1.6.4}$$

the space–time interval will be automatically invariant under this transformation thanks to an identity satisfied by hyperbolic functions. Relate $\tanh\theta$ to the velocity. Suppose a third observer moves relative to the second at a speed v', that is, with *rapidity θ'. Relate his coordinates (x'', t'') to (x, t) going via $(x'.t')$. Show that the rapidity parameter $\theta'' = \theta' + \theta$ in obvious notation. (You will need to derive a formula for $\tanh(A + B)$.) Thus it is the rapidity, and not velocity that really obeys a simple addition rule. Show that if v and v' are small (in units of c), that this reduces to the daily life rule for addition of velocities. (Use the Taylor series for $\tanh \theta$.) This is an example of how hyperbolic functions arise naturally in mathematical physics.*

Problem 1.6.5. *A magnetic moment μ in a magnetic field h has energy $E_{\pm} = \mp \mu h$ when it is parallel (antiparallel) to the field. Its lowest energy state is when it is aligned with h. However at any finite temperature, it has a nonzero probabilities for being parallel or antiparallel given by $P(par)/P(antipar) = \exp[-E_+/T]/\exp[-E_-/T]$ where T is the absolute temperature. Using the fact that the total probability must add up to 1, evaluate the absolute probabilities for the two orientations. Using this show that the average magnetic moment along the field h is $m = \mu \tanh(\mu h/T)$ Sketch this as a function of temperature at fixed h. Notice that if $h = 0$, m vanishes since the moment points up and down with equal probability. Thus h is the cause of a nonzero m. Calculate the susceptibility, $\frac{dm}{dh}|_{h=0}$ as a function of T.*

Problem 1.6.6. *Consider the previous problem in a more general light. According to the laws of Statistical Mechanics if a system can be in one of n states labeled by an index i, with energies E_i, then at temperature T the system will be in state i with a relative probability $p(i) = e^{-\beta E_i}$ where $\beta = 1/T$. Introduce the partition function $Z = \sum_i e^{-\beta E_i}$. First write an expression for $P(i)$, the absolute probability (which must add up to 1). Next write a formula for $\langle V \rangle$, the mean value of a variable V that takes the value V_i in state i, i.e., $\langle V \rangle$ is the average over all allowed values, duly weighted by the probabilities. Show that $< E > = -\frac{d \ln Z}{d\beta}$. Give an explicit formula for Z for the previous problem. Show that $\frac{d \ln Z}{d\beta h}$ gives the mean moment along h. Use the formula for Z, evaluate this derivative and verify that it agrees with the result you got in the last problem.*

Problem 1.6.7. *A wire of length L is used to fence a rectangular piece of land. For a rectangle of general aspect ratio compute the area of the rectangle. Use the rule for finding the maximum of a function to find the shape that gives the largest area. Find this area.*

Problem 1.6.8. *Sketch and locate the maxima and minima of $f(x) = (x^2 - 5x + 6)e^{-x}$.*

Problem 1.6.9. *Find the first and second derivatives of $f(x) = e^{x/(1-x)}$ at the origin.*

Problem 1.6.10. *Imagine a life guard situated a distance d_1 from the water. He sees a swimmer in distress a distance L to his left and distance d_2 from the shore. Given that his speed on land and water are v_1 and v_2 respectively, with $v_1 > v_2$, what trajectory will get him to the swimmer in the least time? Does he rush towards the victim in a straight line joining them, does he first run on land until he is in front of the victim and then swim, does he head for the water first and then swim over, or does he do something else? Pick some trajectory composed of two straight line segments in each medium (why) and show that for the least time $\frac{\sin \theta_1}{\sin \theta_2} = \frac{v_1}{v_2}$ where the angles θ_i are the angles of the segments with respect to the normal to the shoreline.*

This problem has an analog in optics. If light is emitted at a point in a medium where its velocity is v_1 and arrives at a point in an adjacent medium where its velocity is v_2, the route it takes is arrived at in the same fashion since light takes the path of least time. The above equation is called Snell's Law.

Problem 1.6.11. *The volume of a sphere is $V(R) = \frac{4\pi R^3}{3}$. What is the rate of change of the volume with respect to R? Does it make sense?*

Problem 1.6.12. (Implicit Differentiation). *You know how to find the derivative dy/dx when $y(x)$ is given. Suppose instead I tell you that y and x are related by an equation, say $x^2 + y^2 = R^2$ and ask you to find the derivative at each point. There are two ways. The first is to solve for y as a function of x and then let your spinal column take over, i.e., by changing x infinitesimally and computing the corresponding change in y given by the functional relation. The second is to imagine changing x and y infinitesimally while preserving the constraining relation (a circle in our example). The latter condition allows us to relate the infinitesimals Δx and Δy and allows us to compute their ratio in the usual limit. Show that the derivative computed this way agrees with the first method.*

Find the slope at the point $(2,3)$ on the ellipse $3x^2 + 4y^2 = 48$ using implicit differentiation.

Problem 1.6.13. *Find the stationary points of $f(x) = x^3 - 3x + 2$ and classify them as maxima or minima.*

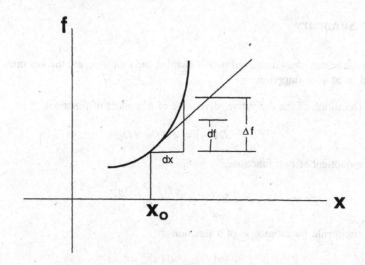

Figure 1.8. The meaning of the differentials df and dx.

1.7. Differentials

Consider a function $f(x)$ shown in Fig. 1.8.

If we change x by Δx at the point x_0, we write the change in f as

$$\Delta f = \left.\frac{df}{dx}\right|_{x_0} \Delta x + \dots \tag{1.7.1}$$

where the dots stand for terms of order $(\Delta x)^2$ and beyond. We expect the latter to be relatively insignificant as $\Delta x \to 0$. Let us now introduce the *differentials df* and dx such that

$$df = \left.\frac{df}{dx}\right|_{x_0} dx \tag{1.7.2}$$

with no approximation or requirement that either differential be small. What this means is that df is the change the function would suffer upon changing x by dx, *if we moved along the tangent to the function at the point x_0 as in Fig. 1.8.*

Note that we always have the option of taking dx vanishingly small, in which case df, which is the change in f to first order in dx, becomes a better and better approximation to Δf, the actual change in f (and not just along its tangent at x_0). This is always how we will use the differentials in this book, although the concept has many other uses. Thus when you run into an equation involving differentials you should say: "I see he is working to first order in the change dx." The advantage of using df will then be that I don't have to keep saying to "to first order" or use the string of dots.

1.8.   Summary

Of the numerous ideas discussed in this chapter, the following are the key ones and should be at your fingertips.

- Definition of the derivative, derivative of a product of functions

$$D(fg) = gDf + fDg$$

  a quotient of two functions

$$D(\frac{f}{g}) = \frac{[gDf - fDg]}{g^2}$$

  chain rule for a function of a function

$$\frac{df(u(x))}{dx} = \frac{df(u)}{du} \cdot \frac{du(x)}{dx}.$$

- The notion of the Taylor series

$$f(x) = f(0) + xf^{(1)}(0) + \frac{x^2}{2}f^{(2)}(0) + \dots$$

  about the origin or about the point a

$$f(a + x) = f(a) + xf^{(1)}(a) + \frac{x^2}{2}f^{(2)}(a) + \dots$$

- The following series to the order shown

$$e^x = 1 + x + \frac{x^2}{2!} + \frac{x^3}{3!} \dots$$

$$e^x = \lim_{N \to \infty} \left[1 + \frac{x}{N}\right]^N.$$

$$\cos x = 1 - \frac{x^2}{2!} + \dots$$

$$\sin x = x - \frac{x^3}{3!} + \dots$$

$$(1 + x)^p = 1 + px + \frac{(p)(p-1)}{2}x^2 + \dots$$

- Definition of the hyperbolic function, in particular

$$\cosh x = \frac{e^x + e^{-x}}{2}$$

$$\sinh x = \frac{e^x - e^{-x}}{2}$$

their symmetry under $x \to -x$ and functional identities, especially

$$\cosh^2 x - \sinh^2 x = 1.$$

If you need a formula for $\cosh 2x$ or any other identity, you can get it from the definition of the hyperbolic functions in terms of exponentials. The power series for these functions can also be obtained from that of the exponential function.

- The $\ln x$ function, the identity

$$x = e^{\ln x}$$

and the series

$$\ln(1 + x) = x - \frac{x^2}{2} + \frac{x^3}{3} + \ldots$$

and its derivative

$$D \ln x = \frac{1}{x}.$$

$$x^a = e^{a \ln x}.$$

- Trigonometric functions, identities and derivatives, radian measure for angles. These will not be listed here since you must have already learned them by heart as a child.

- The definition of the differential, that:

$$df = f' dx$$

is an exact relation which defines df in terms of the derivative at the point x, and that as $dx \to 0$, $df \to \Delta f$, the actual change in f.

INTEGRAL CALCULUS

2.1. Basics of Integration

We have so far focused on the question of finding the derivative of a *given* function $f(x)$. We are familiar with the constructive procedure for finding this derivative: change x by some amount Δx, calculate the change in the function, take the ratio of the two changes, and so on.

Often one faces the reverse question: *find the function given its derivative.* For example in the case of a falling rock, one is given the acceleration a due to gravity, and one wants the velocity $v(t)$. Once the velocity is found, one wants to do this again, to obtain the position $x(t)$. Let us consider this problem in general, using the symbol $F(x)$ to denote the function which is to be determined, given that its derivative is $f(x)$. (As far as possible, we will use upper and lower case symbols to denote a function and its derivative. Occasionally we will have to bow to tradition as in the case of velocity and position.)

Let us now analyze this reverse problem, called the problem of *integration*. We are to find $F(x)$ given its rate of change $f(x)$. This is like saying: I have been putting $50 a month in the bank for the last 5 years, how much money is in my account? The answer cannot be given, unless I reveal how much I started with. Similarly, if I tell you a car has been moving at a velocity of 50 mph for six hours and ask you where it is now, you will want to know where it was at the beginning of the six hour trip. The initial bank balance and initial car position are examples of *initial conditions*. Thus in the general problem too, we must specify not only a rate of change, but also $F(x_0)$, the value of $F(x)$ at some $x = x_0$. Given this, we can proceed as follows to find $F(x)$. Take first the simple case when the function has a steady rate of change, i.e., if $f(x) = f$ is a constant. Clearly

$$F(x) = F(x_0) + (x - x_0)f. \tag{2.1.1}$$

Now for the general case where $f(x)$ varies from point to point, depicted in Fig. 2.1.

To deal with the varying $f(x)$, we divide the interval $x - x_0$ into N tiny parts of width $\Delta x = (x - x_0)/N$. Within any one of these tiny intervals numbered i and

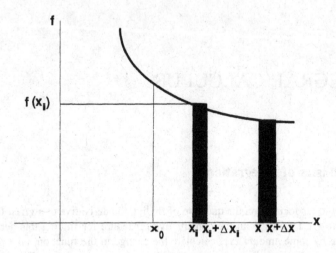

Figure 2.1. Graphical integration of $f(x)$. The definite integral is given by the limit in which the number of intervals goes to infinity, and equals the area under the function between the lower and upper limits x_0 and x.

starting at $x = x_i$, we see that $x \simeq x_i$, and the function f is essentially constant, i.e., $f \simeq f(x_i)$, and the change in F is approximately $\Delta F \simeq f(x_i)\Delta x$. Adding all these changes we get

$$F(x) \simeq F(x_0) + \sum_{i=1}^{N} f(x_i)\Delta x, \qquad (2.1.2)$$

where we use the $\simeq$ symbol to reflect the fact that we are making an approximation within each segment by treating f as a constant. However if we let $N \to \infty$ these errors vanish and we obtain

$$F(x) = F(x_0) + \int_{x_0}^{x} f(x)dx, \qquad (2.1.3)$$

where by definition

$$\int_{x_0}^{x} f(x)dx = \lim_{N \to \infty} \sum_{i=1}^{N} f(x_i)\Delta x. \qquad (2.1.4)$$

One refers to $\int_{x_0}^{x} f(x)dx$ as the *definite integral* of the *integrand* f with respect to the *integration variable* x, between the *lower limit* x_0 and *upper limit* x. If you look at Fig. 2.1 you will see that each term $f(x_i)\Delta x$ in the sum on the right-hand side of the above equation stands for the area of a rectangle of base Δx and height

$f(x_i)$, shown by the shaded region on the left. At any large but finite N, these areas approximate, but do not exactly equal, the area under the function between x_0 and x. For example in the one highlighted segment named i, the height of the function at the left end of the segment, $f(x_i)$, is greater than in the rest of the segment and the area of the rectangle is larger than the area under the curve. The opposite could easily have happened had the function been on the rise. *However in the limit $N \to \infty$, the sum becomes the definite integral as well as the area under the graph between x_0 and x.*[1]

Digression on the limit $N \to \infty$

The limit $N \to \infty$, in which the number of intervals goes to infinity, must be analyzed carefully, but I will not do so here, referring you to books on calculus. Instead a few points will be raised and briefly treated. First let us understand how it is that as we let $N \to \infty$, we get a finite limit. Say we send $N \to \infty$ by doubling it sequentially. Each time we double N, by subdividing each interval into two, we double the number of terms in the sum, but the contribution from each is roughly halved since Δx gets halved, while $f(x_i)$ is essentially same in each of the two segments that evolved from one parent. Thus the decrease in the segment width (going down as $1/N$) is precisely offset by the increase in the number of segments (growing as N) and a finite limit emerges. Next, how about the fact that in each segment we treated f as a constant, when in fact it varies? In other words, was it all right to evaluate f at the left end of each segment, which led to an error in the area (overestimate, in the case of our particular function)? While it is true that as we increase N, the segments get narrower, and the error committed in each segment from treating the function as a constant indeed decreases; it is also true that the number of segments is growing without limit and the sum of these errors could end up being finite. After all, as we reduce the segment size, the contribution from each segment also goes to zero, but we do not drop these, since there are enough of them to build up the area under the graph. However the error due to the variation of f within each segment is not of this type, and does indeed become negligible for the following (simplified) reason. Consider two sums, both approximations to the area, one where we take f at the left end of each segment (as we did) and one where we take f at the right end of each segment, where $x = x_i + \Delta x$. Even though f is not a constant, we assume it is smooth enough for us to say that (for small Δx)

$$f(x_i + \Delta x) \simeq f(x_i) + \frac{df}{dx}\Delta x. \qquad (2.1.5)$$

The *difference* between the two sums is given by $\frac{df}{dx}(\Delta x)^2$ *per segment*. The sum of these differences over the N segments will vanish as $N \to \infty$ (provided the function is not pathological), *because there are N terms in the sum, and each term is of order* $(\Delta x)^2 \propto 1/N^2$. Equivalently, look at the shaded rectangle in Fig. 2.1 at the point x_i and note that whereas the area of the rectangle goes as $f(x_i)\Delta x$, the excess at the top is shaped like an inverted right triangle of base Δx and height $|\Delta f = \frac{df}{dx}\Delta x|$. Thus although the contribution of the segment and the error are both infinitesimal, the latter is an infinitesimal of second order and adds up to nothing even when summed over N segments as $N \to \infty$.[2]

Although we considered two specific schemes which evaluated f at the end points of each segment, similar arguments easily show that in general the limit of the sum is the same no matter where in each segment we evaluate the function.

---

[1] We have assumed f is everywhere positive. If $f < 0$, the contribution to the area will be negative. Thus if $f(x) = \sin x$, and we measure the area from $x_0 = 0$, it will rise until $x = \pi$ (reaching a value of 2) and then start dropping as $\sin x$ turns negative. At the end of a full period ($x = 2\pi$) it will return to zero.

[2] For yet another version of the argument, notice that since the right edge of each sector is the same as the left edge of the next one, the two schemes involve the same sums except at the end points, whose contribution is therefore of order Δx.

So the problem of finding the function given its derivative reduces to finding the area under its derivative between x_0 and x (which gives the change in F) and adding on the initial value $F(x_0)$.

How is the area to be evaluated? There two ways.

The *numerical or graphical way* is to simply plot f on some graph paper and count the squares. (The size of the squares will be determined by the accuracy sought). Thus for example, the distance covered by a car traveling at some variable velocity $v(t)$ between times t_0 and t is given by plotting the velocity and measuring the area under it between the times t_0 and t. If we add to this the initial position, we get the position at time t. The graphical method is a constructive method, where we can follow an algorithm. It is essentially what computers often use to *numerically* evaluate any integral. Of course they cannot send $N \to \infty$ but can make it large enough to reach a point where further increase in N makes changes that are less then some preassigned measure of accuracy. Defined this way, the graphical method works for any function you can draw on graph paper. However, it does not promise exact answers, just answers good to some preassigned accuracy.

In the *analytical* way, one focuses on the *functional form* of the answer. For example in the case of a rock falling under a constant acceleration a, the increase in velocity over a time t, $v(t) - v(0)$, is given by the area of a rectangle of height a and width t. While this area can be numerically determined by counting squares in the graphical scheme, it can also be given analytically by the expression $v(t) - v(0) = at$. Or consider a particle moving at a velocity $v(t) = at$. The distance it travels between $t = 0$ and $t = t$ is the area under the straight line $v(t) = at$ between these limits. Again we don't need to count squares for this, we know the area of this triangle of base t and height at is $at^2/2$, so that $x(t) - x(0) = at^2/2$. *In other words, the area under the curve is itself a function of the upper limit and we are writing down its functional form.* The answers are exact and also solve the problem once and for all, i.e., for all t. In other words, for each given t, we do not draw a new triangle, we simply evaluate $at^2/2$ for the given value of t. This is what we would like to do in general. But the strategy is not to appeal to areas of well- known figures (like rectangles and triangles) since we will quickly run into not-so-well-known shapes, but the following.

Let us consider

$$\int_{x_0}^{x} f(x)dx \qquad (2.1.6)$$

as a function of the upper limit x, holding the lower limit fixed. We know that when $x = x_0$ this function vanishes. Next we ask how the function grows with increasing x. To this end we change the upper limit by Δx. We see from Fig 2.1 (shaded rectangle at the right) that the integral changes by $f(x)\Delta x$ (plus higher order terms.) Thus the rate of change of the integral is simply $f(x)$. This is not surprising since that is how we cooked up the integral in the first place! (We are simply running our analysis backwards.)

Let us suppose we can somehow come up with a function $\bar{F}(x)$, called an *integral of* f.[3] which is *any* function whose derivative is also $f(x)$. We cannot yet say $\bar{F}(x)$ equals the area integral just because the two have the same derivative: they could differ by a constant. So let us write

$$\int_{x_0}^{x} f(x)dx = \bar{F}(x) + c \tag{2.1.7}$$

where the constant c is to be determined. We do so by requiring that when $x = x_0$, both sides vanish:

$$\int_{x_0}^{x_0} f(x)dx = 0 = \bar{F}(x_0) + c \tag{2.1.8}$$

$$c = -\bar{F}(x_0) \tag{2.1.9}$$

$$\int_{x_0}^{x} f(x)dx = \bar{F}(x) - \bar{F}(x_0). \tag{2.1.10}$$

We have thus manufactured a function which grows at the rate $f(x)$ and vanishes at x_0. It must therefore equal the definite integral from x_0 to x.

Let us consider as an example the case $f(x) = x^2$, $x_0 = 1$, $x = 3$, i.e., we are looking for the area under the graph $f(x) = x^2$ between the points 1 and 3. It is clear that $\bar{F}(x) = x^3/3$ is an integral since $D\bar{F}(x) = f(x) = x^2$. So the area is given by

$$\int_{1}^{3} f(x)dx = \frac{x^3}{3}\Big|_3 - \frac{x^3}{3}\Big|_1 = \frac{3^3}{3} - \frac{1^3}{3}. \tag{2.1.11}$$

where $f(x)|_a$ means $f(x)$ evaluated at $x = a$.

There is a problem now since there is more than one choice for the integral of f. What if we had used $\bar{\bar{F}}(x) = x^3/3 + 17$ which also has the same derivative? Nothing changes:

$$\int_{1}^{3} f(x)dx = \left(\frac{x^3}{3} + 17\right)\Big|_3 - \left(\frac{x^3}{3} + 17\right)\Big|_1$$

$$= \left(\frac{3^3}{3} + 17\right) - \left(\frac{1^3}{3} + 17\right) = \frac{3^3}{3} - \frac{1^3}{3}.$$

In other words, the constant that distinguishes one integral from another drops out in the formula for the definite integral, which represents the definite area enclosed between the prescribed limits.

Thus:

To find a definite integral, which denotes the area under a function $f(x)$ between two end points, we take any integral $\bar{F}(x)$, or $\bar{\bar{F}}$, etc., and find its difference between these points.

---

[3] Sometimes this is called the *primitive* or *antiderivative* of f.

Often the end points are denoted by x_1 and x_2 in which case

$$\int_{x_1}^{x_2} f(x)dx = \bar{F}(x_2) - \bar{F}(x_1) \equiv \bar{F}(x)|_{x_1}^{x_2} = \bar{\bar{F}}(x)|_{x_1}^{x_2} \cdots \qquad (2.1.13)$$

Consider the area under f between x_0 and x. This too is one member of the family of integrals of f with the special property that it vanishes when $x = x_0$. It can be related to any other member of the family $\bar{F}(x)$, which does not vanish at $x = x_0$, through the difference $\bar{F}(x) - \bar{F}(x_0)$. This result can be understood algebraically: the above difference has the twin properties of having the right x-derivative $(f(x))$ as well as vanishing at $x = x_0$. While the two terms in the difference will vary as we change from one member of the family to another, the difference will not.

It is sometimes, but not always, possible to associate an area with every integral. Thus if $f(x) = x^2$, then $\bar{F}(x) = x^3/3$ stands for the area from the origin to the point x. We know the area is counted from $x_0 = 0$ since $\bar{F}(x = 0) = 0$. On the other hand $\bar{\bar{F}}(x) = x^3/3 - 1/3$ stands for the area reckoned from the lower limit $x_0 = 1$ since it vanishes when $x = 1$. Consider however $f(x) = \sin x$. The function $\bar{F}(x) = -\cos x + 97$ certainly obeys $D\bar{F}(x) = \sin x$, but it cannot stand for the area under the sine function from any point x_0 to x since the area under this oscillating function is bounded by 2. In those cases where the area interpretation is possible, we can understand Eqn. (2.1.13) as follows. The function $\bar{F}(x)$ equals the area from some point to the point x, while $\bar{F}(x_0)$ equals the area from the *same* point to the point x_0. (Imagine the case where the initial point is to the left of x_0 which is to the left of x.) Thus their difference is the area between x_0 and x. As we go from $\bar{F}$ to $\bar{\bar{F}}$ (assuming the latter also admits an area interpretation) we merely change the lower limit from which the area is measured.

Let us summarize where we stand in our quest for the function $F(x)$ with a known derivative $f(x)$ and initial value $F(x_0)$ at x_0. We have found

$$F(x) - F(x_0) = \int_{x_0}^{x} f(x)dx \qquad (2.1.14)$$

$$= \bar{F}(x)|_{x_0}^{x} \qquad (2.1.15)$$

$$= \bar{\bar{F}}(x)|_{x_0}^{x} \cdots \qquad (2.1.16)$$

At the left we have $F(x) - F(x_0)$, the change in the function we were interested in, over the interval x_0 to x, with a rate of change $f(x)$.

This is then expressed as the definite integral over the same interval. This was clearly progress, since we have identified the definite integral with the area under f between these limits and we have the choice of computing it numerically.

The subsequent equations relate the definite integral to the difference of *any* integral of $f(x)$, $\bar{F}(x)$, $\bar{\bar{F}}$ etc., between these limits. This seems to be a wonderful result since it gives exact results in analytical form. But the catch is of course that the only way to find these integrals is to guess the answer given the derivative is $f(x)$. But that is the problem we started with! All we seem to have learned is that our original function F is a member of the family of functions with f as their derivative and that the change in F is the change in any other member, since the constant that differentiates them drops out.

All this is correct. The analytical scheme does indeed tell us to guess the solution. How can we even call it a scheme? Suppose I told you that in addition to the high school formula $(-(b/2a) \pm \sqrt{b^2 - 4ac}/2a)$ I have found another way to find the roots of a quadratic equation $ax^2 + bx + c = 0$: guess the answer and see if it satisfies the equation. You will laugh. But guessing the integral of a function is not in the same league, as we will see.

First, the guessing is not totally wild. In the example considered, $f(x) = x^2$, it went as follows. We knew that each time we took a derivative of a power of x we would pull down the exponent and reduce the power by one. So to end up with x^2, we had to start with $\bar{F} = Ax^3$, where A is a constant to be fixed. Upon taking the derivative, we get $3Ax^2$. Setting this equal to x^2, we fix $A = 1/3$ and obtain $\bar{F}(x) = x^3/3$. We can similarly guess that

$$f(x) = x^n \leftrightarrow \bar{F}(x) = \frac{x^{n+1}}{n+1} + c, \qquad (2.1.17)$$

where c is any constant. Indeed, every time we take a derivative, we also learn about an integral by running the calculation backwards. Thus for example, knowing that $D \sin x = \cos x$, we can conclude that

$$f(x) = \cos x \leftrightarrow \bar{F}(x) = \sin x + c. \qquad (2.1.18)$$

People have been taking derivatives and keeping records for a long time. Now you can benefit from all this accumulated wisdom, by buying yourself what are called Tables of Integrals, which contain an assortment of such results. Since the guessing game has been going on for centuries, these Tables are quite fat and most sensible people have a copy at home and one at work. The two integrals given above will however be listed as follows in the Tables:

$$\int x^n dx = \frac{x^{n+1}}{n+1} + c \qquad (2.1.19)$$

$$\int \cos x \, dx = \sin x + c \qquad (2.1.20)$$

Notice several things. First no limits are shown. The upper limit is understood to be x and the lower limit is left unspecified, as is the constant c. Often one does not bother to write the c on the right-hand side. One refers to $\frac{x^{n+1}}{n+1} + c$, with c unspecified, as an *indefinite integral* of x^n. Clearly we can't associate with it a numerical value or a geometrical area. What counts here is the functional form or x-dependence which ensures (independent of c) that the derivative is $f(x)$. If however we want an enclosed area, it is easy to get it by evaluating the difference of the indefinite integral between the end points (wherein the unknown constant drops out.)

One does not also use different symbols $\bar{F}$, $\bar{\bar{F}}$ for different integrals of f. In speaking of any one member one simply refers to it as $F(x)$. Thus one writes

$$\int_{x_1}^{x_2} f(x)dx = F(x)|_{x_1}^{x_2}, \qquad (2.1.21)$$

where F is any integral of f.

To make sure you have understood all the ideas, let us work out a simple example. A function $F(x)$ is changing at a rate $f(x) = x^2$. It has the value 12 at $x = 1$. What is its value at $x = 3$? What is its value at some general x? As for the first question, we write

$$F(3) = F(1) + \int_1^3 x^2 dx = 12 + \frac{x^3}{3}\bigg|_1^3 = 12 + [3^3/3 - 1^3/3] = 12 + \frac{26}{3}, \quad (2.1.22)$$

where we have used the integral $x^3/3$ to evaluate the definite integral. Likewise

$$F(x) = 12 + \int_1^x x^2 dx = 12 + \frac{x^3}{3}\bigg|_1^x = 12 + [x^3/3 - 1^3/3] = \frac{x^3}{3} + \frac{35}{3}, \quad (2.1.23)$$

Whereas the integral we used in evaluating the definite integral in both Eqns. (2.1.22-2.1.23) was $x^3/3$, the answer to the second problem we posed is the integral $x^3/3 + 35/3$. They all belong to the same family of integrals of x^2. The former may be interpreted as the area from $x_0 = 0$, the latter from $x_0 = \sqrt[3]{-35} = -\sqrt[3]{35}$.

We have thus far been interested in the area integral as a means of solving a problem of finding a function whose rate of change is given. But the area can be interesting in its own right. Say we want to know the area of an ellipse of semi-major axis a and semi-minor axis b. We could draw it on a piece of graph paper and count squares. This will give a numerical answer, good to any chosen accuracy. Or we could do the following analytic calculation which will give the exact answer in terms of the parameters a and b. Let the ellipse be described by

$$\frac{x^2}{a^2} + \frac{y^2}{b^2} = 1. \qquad (2.1.24)$$

By drawing a sketch, and solving for y as a function of x, one can see that the area in the first quadrant is given by

$$\int_0^a \frac{b}{a}\sqrt{a^2 - x^2}dx, \qquad (2.1.25)$$

In the analytical method one first guesses that in this problem

$$F(x) = \frac{b}{2a}\left[x\sqrt{a^2 - x^2} + a^2 \sin^{-1}(\frac{x}{a})\right] \qquad (2.1.26)$$

is an integral of the given integrand, as you *must* verify by differentiation. (If you are new to this game, you may be amazed how one could make this guess. As you learn more tricks of the trade, you will see how this is done.) In any event, given this fact, the area of the ellipse is readily found (upon remembering to multiply by 4 for the four quadrants):

$$\text{Area} = 4 \cdot \frac{b}{2a}\left[x\sqrt{a^2 - x^2} + a^2 \sin^{-1}(\frac{x}{a})\right]_0^a = \pi ab. \qquad (2.1.27)$$

No matter how many tiny squares you count in the numerical scheme, you are not going to get an analytical, exact formula for the area, in terms of π, a, and b.

It was mentioned earlier that guessing the integrals of functions has been going on for centuries and there exist Tables of Integrals. People's attitude to the use of Tables varies. At one end we have some one like Enrico Fermi who used to challenge his colleagues that before they could look up an integral in the library, he would evaluate it himself. At the other end there is someone like myself who challenges his colleagues with the exact reverse. In any event, it is socially acceptable to look up integrals in the Tables, and in fact this can take considerable skill. This is because not every known integral will be listed there as such; one will have to cast the problem at hand into one of many standard forms using various properties true for all integrals. We begin with an example, and go on to list more such properties. Say you want the integrals of x^2 and x^3. Either you can guess them yourself or look them up under $\int x^n dx$. But you will never find the integral of $3x^2 + 5x^3$ in the Tables, (nor should you buy Tables which list it) because you can relate the latter integral to the former by using the following result.

Linearity property
If the integral of f and g are known (to be F and G), then

$$\int^x (af + bg)dx = aF(x) + bG(x) \qquad (2.1.28)$$

where a and b are constants and where the constant on the right-hand side and lower limit on the left-hand side are suppressed. The correctness of these results follows either by differentiation (which tells us that the derivative of the right-hand side must equal the integrand) or the geometrical definition in terms of areas.

Here are a few more properties of integrals, with more to follow later.

Composition rule

$$\int_a^b f(x)dx = \int_a^c f(x)dx + \int_c^b f(x)dx. \qquad (2.1.29)$$

The truth of this is obvious in the geometrical construction if $a \le c \le b$. We will demand that it be true for other orderings as well. It then follows that

$$\int_a^b f(x)dx + \int_b^a f(x)dx \;=\; \int_a^a f(x)dx = 0 \text{ so that} \qquad (2.1.30)$$

$$\int_a^b f(x)dx \;=\; -\int_b^a f(x)dx. \qquad (2.1.31)$$

In other words the integral changes sign under the exchange of the limits.

Integration by parts
Next we consider the integral of a product of functions. Let us begin with

$$D(FG) = Fg + Gf \qquad (2.1.32)$$

and integrate both sides with respect to x between the limits x_1 and x_2 to obtain:

$$FG\big|_{x_1}^{x_2} = \int_{x_1}^{x_2} Fg\,dx + \int_{x_1}^{x_2} Gf\,dx. \qquad (2.1.33)$$

This is usually written by rearranging terms as follows:

$$\int_{x_1}^{x_2} Fg\,dx = FG\big|_{x_1}^{x_2} - \int_{x_1}^{x_2} Gf\,dx \qquad (2.1.34)$$

and is referred to as *integration by parts*. The GF term, coming from the end points is called a *surface term*. There is a subtle point connected with Eqn. (2.1.34). The left hand side has a definite numerical value for any given F and g. The right-hand side, however, depends on G, which is not uniquely determined by the given g. You should do the following exercise to resolve this point.

Problem 2.1.1. *Show that if we change G to $G + c$, the right-hand side does not change, due to a cancellation between changes in the two terms.*

Here is an example to show how all this works. Say you know how to integrate x and $\cos x$ and want the integral of $x \cos x$. So we choose $F = x$ and $g = \cos x$. We find

$$\int_{x_1}^{x_2} x \cos x\,dx = x \sin x\big|_{x_1}^{x_2} - \int_{x_1}^{x_2} \sin x \cdot 1\,dx = [x \sin x + \cos x]_{x_1}^{x_2} \qquad (2.1.35)$$

Notice that had we chosen the integral of $\cos x$ to be $\sin x + 17$, nothing would have changed; two extra terms $\pm 17(x_2 - x_1)$ would have appeared and canceled in the middle expression above. Note also that had we chosen $F = \cos x$ and

$g = x$, we would have obtained another integral equal to the original one, but not any easier to evaluate, since it would have had $x^2 \sin x$ in the integrand.

Dummy variables
Consider the sum

$$S = c_0 + c_1 + c_2 \ldots c_{10} \qquad (2.1.36)$$

where the c's are a given set of numbers. The sum can be written in many equivalent ways, of which two are shown below:

$$S = \sum_{n=0}^{10} c_n \equiv \sum_{m=0}^{10} c_m \qquad (2.1.37)$$

The indices n and m are dummies since nothing depends on the specific name we use for the index: in both cases the indices m or n run over the same eleven values and the sum is over the same terms c_0 through c_{10}. Likewise the symbol x in the integral on the left-hand side of Eqn. (2.1.13) is called a *dummy variable* in the sense that nothing changes if we replace it by another symbol, say x'. In all cases the integral stands for the area under the same function between the same limits. Thus for example

$$\int_3^{19} f(x)dx = \int_3^{19} f(x')dx' = \int_3^{19} f(\phi)d\phi \qquad (2.1.38)$$

and so on.

After this set of rules, we turn to a few tricks of the trade.

Problem 2.1.2. *Find the integral of* $\ln x$ *using integration by parts by rewriting* $\ln x = 1 \cdot \ln x$. *Make the right choice for which factor is to be* F *and which is to be* g.

Problem 2.1.3. *Consider the function*

$$F(n) = \int_0^\infty x^n e^{-x} dx \qquad (2.1.39)$$

where n *is a non-negative integer. Show using integration by parts that* $F(n) = nF(n-1)$ *and that* $F(n) = n!$. *The* **gamma function** *is defined by* $\Gamma(n) = F(n-1)$. *What is* $0!$ *as defined by this integral?*

Problem 2.1.4. *Consider* $\int_0^\pi \cos^2 x dx$. *Give arguments for why this must equal* $\int_0^\pi \sin^2 x dx$. *By adding the two integrals and averaging, and using a well-known trigonometric identity, show that*

$$\int_0^\pi \cos^2 x dx = \int_0^\pi \sin^2 x dx = \pi/2. \qquad (2.1.40)$$

This is a very useful result.

2.2. Some Tricks of the Trade

The tricks one uses to evaluate integrals are so numerous that we cannot hope to cover them all here. There are however two fundamental ploys that are frequently employed. These are

- Substitution or change of variable.

- Differentiating with respect to a parameter.

As an example of the former consider

$$F(x_1, x_2) = \int_{x_1}^{x_2} f \, dx \qquad (2.2.1)$$

$$= \int_{x_1}^{x_2} \frac{x^2 \, dx}{(x^3 + 4)^2}. \qquad (2.2.2)$$

Let us say we only know to integrate powers of x. The integrand f here does not have that form. We will now bring it to that form by a change of variable. Let

$$u = x^3. \qquad (2.2.3)$$

Our original goal was to plot f as a function of x, chop the region between x_1 and x_2 into segments of width dx, evaluate the products $f(x)dx$ over the segments and add them all up in the appropriate limit. Our hope is that the job will be simpler in terms of u. Suppose we express f in terms of u and plot $f(u)$. Along the u-axis the points $u_1 = x_1^3$ and $u_2 = x_2^3$ define the limits. *Every value that f took as a function of x it will now take as a function of u at the corresponding point $u = x^3$. We do not however want the area under $f(u)$ between u_1 and u_2., i.e., we do not want $\int f(u)du$. We want $\sum_i f(u_i)dx$, which is the old area we began with.* Thus given two nearby points separated by du, we want to form the product $f(u)dx$, where dx is the corresponding separation along the x-axis. It is clear that since x is a function of u

$$dx = \frac{dx}{du} du. \qquad (2.2.4)$$

Thus the integrand for the u-integration is $f(u)dx/du$, where the factor dx/du is to be expressed in terms of u. Going back to our problem

$$u = x^3 \qquad (2.2.5)$$

$$\frac{du}{dx} = 3x^2 \qquad (2.2.6)$$

$$\frac{dx}{du} = \frac{1}{3x^2} = \frac{1}{3u^{2/3}} \qquad (2.2.7)$$

$$f(x)dx \longrightarrow f(x(u)) \cdot \frac{du}{3u^{2/3}} = \frac{du}{3(u + 4)^2} \qquad (2.2.8)$$

$$\int_{x_1}^{x_2} f(x)dx = \int_{u_1=x_1^3}^{v_2=x_2^3} \frac{du}{3(u+4)^2}. \qquad (2.2.9)$$

Switching once more to $v = u + 4$ we finally obtain

$$\int_{x_1}^{x_2} \frac{x^2 dx}{(x^3+4)^2} = \int_{v_1=x_1^3+4}^{v_2=x_2^3+4} \frac{dv}{3v^2}. \qquad (2.2.10)$$

It is now trivial to integrate $1/v^2$ between the given limits.

Let us recall the main point in the above manipulations. Given a difficult integrand, we hope that by going to a new variable, we can end up with the simpler integrand. The new integrand is however not just the old one expressed in terms of the new variable, but that times the *Jacobian*

$$J\left(\frac{x}{u}\right) = \frac{dx}{du}. \qquad (2.2.11)$$

Thus

$$\int_{x_1}^{x_2} f(x)dx = \int_{u(x_1)}^{u(x_2)} f(x(u))J\left(\frac{x}{u}\right) du \qquad (2.2.12)$$

with the limits expressed in terms of the corresponding variables.

Our mission is accomplished only if the new integrand is simple. In our example, the denominator of the original integrand did simplify upon going to $v = x^3 + 4$, but the numerator did not. But fortunately it got canceled by the Jacobian. Conversely, the substitution of v would not have been very effective without the x^2 in the numerator.

Problem 2.2.1. *Evaluate $\int_{x_1}^{x_2} \frac{dx}{\sqrt{a^2-x^2}}$ by switching to θ defined by $x = a\sin\theta$. Assume $0 < x < \pi/2$.*

Problem 2.2.2. *Show $\int_0^\infty \frac{dx}{x^2+a^2} = \frac{\pi}{2a}$ by switching to θ defined by $x = a\tan\theta$. Make sure the change of variables is sensible, namely, that every x in the range of integration be reached by some choice of θ.*

Problem 2.2.3. *Evaluate $\int_0^1 e^{\sqrt{x}}dx$. Show that $\int_0^\infty e^{-x^4}dx = \Gamma(\frac{5}{4})$.*

Problem 2.2.4. *To get some familiarity with what kind of manipulations are legal, let us compute $C(R)$, the circumference of a circle $x^2+y^2 = R^2$. We will find that part that lies in the first quadrant and multiply by 4. Consider two points on the circle separated by $(\Delta x, \Delta y)$. The arc length between them may be approximated by the pythagorean distance*

$$ds = \sqrt{(\Delta x)^2 + (\Delta y)^2} = \Delta x\sqrt{1 + \left(\frac{\Delta y}{\Delta x}\right)^2}.$$

The Δ's are however not independent: If (x, y) satisfies the equation for the circle, so must the displaced point. Show that this implies $\frac{\Delta y}{\Delta x} = -\frac{x}{y}$ to leading order in the infinitesimals. Consequently $C(R) = \int ds = 4 \int_0^R \sqrt{1 + (\frac{x}{y})^2} dx$. Eliminate y in favor of x in the integrand and evaluate it by a trigonometric change of variables. What would be the formula for the arc length of a general curve $y - f(x) = 0$?

Problem 2.2.5. *Evaluate $\int_0^{\pi/2} \frac{\cos x\, dx}{(2+\sin x)}$.*

Problem 2.2.6. *Find the area enclosed between the curve $y = x^2$ and the lines $y = |x|$*

Problem 2.2.7. *Evaluate $\int_0^1 2^x dx$.*

Let us now consider the second trick of *differentiating with respect to a parameter.* Consider

$$I_0(a) = \int_0^\infty e^{-ax^2} dx \qquad (2.2.13)$$

This integral cannot be evaluated by any of the standard means. On the other hand if the integral in question had been

$$I_1(a) = \int_0^\infty e^{-ax^2} x\, dx \qquad (2.2.14)$$

we could have changed to $u = x^2$ and evaluated the integral.

Problem 2.2.8. *Show that $I_1(a) = \frac{1}{2a}$.*

Let us see how far we can get with $I_0(a)$. The notation itself tells us that the integral depends on just the *parameter* a. Its dependence on a can be found by *scaling.* In terms of $u = \sqrt{a}x$,

$$I_0(a) = \frac{1}{\sqrt{a}} \int_0^\infty e^{-u^2} du = \frac{1}{\sqrt{a}}c \qquad (2.2.15)$$

where c is a constant independent of a. In the next chapter we will learn how to do this very important integral and find out that $c = \frac{\sqrt{\pi}}{2}$, so that

$$I_0(a) = \frac{1}{2}\sqrt{\frac{\pi}{a}}. \qquad (2.2.16)$$

Now it turns out that given $I_0(a)$, we can evaluate a whole family of related integrals by differentiating both sides of Eqn. (2.2.13) with respect to a. Doing this once we find

$$\frac{dI_0(a)}{da} = \int_0^\infty \frac{\partial e^{-ax^2}}{\partial a} dx \qquad (2.2.17)$$

$$-\frac{1}{4a}\sqrt{\frac{\pi}{a}} = \int_0^\infty (-x^2)e^{-ax^2} dx \text{ so that} \qquad (2.2.18)$$

$$I_2(a) \equiv \int_0^\infty (x^2)e^{-ax^2} dx = \frac{1}{4a}\sqrt{\frac{\pi}{a}}. \qquad (2.2.19)$$

This trick can be used to evaluate

$$I_n(a) = \int_0^\infty (x^n)e^{-ax^2} dx \qquad (2.2.20)$$

for n even. If n is odd, we can switch to x^2 as the integration variable. The details are left to the following exercise.

Problem 2.2.9. *Evaluate* $I_3(a)$, $I_4(a)$.

This business of going inside an integral and differentiating with respect to a parameter like a seems to make some students a bit uneasy. Let us see why this is perfectly legal in a general case where

$$F(a) = \int f(x,a)dx \qquad (2.2.21)$$

where it is assumed the limits are independent of a. Consider the following sequence of operations

$$F(a + \Delta a) = \int f(x, a + \Delta a)dx \qquad (2.2.22)$$

$$F(a + \Delta a) - F(a) = \int f(x, a + \Delta a)dx - \int f(x, a)dx \qquad (2.2.23)$$

$$= \int (f(x, a + \Delta a) - f(x, a))dx \qquad (2.2.24)$$

Dividing both sides by Δa and taking the limit we get the desired result

$$\frac{dF(a)}{da} = \int \frac{\partial f(x, a)}{\partial a} dx. \qquad (2.2.25)$$

Sometimes we have to be more devious. Recall from Exercise (2.2.2.) that

$$\int_0^\infty \frac{dx}{x^2 + a^2} = \frac{\pi}{2a}. \qquad (2.2.26)$$

By taking the a derivative we can show that

$$\int_0^\infty \frac{dx}{(x^2 + a^2)^2} = \frac{\pi}{4a^3}.$$

(2.2.27)

Suppose we only knew that

$$\int_0^\infty \frac{dx}{x^2 + 1} = \frac{\pi}{2}$$

(2.2.28)

and wanted to evaluate the integral with $(x^2 + 1)^2$ in the denominator. We have no parameter to differentiate. *Then we must introduce one!* First we view the given integral as

$$\int_0^\infty \frac{dx}{x^2 + a^2}$$

evaluated at $a = 1$. Next we can change to $u = ax$ in Eqn. (2.2.28) and deduce Eqn. (2.2.26). Thereafter we proceed as before in taking parametric derivatives. At the end we set $a = 1$. The notion of embedding a zero parameter problem into a parametrized family is very powerful and has many uses.

Problem 2.2.10. *Consider*

$$I = \int_0^1 \frac{t - 1}{\ln t}.$$

Think of the t in $t - 1$ as the $a = 1$ limit of t^a. Let $I(a)$ be the corresponding integral. Take the a derivative of both sides (using $t^a = e^{a \ln t}$) and evaluate dI/da by evaluating the corresponding integral by inspection. Given dI/da obtain I by performing the indefinite integral *of both sides with respect to a. Determine the constant of integration using your knowledge of $I(0)$. Show that the original integral equals $\ln 2$.*

Problem 2.2.11. *Given*

$$\int_0^\infty e^{-ax} \sin kx\, dx = \frac{k}{a^2 + k^2},$$

evaluate $\int_0^\infty x e^{-ax} \sin kx\, dx$ and $\int_0^\infty x e^{-ax} \cos kx\, dx$.

2.3. Summary

Here are the key ideas to remember at all times:

- The equation

$$\int_{x_1}^{x_2} f(x)dx = F(x) - F(x_0)$$

  can be viewed in two ways. First, it gives the change in a function $F(x)$ between x_0 and x given that its rate of change is $f(x)$, in terms of the definite integral over f between the said limits, assuming the integral can be found somehow, say as the area under the graph. Conversely, if we know *any* integral F of f, this equation expresses the definite integral in terms of the latter. The members of the family of integrals differ by constants which drop out in the difference computed in the definite integral.

-

$$\int_{x_1}^{x_2} f(x)dx = -\int_{x_2}^{x_1} f(x)dx$$

$$\int_{x_1}^{x_2} F(x)g(x)dx = FG\big|_{x_1}^{x_2} - \int_{x_1}^{x_2} f(x)G(x)dx$$

$$\int_{x_1}^{x_2} f(x)dx = \int_{u(x_1)}^{u(x_2)} f(x(u))\frac{dx(u)}{du}du,$$

where $\frac{dx(u)}{du} = J(\frac{x}{u})$ is the Jacobian.
If

$$F(a) = \int_{x_1}^{x_2} f(x,a)dx$$

then,

$$\frac{dF(a)}{da} = \int_{x_1}^{x_2} \frac{\partial f(x,a)}{\partial a}dx.$$

CALCULUS OF MANY VARIABLES

We now turn to differential and integral calculus of many variables, in that order. Most of our discussion will be limited to the case of two variables since it illustrates most of the new ideas. We will however occasionally discuss three variables since we live and solve problems in three dimensions.

3.1. Differential Calculus of Many Variables

Let us begin with a function $f(x, y)$ of two variables. For example (x, y) could label points in the plane and f could be some function such as the temperature $T(x, y)$ or $h(x, y)$, the elevation above sea level.

The *partial derivative* with respect to, say x, is defined as

$$\frac{\partial f}{\partial x} \equiv f_x = \lim_{\Delta x \to 0} \frac{f(x + \Delta x, y) - f(x, y)}{\Delta x}. \tag{3.1.1}$$

Thus, to find the partial derivative along x, we imagine moving infinitesimally in just the x-direction and measuring the rate of change. Operationally this means that while taking the x-partial derivative, we treat y as a constant since it is indeed held constant. A similar set of relations exist for f_y, the derivative in the y-direction. Consider for example the following function and its partial derivatives:

$$f(x, y) = x^3 + 2xy^2 \tag{3.1.2}$$
$$f_x = 3x^2 + 2y^2 \tag{3.1.3}$$
$$f_y = 4xy. \tag{3.1.4}$$

Conversely the x-partial derivative determines the change in the x-direction:

$$f(x_0 + \Delta x, y_0) = f(x_0, y_0) + \frac{\partial f}{\partial x}\big|_{x_0, y_0} \Delta x + \cdots \tag{3.1.5}$$

where (x_0, y_0) is some point in the plane and the partial derivative is evaluated there. The ellipsis refer to higher order terms. What if we wanted to move to a point displaced in both the x and y directions? We do this in two stages:

$$f(x_0 + \Delta x, y_0 + \Delta y) - f(x_0, y_0)$$
$$= f(x_0 + \Delta x, y_0 + \Delta y) - f(x_0 + \Delta x, y_0) + f(x_0 + \Delta x, y_0) - f(x_0, y_0)$$
$$= \left.\frac{\partial f}{\partial y}\right|_{x_0 + \Delta x, y_0} \Delta y + \left.\frac{\partial f}{\partial x}\right|_{x_0, y_0} \Delta x + \cdots$$
$$= \left.\frac{\partial f}{\partial y}\right|_{x_0, y_0} \Delta y + \left.\frac{\partial f}{\partial x}\right|_{x_0, y_0} \Delta x + \cdots$$

where in going to the last equation we have ignored the change in the y-partial derivative as we move from (x_0, y_0) to $(x_0 + \Delta x, y_0)$. This is because we wish to work to just first order in the infinitesimals. (That is to say, the y-partial derivative, itself a function of two variables, assumed to be differentiable, will change by an amount proportional to Δx. Due to a prefactor Δy multiplying this derivative, the neglected term is of order $\Delta x \Delta y$.)

What about a Taylor series with more terms as in the case of one variable? For this we need to look at higher derivatives. There are four of them at second order and are listed below, the expressions in parentheses being the values for the example $f = x^3 + 2xy^2$:

$$f_{xx} = \frac{\partial^2 f}{\partial x^2} = \frac{\partial}{\partial x}\frac{\partial f}{\partial x} \quad (= 6x) \tag{3.1.6}$$

$$f_{yy} = \frac{\partial^2 f}{\partial y^2} = \frac{\partial}{\partial y}\frac{\partial f}{\partial y} \quad (= 4x) \tag{3.1.7}$$

$$f_{xy} = \frac{\partial^2 f}{\partial x \partial y} = \frac{\partial}{\partial x}\frac{\partial f}{\partial y} \quad (= 4y) \tag{3.1.8}$$

$$f_{yx} = \frac{\partial^2 f}{\partial y \partial x} = \frac{\partial}{\partial y}\frac{\partial f}{\partial x} \quad (= 4y) \tag{3.1.9}$$

The last two *mixed derivatives* are always equal if they are continuous. This continuity will be assumed henceforth unless stated otherwise.

Problem 3.1.1. *Establish the equality of the two mixed derivatives. Go back to their definitions (given above) and show that both are given by the limits*

$$f_{xy} = f_{yx}$$
$$= \lim_{\Delta x, \Delta y \to 0} \frac{f(x+\Delta x, y+\Delta y) - f(x+\Delta x, y) - f(x, y+\Delta y) + f(x, y)}{\Delta x \Delta y}.$$

The Taylor series now goes as follows:

$$f(x_0 + \Delta x, y_0 + \Delta y) - f(x_0, y_0) = \frac{\partial f}{\partial x}\Delta x + \frac{\partial f}{\partial y}\Delta y$$

$$+ \frac{1}{2}\left[f_{xx}(\Delta x)^2 + f_{yy}(\Delta y)^2 + f_{xy}\Delta x \Delta y + f_{yx}\Delta y \Delta x\right] + \cdots$$

where it is understood that all derivatives are taken at (x_0, y_0) and where we have not used the freedom to lump the last two terms using the equality of the mixed derivatives.

Problem 3.1.2. *Find the partial derivatives up to second order for the function* $f(x, y) = x^3 + x^2 y^5 + y^4$. *Observe the equality of the mixed derivatives.*

Let us now consider the question of maxima and minima for functions of many variables. Let us recall how it goes for the case of one variable. We plot f on one axis and x along the other. We follow the undulations of f as we vary x. A *stationary point* is one where a change in x does not produce a change Δf to *first order in* Δx. Thus the first derivative, which is the ratio of the former with respect to the latter, vanishes at this point. This point could be a maximum, minimum, or a point of inflexion. To decide between these alternatives we compute the second derivative. The three cases listed above correspond to the second derivative being negative, positive, and zero, respectively. We are generally more interested in the case of a maximum or minimum.

Consider now a function of two variables, $f(x, y)$. Let us plot above each point in the $x - y$ plane the value of f measured in the z direction. Consider for example the function $f = \sqrt{R^2 - x^2 - y^2}$, with $x^2 + y^2 \leq R^2$ and the positive branch of the root. It is clear that the profile of f is just the upper hemisphere of the sphere $x^2 + y^2 + f^2 = R^2$. It is also clear that the north pole, situated on top of $(0, 0)$ is a maximum. How does this come out of a calculation?

Once again we look for *stationary points*, defined to be points where a change in x or y produces no change in f to first order in either Δx or Δy. This change is given by

$$\Delta f = f_x \Delta x + f_y \Delta y. \tag{3.1.10}$$

Even though it is the sum which has to vanish at the stationary point, we can argue that each piece must separately vanish. This is because the first variation has to vanish for *any* choice of the Δx or Δy. If we choose just one of them to be nonzero, the corresponding partial derivative must vanish; there is no room for cancellations between the two terms. Consequently both the partial derivatives must vanish at a stationary point:

$$f_x = f_y = 0 \text{ at stationary point.} \tag{3.1.11}$$

In our example $f = \sqrt{R^2 - x^2 - y^2}$, we find the partial derivatives

$$f_x = -\frac{x}{\sqrt{R^2 - x^2 - y^2}} \qquad\qquad f_y = -\frac{y}{\sqrt{R^2 - x^2 - y^2}} \tag{3.1.12}$$

indeed vanish at the origin. To further diagnose this as a maximum, we must clearly look at the second derivatives. A simple calculation shows that at the origin

$$f_{xx} = f_{yy} = -\frac{1}{R} \tag{3.1.13}$$

with the mixed derivatives vanishing. Going back to the Taylor series

$$f(\Delta x, \Delta y) - f(0,0) \equiv \Delta f = -\frac{1}{2R}\left[(\Delta x)^2 + (\Delta y)^2\right] \qquad (3.1.14)$$

to quadratic order. Notice that we have dropped the first derivatives since we are at a stationary point. *It is clear that no matter which direction we move in,* $\Delta f < 0$. *Thus we are at a maximum.* (Had we chosen the negative square root for f earlier, we would have been dealing with the lower hemisphere and the origin would have been a minimum with $\Delta f > 0$.) We had no trouble here since Δf had a unique sign which could be seen by inspection. This in turn was because the mixed derivative vanished, and the infinitesimals $(\Delta x)^2$ and $(\Delta y)^2$ were positive definite: if they had both multiplied positive (negative) second derivatives, we would have had a minimum (maximum).

Consider next a slightly more general stationary point for some f such that the mixed derivatives are still zero, but f_{xx} and f_{yy} have opposite signs, say with $f_{xx} > 0$. This is a case where moving along x causes f to increase, while moving along y causes it to decrease. This is called a *saddle point*. This is because the shape of the function resembles that of a saddle in the vicinity of its center: as we move along the horse the function rises, while as we move transverse to the horse it falls.

Consider finally a function with all three second order partial derivatives nonzero. *It is not possible now to say what is happening by just looking at the signs of the second derivatives, because* Δf *has cross terms* $\Delta x \Delta y$ *in its expansion and this factor is of indefinite sign. Thus there is no simple relation between the sign of* Δf *and the signs of the partial derivatives.* We shall see how to solve this problem when we learn about matrices. For the present let us note the following. Consider the case where near the origin

$$\Delta f = \left[A(\Delta x)^2 + B(\Delta y)^2\right], \qquad (3.1.15)$$

where A and B are both positive. We clearly are at a minimum since f rises for all displacements. Let us rotate our axes by $45°$ and use new coordinates $x_\pm = \frac{x \pm y}{\sqrt{2}}$. In terms of these

$$\Delta f = \left[\frac{(A+B)}{2}[(\Delta x_+)^2 + (\Delta y_-)^2] + (A-B)\Delta x_+ \Delta x_-\right]. \qquad (3.1.16)$$

Now we know that despite the cross terms, this is just a minimum in disguise. To see this, we simply have to go back to the old coordinates in which cross terms vanish. But what about some other problem with cross terms? Will there be nice coordinates in which cross terms disappear so that the nature of the stationary point can be read off by inspection? Matrix theory answers this in the affirmative as you will learn later in Chapter 9.

3.1.1. Lagrange multipliers

The final topic in this section pertains to locating the maxima or minima of a function within a submanifold determined by a *constraint*. For example let

$$f(x,y) = x^2 - xy + y^2 \qquad (3.1.17)$$

be the temperature distribution in the plane. Let some bug be restricted to live in a circle of radius 5 given by the constraint equation

$$g(x,y) = x^2 + y^2 - 25 = 0. \qquad (3.1.18)$$

What is the hottest point in this bug's world? Since this point need not be a real maximum of the function in the unrestricted plane, it is not necessarily characterized by the vanishing of all its first derivatives. The solution to this problem lies in understanding this point in some detail.

As the bug walks around the circle measuring the temperature, it will find that as it approaches a maximum (minimum) on the circle, the temperature will first go up (down) and then start going down (up). At either extremum, the variation to first order will indeed vanish:

$$df = f_x dx + f_y dy = 0. \qquad (3.1.19)$$

We cannot however conclude that each partial derivative vanishes since dx and dy are not independent, they are chosen so that the displacement is along the (tangent to the) circle. In other words, they are chosen such that g remains zero before and after the displacement:

$$dg = g_x dx + g_y dy = 0 \qquad \text{which implies} \qquad (3.1.20)$$
$$dy = -\frac{g_x}{g_y} dx. \qquad (3.1.21)$$

If we feed this into Eqn. (3.1.19) we find the following condition for the stationary point:

$$\frac{f_x}{g_x} = \frac{f_y}{g_y} \qquad \text{at stationary point.} \qquad (3.1.22)$$

We can now find the two coordinates of the stationary point given the above and the constraint equation $g = 0$. Turning to the given temperature distribution, equation (3.1.22) becomes

$$\frac{2x - y}{2x} = \frac{2y - x}{2y} \qquad (3.1.23)$$

which implies

$$x = \pm y. \qquad (3.1.24)$$

This, along with the constraint $x^2 + y^2 = 25$, tells us there are four stationary points:

$$(x, y) = \pm(5/\sqrt{2}, \pm 5/\sqrt{2}) \tag{3.1.25}$$

Of course we must do more work to see which of these stationary points is really a maximum as compared to a minimum or saddle point.

Problem 3.1.3. (Very important). *If you want to really check Eqn. (3.1.25), simply eliminate y in favor of x in f using the constraint. Find the stationary points of this function of just one variable. Alternatively write* $x = 5\cos\theta$, $y = 5\sin\theta$ *and vary with respect to* θ.

There is an equivalent way to write Eqn. (3.1.22) for the stationary point:

$$f_x = \lambda g_x \tag{3.1.26}$$
$$f_y = \lambda g_y \tag{3.1.27}$$
$$g(x, y) = 0, \tag{3.1.28}$$

where λ is called the *Lagrange multiplier*. The three equations above determine the two coordinates and one Lagrange multiplier. (If there are several stationary points, there can be a different λ for each.) In this form the result generalizes to more variables.

Notice that whether we arrive at this result by eliminating one of dx or dy in favor of the other, the final equations above are symmetric between the variables. *Lagrange invented a clever trick by which the symmetry between the constrained variables is retained at all stages in the calculation and the procedure for finding these equations is reduced to finding the minimum of some other function* $\mathcal{F}$ *with no constraints.* Here is how it works. Consider the bug problem. Let us use polar coordinates r and θ related to x and y by

$$x = r\cos\theta \tag{3.1.29}$$
$$y = r\sin\theta. \tag{3.1.30}$$

These coordinates are chosen because the constraint is simply $g = r^2 - 25 = 0$. Moving on this fixed r curve, we satisfy the constraint. *The other coordinate* θ *describes variations within the constrained space.* We can write in these coordinates

$$df = f_r dr + f_\theta d\theta \tag{3.1.31}$$

for a general displacement. At the maximum on the circle $r = 5$, $df = 0$, but this only implies $f_\theta = 0$ but says nothing about f_r since the variation $dr = 0$. Let us now introduce a new function

$$\mathcal{F} = f - \lambda g, \tag{3.1.32}$$

where λ is a constant to be chosen shortly. Its variation is

$$
\begin{aligned}
d\mathcal{F} &= df - \lambda dg & \text{(3.1.33)} \\
&= (f_\theta - \lambda g_\theta)d\theta + (f_r - \lambda g_r)dr & \text{(3.1.34)} \\
&= (f_\theta)d\theta + (f_r - \lambda g_r)dr, & \text{(3.1.35)}
\end{aligned}
$$

where in the last step we have used the fact that g does not vary with θ since the latter is chosen to parametrize the constrained surface $g = r^2 - 25 = 0$. *Thus f and $\mathcal{F}$ have the same derivatives along the constraint surface no matter λ is.* In particular when f_θ vanishes so does $\mathcal{F}_\theta$. At this point the derivative of $\mathcal{F}$ normal to the surface equals $f_r - \lambda g_r$. Let us now use our freedom in choosing λ to ensure that this combination vanishes. *For this choice, the point in question becomes stationary with respect to all variations (radial and tangential).* This statement is of course true even in the cartesian coordinates. Thus the point we are looking for is the solution to

$$
\begin{aligned}
d\mathcal{F} &= 0 & \text{(3.1.36)} \\
&= (f_x - \lambda g_x)dx + (f_y - \lambda g_y)dy & \text{(3.1.37)}
\end{aligned}
$$

with no constraints on the variations. Consequently the following three equations locate the stationary point and the lagrange multiplier:

$$
\begin{aligned}
f_x &= \lambda g_x & \text{(3.1.38)} \\
f_y &= \lambda g_y & \text{(3.1.39)} \\
g(x,y) &= 0 & \text{(3.1.40)}
\end{aligned}
$$

which is what we had earlier in Eqns. (3.1.26-3.1.28).

To summarize:

To find the stationary points of f, a function of 2 variables, subject to the constraint $g = 0$, find the extrema of $\mathcal{F} = f - \lambda g$ with no constraint. The extremal point and λ are determined by the resulting 2 equations (the vanishing partial derivatives of $\mathcal{F}$) and the equation $g = 0$.

As an example consider the extremization of

$$
f(x,y) = x^2 + 2xy \qquad \text{(3.1.41)}
$$

subject to the constraint

$$
g(x,y) = x^2 + y^2 - 4 = 0. \qquad \text{(3.1.42)}
$$

Thus

$$
\mathcal{F}(x,y) = x^2 + 2xy - \lambda(x^2 + y^2 - 4) \qquad \text{(3.1.43)}
$$

and the stationary points obey

$$\frac{\partial \mathcal{F}}{\partial x} = 2x + 2y - 2\lambda x = 0 \tag{3.1.44}$$

$$\frac{\partial \mathcal{F}}{\partial y} = 2x - 2\lambda y = 0 \tag{3.1.45}$$

$$x^2 + y^2 = 4. \tag{3.1.46}$$

The middle equation tells us $x = \lambda y$, which upon feeding into the first we find

$$\lambda^2 - \lambda - 1 = 0 \tag{3.1.47}$$

$$\lambda_{\pm} = \frac{1 \pm \sqrt{5}}{2}. \tag{3.1.48}$$

It is now straightforward to plug these values in and to obtain the stationary points

$$\frac{\pm 1}{\sqrt{5 + \sqrt{5}}} \left[\sqrt{2}(1 + \sqrt{5}), \sqrt{8} \right] \qquad \lambda_+ \tag{3.1.49}$$

$$\frac{\pm 1}{\sqrt{5 - \sqrt{5}}} \left[\sqrt{2}(1 - \sqrt{5}), \sqrt{8} \right] \qquad \lambda_-. \tag{3.1.50}$$

Shown in Fig. 3.1 is the contour plot of f and the constraint surface. Note that at the extremal points, the contours of constant f are tangential to the constraint circle. This will be taken up in Chapter 7.

Problem 3.1.4. *Repeat the temperature problem done earlier using a Lagrange multiplier. This time find the value of λ as well for each stationary point.*

If you have followed these arguments, you will see that the method generalizes in an obvious way to more coordinates and more constraints.

For example in the case where f and g depend on (x, y, z), you may show that by eliminating say dz, that the following four equations determine the three coordinates of the stationary point and one Lagrange multiplier:

$$f_x = \lambda g_x \tag{3.1.51}$$

$$f_y = \lambda g_y \tag{3.1.52}$$

$$f_z = \lambda g_z \tag{3.1.53}$$

$$g(x, y, z) = 0. \tag{3.1.54}$$

Or you could more easily get the equations by minimizing $\mathcal{F} = f - \lambda g$.

Problem 3.1.5. *Find the shortest distance from the origin to any point on the line $x + 2y = 4$ by using Lagrange multipliers. Check this by more elementary means: by first finding the equation for the line which is perpendicular to the given line and passing through the origin.*

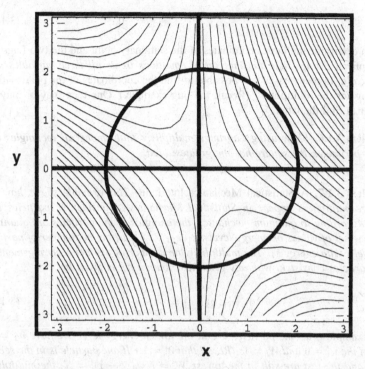

Figure 3.1. Contours of constant $f(x,y) = x^2 + 2xy$ and the constraint circle $x^2 + y^2 = 4$ on which we want to extremize f. The density of lines reflect the height of the function. Thus the stationary points in the first and third quadrants are maxima, while the other two are minima. Note the tangency of the constraint circle and the constant f curves at the extrema.

Finally let us ask what happens if there are three independent variables x, y, z and two constraints $g_1(x, y, z) = 0$ $g_2(x, y, z) = 0$. If we want, we can still eliminate two of the differentials, say dz and dy in favor of the remaining one. It is however wiser to follow Lagrange's idea and use

$$d\mathcal{F} = (f_x - \lambda_1 g_{1x} - \lambda_2 g_{2x})dx + (f_y - \lambda_1 g_{1y} - \lambda_2 g_{2y})dy + (f_z - \lambda_1 g_{1z} - \lambda_2 g_{2z})dz = 0$$
$$(3.1.55)$$

to obtain

$$f_x - \lambda_1 g_{1x} - \lambda_2 g_{2x} = 0 \qquad (3.1.56)$$
$$f_y - \lambda_1 g_{1y} - \lambda_2 g_{2y} = 0 \qquad (3.1.57)$$
$$f_z - \lambda_1 g_{1z} - \lambda_2 g_{2z} = 0 \qquad (3.1.58)$$
$$g_1 = 0 \qquad (3.1.59)$$

$$g_2 = 0, \qquad\qquad (3.1.60)$$

which determine the three coordinates of the stationary point and the two Lagrange multipliers. *Notice that the final set of equations treats all three variables symmetrically.* In many cases it is possible to find the coordinates of the point without solving for the multipliers. (Recall the bug example.) One then may or may not bother to find the latter.

Problem 3.1.6. *Show using Lagrange multipliers that among all rectangles of a given perimeter, the square has the greatest area.*

Problem 3.1.7. A Statistical Mechanics Interlude. *This example shows how Lagrange multipliers appear in Statistical Mechanics. Consider N particles in a box. According to quantum mechanics, the energies of the particles are quantized to some set of values or energy levels, $\varepsilon_1, \varepsilon_2, \dots$. Let n_i be the number of particles in level i with energy ε_i. The multiplicity or number of distinct rearrangements of the particles consistent with any given distribution n_i, is given by*

$$W(n_1, n_2, \dots) = \frac{N!}{n_1! n_2! \dots} \qquad\qquad (3.1.61)$$

For example if all the particles are in the lowest energy level, we have $n_1 = N$, rest of the $n_i = 0$ and $W = 1$. (Recall that $0! = 1$.) If one particle is in the second level and the rest are still in the lowest, $W = N!/(N-1)! = N$, the multiplicity reflecting the N ways to choose the one who goes up. The question is this: which distribution of particles, subject to the constraint that the total number equal N and the total energy equal E, gives the biggest W? Proceed to find this as follows:

- *Work with $S = \ln W$. Argue that*

$$\ln n! \simeq n \ln n - n$$

   for large n by approximating the sum involved in $\ln n!$ by an integral.

- *Write the constraints on the n_i's due to total number N and energy E.*

- *Treat all n_i as continuous variables, introduce Lagrange multipliers α and β for N and E and maximize S.*

- *Derive the Boltzmann distribution $n_i = e^{-\alpha - \beta \varepsilon_i}$.*

The multipliers may then be found by setting the total number and energy coming from this distribution to N and E, respectively. But this is not our problem here.

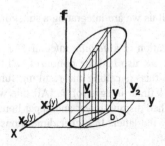

Figure 3.2. Integration of a function of two variables. The domain D is shown on the $x - y$ plane. The integral equals the volume of the cylindrical object standing on top of the domain. We have chosen to slice the volume in planes parallel to the $x - f$ plane.

3.2. Integral Calculus of Many Variables

We begin by recalling how we integrate a function of just one variable. First we plot $f(x)$ along an axis perpendicular to the x-axis, then we select a domain of integration bounded by the limits x_1 and x_2, then we take a thin interval of width Δx at x, form the product $f(x)\Delta x$, which corresponded to the area of a rectangle "standing" on top of that interval and sum over such areas in the limit of vanishing Δx. Geometrically this corresponded to the area of the figure bounded by the two vertical lines $x = x_1$, and $x = x_2$, the x-axis, and the function f. There was however an algebraic way to do it, which did not involve actually measuring and adding areas. For example in the case $f(x) = x^2$, we simply evaluated $\left[F(x) = x^3/3\right]_{x_1}^{x_2}$. The equivalence of the two schemes stems from the fact that in the algebraic scheme we are looking for an $F(x)$ whose derivative is the given $f(x)$, and area construction is the geometric solution to this problem.

Consider now the extension of this notion to a function $f(x, y)$, assumed continuous in a domain D in the $x - y$ plane as shown in Fig. 3.2.

We want to calculate the volume bounded below by the $x - y$-plane, bounded above by the function $f(x, y)$, and at the sides by a cylindrical surface whose cross section in the $x - y$-plane is the domain D. We denote this by

$$F(D) = \int\int_D f(x, y)dx \, dy. \qquad (3.2.1)$$

The notation also tells how we plan to go about computing this. First we take a little patch in D, of size Δx by Δy at the point (x, y), multiply it by the value of f there to obtain the volume of the rectangular solid with base given by the selected patch and height f. We then do a sum over such volumes in the limit of vanishingly small patches and call it $F(D)$. The subscript D in the right-hand side tells us that the little patches must cover D no more, no less. We use two integration symbols to tell us we are integrating a function over a two dimensional domain.

This is the generalization of a *definite* integral $F(x_1, x_2)$ in one variable. In the case of one variable, we also had the option of calling x_2 as x and studying the dependence of F on this x, calling the resulting function (defined up to an additive constant) the indefinite integral of f. All this was possible because the one-dimensional domain had disjoint lower and upper boundaries so we could hold the former fixed and vary the latter. In higher dimensions a natural generalization of this idea does not exist.

To find the integral algebraically we proceed as follows. First we decide to find the volume of the slab that lies between the two vertical planes at distances y and $y + \Delta y$ from the $f - x$ plane, as shown in Fig. 3.2. This is clearly given by the area of either sheet lying within the solid times Δy. As for the former, it is just the integral of $f(x, y)$ with respect to x, between the limits $x_1(y)$ and $x_2(y)$ shown in the figure. The limits thus depend on the domain D and the present value of y. The variable y is treated as a constant during this x-integration, as it indeed is within the slab. The volume of this slab is a function of y. We finally do the sum over such slabs, or equivalently the integral over y between the limits y_1 and y_2 as shown in the figure. In summary

$$F(D) = \int_{y_1}^{y_2} \left[\int_{x_1(y)}^{x_2(y)} f(x, y) dx \right] dy. \tag{3.2.2}$$

We thus see that the two-dimensional integral can be reduced to a sequence of one-dimensional integrals. Note that we could just as easily have sliced the volume into slabs bounded by planes of constant x.

Let us try a simple example: find $V(R)$, the volume of a sphere of radius R, centered at the origin. Let us find $H(R)$, the volume of the hemisphere above the x–y-plane and double the answer. The domain of integration is clearly a circle:

$$D = (x, y, \text{ such that } x^2 + y^2 \leq R^2). \tag{3.2.3}$$

On top of the point (x, y), the height of the hemisphere is $f(x, y) = \sqrt{R^2 - x^2 - y^2}$. At a given y, the range for x is $x_1 = -\sqrt{R^2 - y^2}$, $x_2 = \sqrt{R^2 - y^2}$, as you should convince yourself by drawing a sketch of D or by examining the above equation for D. Thus

$$H(R) = \int_{-R}^{R} \left[\int_{-\sqrt{R^2 - y^2}}^{\sqrt{R^2 - y^2}} \sqrt{R^2 - x^2 - y^2} dx \right] dy. \tag{3.2.4}$$

Next we argue that the integration over the full circle is four times what we get from the first quadrant where both x and y are positive. This is clear if you imagine the sphere with its symmetries. It is also clear even if you have no imagination but can see that the integrand is an even function of both variables and the integration region is symmetric about the origin of coordinates. Thus every patch inside the chosen quadrant has identical counterparts in the other three quadrants related by changing the sign of x, of y, or both. Thus

$$H(R) = 4 \int_0^R \left[\int_0^{\sqrt{R^2 - y^2}} \sqrt{R^2 - x^2 - y^2} dx \right] dy. \qquad (3.2.5)$$

Let us now do the x-integral by the following change of variables:

$$x = \sqrt{R^2 - y^2} \sin\theta \qquad (3.2.6)$$

$$0 \leq \theta \leq \pi/2, \qquad (3.2.7)$$

which leads to

$$\int_0^{\sqrt{R^2 - y^2}} \sqrt{R^2 - x^2 - y^2} dx = \int_0^{\pi/2} \sqrt{R^2 - y^2} \cos\theta \sqrt{R^2 - y^2} \cos\theta d\theta$$

$$= (R^2 - y^2)\pi/4. \qquad (3.2.8)$$

where the repeated factors in the integrand come once from rewriting f in terms of θ and once from changing from dx to $d\theta$. It is now easy to do the y-integral and to multiply the answer by 2 to get $V(R) = \frac{4}{3}\pi R^3$.

Problem 3.2.1. *Fill in the missing steps.*

Problem 3.2.2. *By doing the integral of $f = 1$ over the same domain obtain the area of a circle.*

Consider now the following problem: to integrate $f(x, y)$ over a square of side $2a$ centered at the origin. The answer is

$$I(D) = \int_{-a}^a \left[\int_{-a}^a f(x, y) dx \right] dy. \qquad (3.2.9)$$

The simplification we notice is that the range of x-integration is independent of y. Thus cartesian coordinates are very convenient to use if D is bounded by lines of constant x or y coordinates. Things get even easier if

$$f(x, y) = X(x)Y(y), \qquad (3.2.10)$$

that is, if f factorizes into a product of a function $X(x)$ that depends on just x and $Y(y)$ that depends on just y. Let D still be rectangular in shape so that limits on x are independent of y. Then

$$I(D) = \int_{y_1}^{y_2} Y(y) dy \int_{x_1}^{x_2} X(x) dx, \qquad (3.2.11)$$

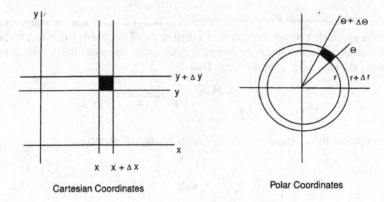

Figure 3.3. Cartesian and polar coordinate systems.

that is, $I(D)$ becomes a product of two one-dimensional integrals. This is as easy as the problem in higher dimensions can get: factorize into a product of one-dimensional integrals. But notice that the very same problem can look very nasty in some other coordinate systems. For example suppose we switched to polar coordinates $(r, \theta)^1$ shown in Fig. 3.3

They are defined by

$$r = \sqrt{x^2 + y^2} \qquad \theta = \tan^{-1}\frac{y}{x} \qquad (3.2.12)$$

or the inverse relations

$$x = r\cos\theta \qquad y = r\sin\theta. \qquad (3.2.13)$$

In this coordinate system, neither will the integrand factorize nor will the limits on each variable be independent of the value taken by the other.

By the same token, a problem that looks nasty in one coordinate system could possibly be simplified by going to a different system. Let us go back to the volume of the sphere. The integrand and domain were

$$f(x, y) = \sqrt{R^2 - x^2 - y^2} \text{ integrand} \qquad (3.2.14)$$

$$x^2 + y^2 \leq R^2 \text{ domain}. \qquad (3.2.15)$$

It is very clear that the problem will be a lot simpler if we use polar coordinates. Indeed we find:

$$f(r, \theta) = \sqrt{R^2 - r^2} \text{ integrand} \qquad (3.2.16)$$

$$r \leq R \text{ domain}. \qquad (3.2.17)$$

---

[1] Sometimes the polar coordinates are referred to as (ρ, θ), (ρ, ϕ), (r, θ), or (r, ϕ).

Notice how f in the new coordinates has simplified: not only is it a product of a function of r times a function of θ, the latter is just unity! Next, the domain of integration is described entirely by one of the coordinates, so that no matter what θ is, r will go over this range. How do we actually do the integral in these coordinates? As in one dimension, it is not enough to just rewrite the integrand and limits in the new system, we must include the Jacobian. In other words, the area element $dxdy$ does not simply get replaced by $dr d\theta$. Indeed the latter does not even have the dimensions of an area. What we need is a Jacobian $J(r, \theta)$ that needs to be inserted to ensure that we are still computing the same object (the volume) as before.

The first step is to better understand the cartesian system. Given a plane, we pick an origin. Then we draw lines of constant x and y, the former are parallel to the y-axis and the latter are parallel to the x-axis, as shown in Fig. (3.3). The intersection of two such lines of constant coordinates defines a point. For example the lines $x = 3$ and $y = 4$ meet at the point we call $(3, 4)$. To perform a two-dimensional integral, we draw two constant-x lines with coordinates x and $x + \Delta x$ and two constant-y lines with coordinates y and $y + \Delta y$. Their intersection encloses a region of size $\Delta x \Delta y$ which we then multiply by $f(x, y)$ in doing the integration. We finally take the limit $\Delta x, \Delta y \to 0$ in defining the integral. All this is understood when we write $\int dxdy$. Figure 3.3 summarizes all this. Let us repeat this with polar coordinates. To locate a point, we first draw a curve of constant coordinate r, which is just a circle of radius r. Then we draw a curve of constant θ, which is just a ray going from the origin to infinity at an angle θ with respect to the x-axis. The intersection of these two curves defines the point called (r, θ). To do integrals, we draw another constant r curve at $r + \Delta r$ and another constant θ curve at $\theta + \Delta\theta$ and multiply the enclosed area by $f(r, \theta)$. What is the enclosed area? It is clear from the figure that the enclosed area has the shape of a rectangle of sides $r\Delta\theta$ and Δr. (All right, it is not quite a rectangle since the constant θ lines are not parallel and the constant r lines are curved. But these objections vanish in the limit of infinitesimal patches.) Notice that the area is simply given by the product of the two lengths Δr and $r\Delta\theta$ because r, θ are *orthogonal* coordinates: curves of constant r and θ meet at right angles. To summarize, what we have learned is that

$$dx\,dy \quad \to \quad r\,dr\,d\theta \tag{3.2.18}$$

$$J(r, \theta) \quad = \quad r. \tag{3.2.19}$$

Let us notice that with the Jacobian in place, the integration measure has the right dimensions. *The factor r which converts the infinitesimal change in θ (at fixed r) to the corresponding displacement $rd\theta$ is called the* scale factor corresponding to θ, *and is often denoted by h_θ.* The scale for r is unity since dr is itself the actual displacement. Likewise the scale factors for the cartesian coordinates is unity. *The Jacobian is the product of the scale factors in orthogonal coordinates.*

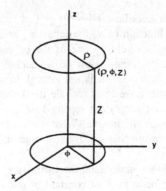

Figure 3.4. The cylindrical coordinate system.

If we go to three dimensions, there are two popular coordinate systems besides the cartesian. One is called the *cylindrical coordinate system* with coordinates:

$$\rho = \sqrt{x^2 + y^2} \tag{3.2.20}$$

$$\phi = \tan^{-1} y/x \tag{3.2.21}$$

$$z = z. \tag{3.2.22}$$

The system is shown in Fig. 3.4.

The surface of constant ρ is a cylinder of radius ρ with the z axis as its axis since ρ measures the distance from the z axis. The surface of constant ϕ is like a door hinged on the z axis, making an angle ϕ with the $x - z$ plane. These two surfaces intersect on a vertical line where the door penetrates the cylinder. Finally the surface of constant z is a plane a distance z above the $x - y$ plane. It intersects this vertical line at one point, called the point (ρ, ϕ, z). Note that this too is an orthogonal coordinate system. You should prove that the scale factors (factors which convert a change in one coordinate with the other held fixed) and Jacobian (product of scale factors for an orthogonal system such as this one) are

$$h_\rho = 1 \quad h_\phi = \rho \quad h_z = 1 \tag{3.2.23}$$

$$J(\rho, \phi, z) = 1 \cdot 1 \cdot \rho \tag{3.2.24}$$

This means for example that at fixed ρ and z, an infinitesimal change $d\phi$ causes the point to move by $h_\phi d\phi = \rho d\phi$. Consequently $\rho d\rho d\phi dz$ is the volume trapped between surfaces infinitesimally separated in all three coordinates.

The other popular coordinates are *spherical coordinates*, shown in Figure (3.5). They are related to cartesian coordinates as follows:

$$r = \sqrt{x^2 + y^2 + z^2} \tag{3.2.25}$$

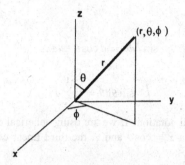

Figure 3.5. Spherical coordinates.

$$\theta = \cos^{-1} z/r \qquad (3.2.26)$$
$$\phi = \tan^{-1} y/x. \qquad (3.2.27)$$

The inverse of the above is

$$z = r\cos\theta \qquad (3.2.28)$$
$$y = r\sin\theta\sin\phi \qquad (3.2.29)$$
$$x = r\sin\theta\cos\phi. \qquad (3.2.30)$$

Since r is the distance from the origin, the constant-r surface is a sphere of radius r. The constant-ϕ surface is once again the door and it cuts the sphere along a great circle of definite longitude. Finally the constant-θ surface is a cone of opening angle 2θ measured from the z axis. It cuts the sphere along a circle of fixed latitude. (Thus θ is zero on the north pole, $\pi/2$ at the equator and π at the south pole.) In the spherical coordinate system we label a point by picking a sphere of radius r and then picking a point on it using latitudes and longitudes.

Problem 3.2.3. *Show with the aid of figures that*

$$h_r = 1 \qquad (3.2.31)$$
$$h_\theta = r \qquad (3.2.32)$$
$$h_\phi = r\sin\theta, \qquad (3.2.33)$$

so that

$$J(r,\theta,\phi) = r^2\sin\theta. \qquad (3.2.34)$$

Volume integrals (over all of space) are done as follows:

$$\int_{-\infty}^{\infty} dx \int_{-\infty}^{\infty} dy \int_{-\infty}^{\infty} dz\, f(x,y,z) = \int_{0}^{\infty} r^2 dr \int_{0}^{2\pi} d\phi \int_{0}^{\pi} \sin\theta d\theta f(r,\theta,\phi).$$
$$(3.2.35)$$

Using the fact that

$$\sin\theta d\theta = d\cos\theta \equiv dz, \tag{3.2.36}$$

we may write

$$\int_0^\pi \sin\theta d\theta = \int_{-1}^1 dz. \tag{3.2.37}$$

This is a commonly used notation. If we are using spherical coordinates, there can be no confusion between $z = \cos\theta$ and z, the third linear coordinate in cartesian or cylindrical coordinates.

Problem 3.2.4. *Show that the volume of a sphere is $V(r) = \frac{4}{3}\pi r^3$ by integrating $f = 1$ over a sphere.*

One can easily argue that if the radius of the sphere is increased from r to $r + dr$, its volume must increase by $S(r)dr$, where $S(r)$ is the area of the sphere of radius r. Conversely, the area of the sphere is

$$S(r) = \frac{d(V(r))}{dr} = 4\pi r^2. \tag{3.2.38}$$

3.2.1. Solid angle

Recall that on a circle of radius r, the arc length ds divided by r gives the angle subtended by the arc at the center: $d\theta = ds/r$. Consider next some closed curve, not necessarily a circle, that encloses the origin as in Fig. 3.6.

Next let us divide it into tiny segments small enough that each is approximately a straight line. If we join the beginning and end of any one segment to the origin by straight lines, these lines will have an angular separation $d\theta$, which is the angular coordinate difference between the beginning and end of the segment. If ds is the length of the segment, $d\theta$ will generally not be equal to ds/r, instead

$$d\theta = \frac{ds\cos\gamma}{r} \equiv \frac{ds_\theta}{r} \tag{3.2.39}$$

where γ is the angle between the normal to line segment and the radial direction.

(Approximating the segment by a straight line and decomposing it vectorially into radial and tangential part, we see that only the tangential part ds_θ contributes to change in θ.) Obviously the sum of all these $d\theta$'s equals 2π. *Thus every closed curve surrounding the origin subtends an angle 2π.* We wish to generalize these notions to three dimensions.

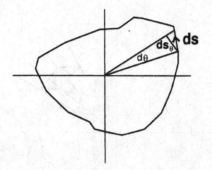

Figure 3.6. Angle subtended by a part of a closed curve.

The obvious generalization of the circle centered at the origin is a sphere centered at the origin. Consider a (roughly rectangular) tiny patch on a sphere of radius r bounded by lines at θ, $\theta + d\theta$, ϕ and $\phi + d\phi$ as shown in Fig. 3.7.

The linear dimensions of this patch are $rd\theta$ and $r\sin\theta d\phi$ upon invoking the scale factors for θ and ϕ. Thus the area of the patch is

$$
\begin{aligned}
dS &= rd\theta r\sin\theta d\phi & (3.2.40) \\
&= r^2[d\cos\theta][d\phi] & (3.2.41) \\
&= r^2 dz d\phi. & (3.2.42)
\end{aligned}
$$

If we integrate this over both angles or, equivalently, over ϕ and $z = \cos\theta$ we find that the area of the sphere is

$$
r^2 \int_0^{2\pi} d\phi \int_{-1}^{1} dz = 4\pi r^2. \qquad (3.2.43)
$$

Consider now our patch of area dS on a sphere of radius r. We say it subtends a *solid angle* $d\Omega$ at the center of the sphere where

$$
d\Omega = dS/r^2. \qquad (3.2.44)
$$

Conversely a patch on a sphere that subtends a solid angle $d\Omega$ has an area $r^2 d\Omega$.

We just saw that the area of a patch of angular width $d\phi$ and $d\theta$ on a sphere of radius r is

$$
dS = r^2[d\cos\theta][d\phi] \qquad (3.2.45)
$$

from which it follows that

$$
d\Omega = [d\cos\theta][d\phi] = dz d\phi. \qquad (3.2.46)
$$

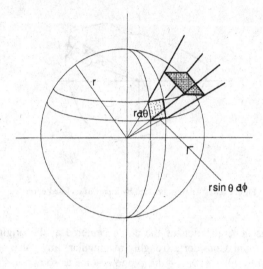

Figure 3.7. Solid angle.

A very useful result is that the total solid angle is 4π:

$$\int d\Omega = \int_0^{2\pi} d\phi \int_{-1}^{1} dz = 4\pi. \tag{3.2.47}$$

Let us next surround the origin by any closed surface and divide it into tiny patches of area dS. Choose the areas so that they are bounded by constant θ and ϕ curves. Then each area subtends a solid angle $d\Omega = d\cos\theta d\phi$ about the origin. One such patch from a surrounding surface is shown in Fig. 3.7. It subtends the same solid angle as the smaller one on the sphere. Note that this area will not generally lie in a plane that is perpendicular to the radial direction, and $d\Omega$ will be less than dS/r^2 by factor which is the cosine of the angle between the radial direction and the normal to the area. In other words

$$d\Omega = \frac{dS \cos\gamma}{r^2}, \tag{3.2.48}$$

where γ is the angle between the normal to the surface (pointing outward if it is part of a closed surface) and the radial direction.

The sum over all the patches of such solid angles clearly adds up to 4π. We shall return to this point in Chapter 7.

Problem 3.2.5. *Find the moment of inertia I of a uniform (constant density) disc of radius R and mass M about an axis through its center. Recall that for any* *object*

$$I = \int_{object} \rho(r,\theta) r^2 r \, dr \, d\theta \qquad (3.2.49)$$

where ρ is the density or mass per unit area. Repeat for the case where the density increases linearly with r, starting at 0 at the center, but the object has the same mass. Before doing any calculations, explain which will be larger and why.

Evaluation of the gaussian integral

Now we provide the method for the evaluation of the *gaussian integral*

$$I(a) = \int_0^\infty e^{-ax^2} dx. \qquad (3.2.50)$$

First let us calculate $2I$ by extending the lower limit to $-\infty$. Then we manipulate as follows:

$$2I(a) \cdot 2I(a) = \int_{-\infty}^\infty e^{-ax^2} dx \cdot \int_{-\infty}^\infty e^{-ay^2} dy \qquad (3.2.51)$$

$$= \int_0^{2\pi} d\phi \int_0^\infty r \, dr \, e^{-ar^2} \qquad (3.2.52)$$

$$= 2\pi \frac{1}{2} \int_0^\infty ds \, e^{-as} \qquad (s = r^2) \qquad (3.2.53)$$

$$= \frac{\pi}{a} \qquad (3.2.54)$$

$$I(a) = \frac{1}{2}\sqrt{\frac{\pi}{a}}. \qquad (3.2.55)$$

The key step consists of recognizing the product of two one-dimensional integrals as a two dimensional integral over the plane and then switching to polar coordinates. The Jacobian r now allows us to change variables to $s = r^2$. You should convince yourself that this trick only works if the limits of the gaussian integration is from 0 to ∞. What goes wrong if the upper limit is finite?

Problem 3.2.6. *Find the volume of a cone of base radius R and height h.*

Problem 3.2.7. (Advanced). *Consider the change from (x,y) to some more general coordinates $u(x,y), v(x,y)$. Thus each point (u,v) corresponds to the intersection of constant-u and constant-v curves. These curves need not even meet at right angles, i.e., the coordinates need not be orthogonal. What is the Jacobian now? Our job is still to find the area of the patch bounded by the curves labeled $u, u + \Delta u, v, v + \Delta v$. We are now looking at the area of a parallelogram rather than a rectangle since the coordinates are not necessarily orthogonal.*

Now, the area of a parallelogram whose adjacent sides are the vectors $\mathbf{a}, \mathbf{b}$ *is* $A = |\mathbf{a} \times \mathbf{b}| = |ab \sin \theta|$. *By drawing a sketch of the constant* u, v *curves show that in this case* $\mathbf{a} = (\mathbf{i} \frac{\partial x}{\partial u} + \mathbf{j} \frac{\partial y}{\partial u}) \Delta u$ *and* $\mathbf{b} = (\mathbf{i} \frac{\partial x}{\partial v} + \mathbf{j} \frac{\partial y}{\partial v}) \Delta v$. *Using the fact that the cross product may also be written as* $a_x b_y - a_y b_x$ *show the general result*

$$J\left(\frac{xy}{uv}\right) = \left| \frac{\partial x}{\partial u} \frac{\partial y}{\partial v} - \frac{\partial x}{\partial v} \frac{\partial y}{\partial u} \right|. \tag{3.2.56}$$

Show that this formula reproduces the Jacobian for polar coordinates derived earlier.

Problem 3.2.8. *Find the volume generated by revolving around the* y-axis, *the figure bounded by the* x-axis, *the line* $x = 4$, *and the curve* $y = x^2$.

3.3. Summary

Here are the main results from this chapter.

- You must know the definitions of $f_x, f_y, f_{xy} = f_{yx}$, etc., and how to evaluate them.

- The change in $f(x, y)$ due to a change in its arguments is

$$f(x_0 + \Delta x, y_0 + \Delta y) - f(x_0, y_0) = \frac{\partial f}{\partial x} \Delta x + \frac{\partial f}{\partial y} \Delta y +$$

$$\frac{1}{2} \left[f_{xx}(\Delta x)^2 + f_{yy}(\Delta y)^2 + f_{xy} \Delta x \Delta y + f_{yx} \Delta y \Delta x \right] + \cdots$$

There is an obvious generalization to functions of more variables.

- A stationary point of f is where df, the first order change, is zero. At this point $f_x = f_y = 0$. This could be a maximum, minimum, or saddle point.

- To extremize a function $f(x, y)$ subject to the condition $g(x, y) = 0$, minimize $\mathcal{F} = f - \lambda g$, where λ is the Lagrange multiplier. The resulting equations are

$$f_x = \lambda g_x$$

$$f_y = \lambda g_y,$$

$$g(x, y) = 0$$

The generalization to more variables and more constraints is obvious.

$$F(D) = \int_{y_1}^{y_2} \left[\int_{x_1(y)}^{x_2(y)} f(x,y)dx \right] dy$$

which means that to integrate a function $f(x,y)$ over a domain D, first fix y, integrate f at this fixed y as a function of just x (treating y as a constant) between limits that depend on y and D; and then do the integral over y between limits set by D. You are also free to integrate in the opposite order.

- Know the relation between polar coordinates (r,θ), also called (ρ,θ), or (ρ,ϕ) and cartesian coordinates in the plane

$$x = r\cos\theta \qquad\qquad y = r\sin\theta$$

$$r = \sqrt{x^2 + y^2} \qquad\qquad \theta = \arctan\frac{y}{x}.$$

Know that in three dimensional cylindrical coordinates one introduces in addition a z-coordinate and invariably refers to the coordinates in the $x-y$ plane as ρ and ϕ.

- Know the spherical coordinates in three dimensions

$$r = \sqrt{x^2+y^2+z^2} \qquad \theta = \arccos\frac{z}{\sqrt{x^2+y^2+z^2}} \qquad \phi = \arctan\frac{y}{x}$$

$$x = r\sin\theta\cos\phi \qquad y = r\sin\theta\sin\phi \qquad z = r\cos\theta.$$

- If you go to noncartesian but orthogonal coordinates $u_1, u_2, ...,$ first identify scale factors h_i such that $h_i du_i$ is the displacement under a change in just u_i. Then the Jacobian is $J = h_1 h_2 \cdots$. In 2-dimensional polar coordinates $h_\rho = 1$, $h_\theta = \rho$, and in 3-dimensions, $h_\rho = 1 = h_z$, $h_\phi = \rho$ in cylindrical coordinates, and $h_r = 1$, $h_\theta = r$, $h_\phi = r\sin\theta$ in spherical coordinates. Thus for example

$$\int_{-\infty}^{\infty} dx \int_{-\infty}^{\infty} dy \int_{-\infty}^{\infty} dz f(x,y,z)$$
$$\rightarrow \int_{0}^{\infty} r^2 dr \int_{0}^{\pi} d\theta \sin\theta \int_{0}^{2\pi} d\phi f(r,\theta,\phi)$$

Note that often one uses $z = \cos\theta$ while working with spherical coordinates.

- One calls $d\cos\theta\, d\phi \equiv dz\, d\phi \equiv d\Omega$ an element of solid angle. The total solid angle is

$$\int d\Omega = \int_{-1}^{1} dz \int_{0}^{2\pi} d\phi = 4\pi.$$

- The solid angle associated with a tiny patch of area is

$$d\Omega = \frac{dS \cos \gamma}{r^2}$$

where γ is the angle between the normal to the area (pointing away from the origin if it is part of a closed surface) and the radial direction.

-

$$\int_{-\infty}^{\infty} e^{-ax^2} dx = \sqrt{\frac{\pi}{a}}.$$

INFINITE SERIES

In Chapter 1, we made a passing reference to series when we discussed the Taylor expansion. We now return to the subject of infinite series for a fuller treatment.

4.1. Introduction

A series is a sum of the following form:

$$S_N = \sum_{n=0}^{N} a_n \qquad (4.1.1)$$

where a_n is called the n-th term. It is assumed that the algorithm for generating the term for each n is provided. An example is $a_n = \frac{1}{n^2+1}$. (In some cases the series may begin with $n = 1$, for example if $a_n = 1/n^2$.)

We will assume unless stated otherwise that, like in our example, the terms are all positive and decreasing with n.

If N is finite, so is the sum, since the individual terms are finite. *The question is what happens as $N \to \infty$?* When will the sum of infinitely many numbers be finite? More precisely, will the sum converge, as per the following definition?

Definition 4.1. *The series converges to S if*

$$|S - S_N| < \varepsilon \quad for \quad N > N(\varepsilon), \qquad (4.1.2)$$

where S_N is the sum to N terms, and where ε is an arbitrarily small, positive, number.

For the most part we will simply like to know if the infinite sum is finite and not care about what the finite value of the sum is.

Let us begin with a simple example that sheds light on many of these questions. Imagine that we have invited an infinite number of guests for a pizza party and we have just two pizzas. So we make up the following rule for the guests: they enter one by one and each person gets to eat half of whatever is left. This way no one goes empty handed. The total consumption is given by the following infinite series

$$S \quad = \quad 1 + 1/2 + 1/4 + 1/8 + \cdots + \tag{4.1.3}$$

$$= \quad \sum_0^\infty 2^{-n}. \tag{4.1.4}$$

We know that the sum is bounded by 2 since it represents the total amount of pizza eaten and we started with just 2. We also know that if we stop the sum at any finite number of terms, say N, the sum will be less than 2, but this difference can be made as small as we want by choosing a large enough N. Thus we say that the series *converges to the limit* 2. This means that given some ε however small, we can find an $N(\varepsilon)$ such that the partial sum

$$S_N = \sum_0^N 2^{-n} \tag{4.1.5}$$

differs from the limit by less than ε. To analyze this problem more carefully, let us recognize that it is a special case of a *geometric series* in which each term is some positive ratio r times the previous term, $r = 1/2$ being the value in our example. Let us choose the first term of the geometric series to be 1, since any other case can be found by rescaling the answer. Let us determine the partial sum of the geometric series as follows:

$$S_N \quad = \quad 1 + r + r^2 + r^3 + \cdots + r^N \tag{4.1.6}$$

$$r S_N \quad = \quad r + r^2 + r^3 + \cdots r^{N+1} \tag{4.1.7}$$

$$\text{Subtracting,} \quad S_N(1-r) \quad = \quad 1 - r^{N+1} \tag{4.1.8}$$

$$S_N \quad = \quad \frac{1 - r^{N+1}}{1 - r} \tag{4.1.9}$$

$$= \quad \frac{1}{1-r} - \frac{r^{N+1}}{1-r}. \tag{4.1.10}$$

The last form tells us the whole story. As long as $r < 1$, the sum approaches the *limit* $S_\infty \equiv S = \frac{1}{1-r}$, given by the first term, whereas at any finite N the sum differs from the limit by the second term. In our example this sum is $\frac{1}{1-1/2} = 2$ as anticipated. As for the approach to the limit, the requirement

$$\frac{r^{N+1}}{1-r} \quad < \quad \varepsilon \text{ implies} \tag{4.1.11}$$

$$N > N(\varepsilon) \quad = \quad \frac{\ln \varepsilon (1-r)}{\ln r} - 1. \tag{4.1.12}$$

Notice that as long as $r < 1$, such an N can always be found and the series converges.

Problem 4.1.1. *In our example, if we want 2^{-55} of pizza to be left over, show that $N.(2^{-55}) = 55$. (Choose the base of the logarithm wisely.)*

Not all series can be handled as completely as the geometric series where the convergence and the approach to the limit are so clearly calculable. We now turn to more general series and ask how we are to know if they are convergent.

4.2. Tests for Convergence

Consider the generic series

$$S = \sum_{n=1}^{\infty} a_n. \qquad (4.2.1)$$

There are many ways to tell if this series converges. We study a few common tests.

The comparison test

If b_n is a convergent series, so is our series if $a_n \leq b_n$ for all n.

We will not attempt to prove this result which is intuitively obvious. It is also clear that even if $a_n > b_n$ for a finite number of terms, the result is still the same since we can do the sum by first setting these terms to zero (thereby restoring the inequality) and adding back their *finite* contribution at the end. *The main point is that the convergence of the series is not decided by any finite number of finite terms at the beginning, but only by the endless string of terms out to infinity.* For example the series with $a_n = \frac{1000}{n^2}$ converges while $a_n = \frac{1}{1000n}$ diverges even though earlier terms in the former dominates their counterparts in the latter.

The converse of the above result is that if $a_n \geq b_n$ and the latter diverges, so does the former. Nothing definite can be said of a series which exceeds term by term a convergent series or one that is bounded term by term by a divergent series.

Problem 4.2.1. *Consider the series $\sum_{n=1}^{\infty} 1/n^2$ which goes as follows*

$$1 + \underbrace{1/4 + 1/9}_{2 \ terms} + \underbrace{1/16 + \cdots 1/49}_{4 \ terms} + \underbrace{1/64 + \cdots + 1/15^2}_{8 \ terms} + \cdots$$

Bound the sum of each bracketed set of terms by replacing each member in the set by the largest member in the set. Show convergence by comparing to a geometric series.

Problem 4.2.2. *It was stated that $a_n = 1/n$ is a divergent series. Show this by comparing it to a series with $a_n = 1/2$ for all n, by grouping terms so that upon replacing terms in any group by the smallest member of the group we get $1/2$.*

Problem 4.2.3. *Given that* $a_n = 1/n$ *diverges what can you say about* $a_N =$ $\frac{1}{1000\sqrt{n}}$?

Ratio test
The series converges if

$$r = \lim_{n \to \infty} \frac{a_{n+1}}{a_n} < 1. \tag{4.2.2}$$

The logic of this test is that if the above condition is satisfied, we may treat the far end of the series as a convergent geometric series as follows. First note that the ratio of successive terms approaching the limit r means that it will not deviate from r, and in particular will not *exceed* r, by more than some arbitrarily small ε after a corresponding $n = N(\varepsilon)$ is crossed. The remaining terms are clearly bounded by a geometric series with ratio $r + \varepsilon < 1$. (Given an $r < 1$ we can always find an $\varepsilon > 0$ such that $r + \varepsilon < 1$ to proceed with this argument.)

It follows from this argument that if $r > 1$, the series diverges. If $r = 1$, more careful analysis is needed.

Consider as an example the series with $a_n = n/3^n$. The limiting ratio is

$$r = \lim_{n \to \infty} \frac{n+1}{3n} = \frac{1}{3} \tag{4.2.3}$$

showing the series converges. On the other hand in the case of $a_n = 1/n$

$$r = \lim_{n \to \infty} \frac{n}{n+1} = 1. \tag{4.2.4}$$

We cannot say using this test if the series converges. Note an important fact: although for any finite n, the ratio is less than 1, this is not enough for convergence — it is the *limit* as $n \to \infty$ that matters.

The integral test
The sum $\sum_{n=1}^{\infty} f(n)$, where f is monotonically decreasing, converges or not according as $\int^L f(x)dx$ converges or not, as the upper limit L is sent to ∞.

The logic of this test is apparent in Fig. 4.1.

It is clear from the figure that $\sum_1^{\infty} f(n)$ equals the total area of rectangles, of height $f(n)$ and width 1, the first two of which are shown. It is obvious that the integral, between the same limits, is less than the sum. Consequently the divergence of the integral ensures the divergence of the sum.

It is also clear from the figure that the sum from $n = 2$ onwards, is given once again by rectangles, but these begin with what used to be the *second* rectangle in the sum from $n = 1$ onwards. Mentally shift the rectangles to the left by one unit and you can see that the sum is now bounded by the integral from $x = 1$ to ∞. It follows that if the integral converges, so must the sum from $n = 2$ onwards. Since

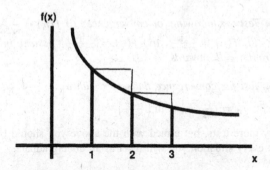

Figure 4.1. Integral test for convergence.

the $n = 1$ term makes a finite contribution, it is clear that the sum and the integral
(as the upper limit goes to infinity) converge or diverge together.

Whereas the ratio test was inconclusive for $a_n = 1/n$ and $a_n = 1/n^2$ since
$r = 1$ in both cases, the integral test tells us that the former is divergent and the
latter is not since

$$\int^L \frac{dx}{x} \to \infty \text{ as } L \to \infty \qquad (4.2.5)$$

while

$$\int^L \frac{dx}{x^2} \simeq 1/L \text{ as } L \to \infty. \qquad (4.2.6)$$

In applying the test you can save yourself some trouble by modifying the
integrand without changing the outcome. For example if

$$f(n) = \frac{3n^2 + 4n + 6}{n^4 + 12n^3 + 4n + 2},$$

you need just look at the integral of the large n limit of the integrand, $f \simeq 3/n^2$.
Likewise you can also replace the given f by another which is everywhere smaller
(larger) and argue that if the latter diverges (converges), so does the given one. For
example you may not be able to integrate $f(n) = e^{-n^2}/(1+n^2)$, but you know it
converges because the larger integral over just $1/(1+n^2)$ does. In this example,
as well as in everything else you do, you get a lot more out of what you learn if
you understand why things work.

Problem 4.2.4. *Test for convergence the series with (i) $a_n = \frac{n^2}{n^3+1}$, (ii) $a_n =$
$\frac{(n!)^3 e^{3n}}{(3n)!}$, (iii) $a_n = 1/(n \ln n)$, (iv) $a_n = \ln(1 + 1/n)$, (v) $a_n = \frac{e^n}{\sqrt{n}}$ and (vi)
$a_n = 1/(n n^{1/n})$. For the last part use $x^n = e^{n \ln x}$. Assume all the series begin
with $n = 2$.*

Problem 4.2.5. *Test the following for convergence:* (i) $f(n) = \frac{1}{\sqrt{3n^4+6n^3+98}}$, (ii) $f(n) = \frac{e^n}{e^{4n}+5}$, (iii) $f(n) = \frac{1}{\cosh n}$, (iv) $f(n) = \frac{1}{n^2 \ln n}$. *Assume in all cases that the sum goes from $n = 2$ onwards.*

Problem 4.2.6. *Test for convergence the series with* $a_n = \frac{1}{\sqrt{n^3+1}}$, $a_n = \frac{1}{(\ln 3)^n}$, $a_n = (1 - 1/n)^n$, $a_n = \frac{1}{\sqrt[n]{2}}$, $a_n = \ln\left(\frac{n+2}{n+3}\right)$.

There are many more tests, but armed with the above you should be able to deal with just about every situation you will meet as an undergraduate.

Absolute convergence
We will discuss series with nonpositive terms very briefly, offering no proofs.

Definition 4.2. *A series with terms of both signs* converges absolutely *if the sum with all terms replaced by their absolute values converges.*

Thus, for example, of $\sum_1^\infty (-1)^{n+1}/n^2$ converges to $\pi^2/12$ as it stands, and to $\pi^2/6$ if we use the absolute values. Thus it is absolutely convergent. On the other hand the series

$$S = 1 - 1/2 + 1/3 - 1/4 + \cdots \tag{4.2.7}$$

certainly diverges if we use the absolute values whereas the sum given above converges to $\ln 2$.

Definition 4.3. *A series which converges with the minus signs in place, but not without, is said to be* conditionally convergent.

In this book we will be only interested in absolutely convergent series. The adjective absolute may be omitted, but is to be understood.

A result you may find useful is that if the signs alternate and the terms are monotonically decreasing in magnitude, the series converges. We will not prove this.

4.3. Power Series in x

The series considered so far had the feature that a_n depended just on n. Thus the series, if convergent, represented just a number, say 2, in the pizza example. We now introduce a variable x into the series by considering a *power series*

$$S(x) = \sum_{n=0}^\infty a_n x^n. \tag{4.3.1}$$

We can think of all previous cases as the above restricted to $x = 1$. But with x in place, a new feature appears: the series may or may not converge depending on the value of x. We deal with this series exactly as before. Let us take the ratio, which now depends on x and ask when it will imply absolute convergence:

$$r(x) \;=\; \lim_{n \to \infty} \left| \frac{a_{n+1} x^{n+1}}{a_n x^n} \right| \qquad (4.3.2)$$

$$=\; |x| \lim_{n \to \infty} \left| \frac{a_{n+1}}{a_n} \right| \equiv |x| r. \qquad (4.3.3)$$

It follows that the series converges for $|x| r < 1$ or

$$|x| < R = \frac{1}{r} = \lim_{n \to \infty} \left| \frac{a_n}{a_{n+1}} \right| \qquad (4.3.4)$$

where R is called the *interval of convergence*, $-R < x < R$.[1]

Since we can associate with every point in this interval a limit, we have defined a function in this interval. In particular, if the coefficients a_n form a Taylor series, $a_n = f^n(0)/n!$, the series defines a function whose n-th derivative at the origin is $f^n(0)$. It should be clear that if x were replaced by $x - 2$ say, in all of the above, r would still be the same, but the interval of convergence would be centered at $x = 2$.

Let us recall how the ratio test went for the exponential series:

$$e^x = \sum_0^\infty \frac{x^n}{n!}. \qquad (4.3.5)$$

Since $r = \lim_{n \to \infty} 1/(n+1) = 0$, we see $R = \infty$, and the series converges for all finite x. The same was true for the sine, cosine, sinh, and cosh functions.

On the other hand in the case of

$$\frac{1}{1 - x} = 1 + x + x^2 + x^3 + \cdots \qquad (4.3.6)$$

$R = 1$, and we need $|x| < 1$ for the series to converge to the function on the left-hand side. Beyond this, the function has a life of its own, but the series is meaningless. In Chapter 6 we will learn how to get around this.

Problem 4.3.1. *Find R for the following series: (i) $\sum_1^\infty \frac{x^{2n}}{2^n n^2}$, (ii) $\sum_1^\infty \frac{(x-1)^n}{n}$, (iii) $\sum_1^\infty \frac{x^{3n}}{n}$, (iv) $\sum_1^\infty \frac{x^n}{n(n+1)}$, (v) Consider $\sum_1^\infty \frac{n^{1/n}}{x^n}$ and think of $1/x$ as the variable. (vi) Use a similar trick to find the interval of convergence of $\sum_1^\infty \frac{(x-1)^n}{n x^n}$. (vii) Show that $\sum_1^\infty n! x^n$ has $R = 0$.*

---

[1] This formula needs to be modified if the series does not contain every power of x. For example, an even function would contain only even powers of x. In this case you must view it as a series in x^2, and R computed from the ratio of successive terms would give the interval of convergence in x^2.

Let us now imagine that we have series for some function $f(x)$ within some interval of convergence. Here are some properties of such an expansion.

- If $f(x) = g(x)$, they both have the same series. (For example, take $f(x) = 2\sin x \cos x$ and $g(x) = \sin 2x$.) In other words if

$$f(x) = \sum_1^\infty a_n x^n \text{ and} \tag{4.3.7}$$

$$g(x) = \sum_1^\infty b_n x^n \text{ then} \tag{4.3.8}$$

$$f(x) = g(x) \text{ implies} \tag{4.3.9}$$

$$a_n = b_n. \tag{4.3.10}$$

Is this reasonable? In general, given that a sum of terms equals another sum, we cannot equate the constituents of either sum to their counterparts, as is known from the well-known case $4 + 5 = 6 + 3$. *The point is that here the sums are equal at every point within the interval of convergence and there is an infinite number of such points, and hence an infinite number of equations subsumed by the statement $f(x) = g(x)$.* It is reasonable that these in turn lead to an infinite number of related conditions, namely $a_n = b_n$. As to exactly how this happens, note that the Taylor series is found by evaluating all the derivatives of the given function at the origin and the numerical equality of f and g ensures the equality of all the derivatives: $a_n = f^{(n)}(0)/n! = g^{(n)}(0)/n! = b_n$.

Conversely, is it true that two different functions cannot have the same Taylor series? Let us first try to manufacture an example that violates this. Let f be some function with a Taylor series. Let us define $g = f + e^{-1/x}$. Consider the derivative of the added term at the origin. It is given by $\frac{1}{x^2}e^{-1/x}$ which goes to zero as x is reduced to zero since the exponential vanishes faster than the power grows. The same goes for all higher derivatives. Thus it appears that f and g have the same series despite being different functions. The error in the argument is that we have been approaching the origin from just the right. Had we approached from the left, we would have found that all the derivatives were infinite! In other words the added function does not have a unique derivative at the origin and hence has no Taylor expansion there. What if we had added e^{-1/x^2}, for which all the right and left derivatives are equal to zero? To weed out this possibility, and to make the converse relation hold, we must view this $f(x)$ as the value of a function of a "complex variable" $z = x + iy$, evaluated on a limited region where $y = 0$. It will then turn out that there are "other" directions (of nonzero y), besides left and right, along which the derivative is infinite. This explanation is not meant to make too much sense at this point. All you have to know now is that Taylor series and functions will have a one-to-one relation only when real variables are extended to complex variables.

- Sometimes we will truncate the infinite series for a function after some number of terms. If we stop at the x^5 term, we say *we are working to fifth order in the expansion*. We may choose to keep just the first few terms if we just want a good approximation to the function for small x. For example we can approximate $\sin x$ by $x - x^3/6$ if x is small (in radians). Consider the not so small case of 45°, which is $\pi/4$ radians. The exact answer is $1/\sqrt{2} = .70710$ to five places, while the above approximation gives .70465. How small an x is small enough? One common criterion is to add one more term to the approximation and see by how much the sum changes. If the change is small enough, we stop; if not, we add one more term and so on.

Consider for example the case of relativistic kinematics wherein $P = \frac{mv}{\sqrt{1-v^2}}$. The leading term gives $P/m = v$. The next term adds $v^3/2$. For an object moving at 1/100-th the velocity of light, this is a relative correction of roughly one part in 10^4. If this is the desired accuracy, we stop here. Adding one more term changes the answer by roughly one part in 10^6 and so on. It is clear that at small velocities we are better off using a polynomial approximation that is so much easier to handle and yet numerically indistinguishable (for a given tolerance.)

- To obtain the first few terms in the series for a function, we will generally not take the first few derivatives, but follow a shortcut as is illustrated by the following example.

Consider the series for the function $\exp\left[\frac{x}{1-x}\right]$ at the origin.[2] Suppose we want to go up to x^3. The brute force way is to take three derivatives of this object. (Try it!) An easier way is as follows:

$$\exp\left[\frac{x}{1-x}\right]$$
$$= 1 + \frac{x}{1-x} + \frac{1}{2}\left[\frac{x}{1-x}\right]^2 + \frac{1}{3!}\left[\frac{x}{1-x}\right]^3 + O(x^4) \qquad (4.3.11)$$
$$= 1 + x(1 + x + x^2 + \cdots) + \frac{1}{2}x^2(1 + x + \cdots)^2 + \frac{1}{3!}x^3(1)^3 + \cdots$$
$$= 1 + x + \frac{3}{2}x^2 + \frac{13}{6}x^3 + O(x^4). \qquad (4.3.12)$$

Thus we first expanded the exponential to the desired order, then we expanded $x/(1-x)$ to the desired order, dropping all terms of order x^4 and beyond. The series is good for $|x| < 1$.

---

[2] The exponential function e^x is sometimes written as $\exp x$, especially if the exponent is complicated, as in this case.

This is a general strategy: in computing a function which itself is composed of other functions, we can replace each constituent function by its series within the common interval of convergence. These series may be truncated at the desired order to obtain the series for the composite function. Thus to find the series for $f(x) = \sin(\cos x)$ to order x^4, we can replace $\cos x$ by $1 - x^2/2! + x^4/4!$ inside the sine and obtain

$$f(x) = \sin(1 - x^2/2! + x^4/4!) \tag{4.3.13}$$

How many terms should we keep in the expansion of the sine to go to order x^4? An infinite number: if you expand the sine out in powers of its argument, you will see that contributions to x^2 and x^4 terms come from all terms. For example the fifth-order term in the sine, $(1 - x^2/2! + x^4/4!)^5/5!$, contributes $-5x^2/2!5!$ to the x^2 piece of f. So we isolate the 1 in the expansion of the cosine to write (to order x^4)

$$
\begin{aligned}
\sin(1 - \frac{x^2}{2!} + \frac{x^4}{4!}) &= \sin 1 \cos(-\frac{x^2}{2!} + \frac{x^4}{4!}) + \cos 1 \sin(-\frac{x^2}{2!} + \frac{x^4}{4!}) \\
&= (1 - (-\frac{x^2}{2!} + \frac{x^4}{4!})^2/2!) \sin 1 + (-\frac{x^2}{2!} + \frac{x^4}{4!}) \cos 1 \\
&= (1 - \frac{x^4}{(2!2!2!)}) \sin 1 + (-\frac{x^2}{2!} + \frac{x^4}{4!}) \cos 1 \\
&= \sin 1 - \frac{\cos 1}{2} x^2 + (\frac{\cos 1}{24} - \frac{\sin 1}{8}) x^4 \tag{4.3.14}
\end{aligned}
$$

• Consider now an exact relation between functions such as

$$\sin^2 x + \cos^2 x = 1. \tag{4.3.15}$$

Now the sine and cosine have expansions:

$$\sin x = x - \frac{x^3}{3!} + \frac{x^5}{5!} + \cdots \tag{4.3.16}$$

$$\cos x = 1 - \frac{x^2}{2!} + \frac{x^4}{4!} + \cdots \tag{4.3.17}$$

If we insert these series into the left-hand side of Eqn. (4.3.15), and compare it to the right-hand side, where the function has just on term, namely a zeroth order term equal to 1, we should find that all nonzero powers of x must exactly cancel *order by order*. For instance if we work to order x^2 and set

$$\sin x = x \tag{4.3.18}$$

$$\cos x = 1 - \frac{x^2}{2} \tag{4.3.19}$$

we find

$$\sin^2 x + \cos^2 x = x^2 + (1 - \frac{x^2}{2})^2 = x^2 + 1 - x^2 + \text{ higher order terms.}$$

$$(4.3.20)$$

(What about the x^4 term that comes when we square the cosine? We cannot worry about this term, having neglected similar terms earlier in the calculation. If you want the x^4 term to work out correctly, you must keep such a term at *all* stages.)

Problem 4.3.2. *Verify the identity Eqn. (4.3.15) to order x^6.*

Problem 4.3.3. *Consider the Taylor series for $(1+x)^p$, for any real p. What is the coefficient of x^n? What is R, the interval of convergence? Verify that $(1 + x)^p (1 + x)^q = (1 + x)^{p+q}$ to order x^3.*

Problem 4.3.4. *In relativistic mechanics the energy E and momentum P of a particle are given as a function of velocity v (in units where the velocity of light is unity) by*

$$E = \frac{m}{\sqrt{1 - v^2}} \qquad P = \frac{mv}{\sqrt{1 - v^2}}$$

m being the rest mass. Verify that $E^2 = P^2 + m^2$. Expand E and P in a Taylor series keeping up to v^4 term. Verify that this identity is satisfied to this order.

- We have seen that as far as functional relations are concerned we may replace the function by its series. The same goes for derivatives and integrals: if $f(x)$ has a convergent Taylor series within an interval, we may differentiate it term by term to get the series for the derivative and integrate it term by term to obtain the series for the integral, always within the same interval. As an example, consider the series for $\ln(1 + x)$:

$$\ln(1 + x) = x - x^2/2 + x^3/3 + \cdots + (-1)^{n+1}\frac{x^n}{n} + \cdots \qquad (4.3.21)$$

which converges for $|x| < 1$ (as you should verify). Taking derivatives of both sides, we get the series for $1/(1 + x)$, also in the same interval:

$$\frac{1}{1 + x} = 1 - x + x^2 - x^3 + \cdots \qquad (4.3.22)$$

We could just as well start with Eqn. (4.3.22) integrate term by term from 0 to x to get the series for $\log(1 + x)$.

Problem 4.3.5. *Start with*

$$\int_0^x \frac{du}{1+u^2} = \tan^{-1} x.$$

Develop the integrand in a Taylor series at the origin, integrate it term by term by and show $\pi/4 = 1 - 1/3 + 1/5 - 1/7 + \cdots$

Problem 4.3.6. *The exact formula for the period of a simple pendulum of length* L *is*

$$T = 4\sqrt{L/g} \int_0^{\pi/2} \frac{d\phi}{\sqrt{1 - k^2 \sin^2 \phi}}$$

where g *is the acceleration due to gravity and* $k = \sin \frac{\alpha}{2}$, α *being the initial angular displacement. Show that in the limit of very small* α *this reduces to the formula we all learned as children:* $T = 2\pi \sqrt{L/g}$. *By expanding the integrand in a series in* $k^2 \sin^2 \phi$, *develop the next correction. What is the fractional change in* T *due to this term for* $\alpha = \pi/3$? *The integral in question is called an* elliptic integral. *But we do not have to know anything about this fancy object if we just want to get an answer that is numerically very good for small oscillations. In general we can deal with many difficult integrals this way. For example, we do not know how to integrate* e^{-x^2} *between finite limits. However if we want a good approximation to* $\int_0^a e^{-x^2} dx$, *for small* a, *we simply expand the exponential and integrate term by term.*

Problem 4.3.7. *Expand out to* x^4 *the following functions: (i)* $\sin(x + \frac{\pi}{4})$ *(ii)* e^{x-1}, *(iii)* $\ln(2 + x)$.

Sometimes we need to determine the convergence of series of the form:

$$f(x) = \sum_1^\infty \frac{\sin nx}{n^2}. \tag{4.3.23}$$

In this case we can argue that since $|\sin x| \leq 1$, and the series with all the sines set equal to unity converges, so does $f(x)$. There are however more complications in such series where the terms are not just powers of x; there are various notions like absolute versus uniform convergence. We do not discuss them here but simply alert you that in such cases you may not be able to differentiate the series to get the series for the derivative of f.

4.4. Summary

Here is the summary of this chapter.

- The series $\sum_{n=0}^{\infty} a_n$ converges to S if $|S - S_N| < \varepsilon$ for $N > N(\varepsilon)$ where S_N is the sum of the first N terms.

- The series

$$\sum_{n=0}^{\infty} a_n$$

  converges absolutely if the series with a_n replaced by $|a_n|$ converges. If a series converges but not if the terms are replaced by absolute values, then it converges conditionally.

- Know the following about the geometric series

$$1 + r + r^2 + \cdots + r^n = \frac{1 - r^{n+1}}{1 - r}$$

$$\sum_{n=0}^{\infty} r^n = \frac{1}{1 - r} \qquad |r| < 1$$

- The series

$$\sum_{n=0}^{\infty} a_n$$

  of positive terms converges if $a_n \leq b_n$ for n beyond some value and if the sum over the positive numbers b_n converges. This is called the comparison test.

- The series

$$\sum_{n=0}^{\infty} a_n$$

  converges absolutely if

$$\lim_{n \to \infty} \left| \frac{a_{n+1}}{a_n} \right| < 1.$$

  This is the ratio test.

- The sum of positive monotonic terms $\sum_{n=0}^{\infty} f(n)$ converges or diverges along with $\int^L f(x)\,dx$ as $L \to \infty$. To do the integral it is enough to use an integrand that agrees with f for large x. You can also trade the given f for another smaller one (while showing the sum diverges) or a larger one (to show it converges) if that makes it easy to integrate.

- The power series $\sum_{n=0}^{\infty} a_n x^n$ converges absolutely within the interval of convergence $|x| < R$, where

$$R = \lim_{n \to \infty} \left| \frac{a_n}{a_{n+1}} \right|.$$

- Relations between functions will be satisfied order by order when they are replaced by their power series. You must know how to expand functions of functions, out to some desired order within the common interval of convergence.

- The power series representing a function may be integrated term by term and differentiated term by term within the interval of convergence to obtain the series for the integral or derivative of the function in question.

COMPLEX NUMBERS

5.1. Introduction

Let us consider the quadratic function $f(x) = x^2 - 5x + 6$ and ask where it vanishes. If we plot it against x, we will find that it vanishes at $x = 2$ and $x = 3$. This is also clear if we write f in factorized form as $f(x) = (x - 2)(x - 3)$. We could equivalently use the well-known formula for the roots $x_\pm$ of a quadratic equation:

$$ax^2 + bx + c = 0 \qquad (5.1.1)$$

namely

$$x_\pm = \frac{-b \pm \sqrt{b^2 - 4ac}}{2a} \qquad (5.1.2)$$

to find the roots $x_\pm = 2, 3$. Suppose instead we consider

$$x^2 + x + 1 = 0. \qquad (5.1.3)$$

A plot will show that this function is always positive and does not vanish for any point on the x-axis. We are then led to conclude that this quadratic equation has no roots. Let us pass from the graphical procedure which gives no solution to the algebraic one which does give some form of answer even now. It says

$$x_\pm = \frac{-1 \pm \sqrt{-3}}{2}. \qquad (5.1.4)$$

The problem of course is that we do not know what to make of $\sqrt{-3}$ since there is no real number whose square is -3. Thus if we take the stand that a number is not a number unless it is a real number, we will have to conclude that some quadratic equations have roots and other do not.

5.2. Complex Numbers in Cartesian Form

This is how it was for many centuries until the rather bold suggestions was made that we admit into the fold of numbers also those of the form $\sqrt{-3}$. *All we need to know to do this is a set of consistent rules for manipulating such numbers; being able to visualize them is no prerequisite.* It is really up to us make up the rules since these entities have come out of the blue. The rules must, however, be free of contradictions. Of course, all this is pointless if the whole enterprise does nothing but merely exist. In the present case the idea has proven to be a very seminal one and we will see some of the evidence even in this elementary treatment.

We are dealing here with a case of mathematical abstraction or generalization, an example of which you have already seen, when we extended the notion of powers from integers to all real values, and examples of which you will see more than once in this course, say when we extend the notion of vectors in three dimensions with real components (which we can readily imagine) to vectors in any number of dimensions, or vectors with complex components (like $\sqrt{-3}$) which we cannot visualize. A general guideline when embarking on such generalizations is that we impose on the new entities as many properties of the more familiar entities as is possible. For example, when we passed from integer powers a^n to arbitrary powers a^x (whatever that meant) we demanded that noninteger powers obey the same rule of composition, i.e., $a^x a^y = a^{x+y}$ for all x and y.

Returning to our problem we will first demand that square roots of negative numbers (whatever they mean) still obey the rule that $\sqrt{ab} = \sqrt{a}\sqrt{b}$. Thus $\sqrt{-3} = \sqrt{3}\sqrt{-1}$. *The point of this is that the problem of taking the square root of any negative number is reduced to taking the root of -1.* Thus the basic building block we need to introduce, called the unit imaginary number, is

$$i = \sqrt{-1}. \tag{5.2.1}$$

In terms of i, the answer to Eqn. (5.1.3) is

$$x_\pm = -\frac{1}{2} \pm \frac{\sqrt{3}}{2}i. \tag{5.2.2}$$

We will postulate that i will behave like a real number in all manipulations involving addition and multiplication and that the only new feature it will have is the one that defined it, namely that its square equals -1.

We now introduce a general complex number

$$z = x + iy \tag{5.2.3}$$

and refer to x and y as its *real and imaginary parts* and denote them by the symbols Re z and Im z. *A number with just $y \neq 0$ is called a pure imaginary number.* The solution to our quadratic equation has a real part $x = -\frac{1}{2}$ and an imaginary part

$y = \pm\sqrt{3}/2$. We think of $x + iy$ as a *single number*. Indeed, if the number inside the radical had been 3 instead of -3, surely we would have treated say $-\frac{1}{2} + \sqrt{3}/2$ as a single number, as one of the roots. The same goes for $-\frac{1}{2} + i\sqrt{3}/2$.

The rules obeyed by complex numbers are as follows. Given two of them z_1 and z_2,

$$z_1 = x_1 + iy_1 \tag{5.2.4}$$

$$z_2 = x_2 + iy_2, \quad \text{we define,} \tag{5.2.5}$$

$$z_1 + z_2 = (x_1 + x_2) + i(y_1 + y_2) \quad \text{addition rule} \tag{5.2.6}$$

$$z_1 z_2 = (x_1 x_2 - y_1 y_2) + i(x_1 y_2 + x_2 y_1) \quad \text{multiplication rule} \tag{5.2.7}$$

where in the last equation we have opened out the brackets as we do with real numbers and used $i^2 = -1$. If we use these rules, we can verify that $x_\pm = -\frac{1}{2} \pm \frac{\sqrt{3}}{2}i$ indeed satisfies Eqn. (5.1.1). Note that $z_1 z_2 = z_2 z_1$.

Problem 5.2.1. *Verify that this is so.*

It was emphasized that we must think of $z = x + iy$ as a single number. *However it is a single number which has two parts, which can be uniquely identified.* Thus although $7 = 5 + 2$ is a single number, the decomposition of 7 into 2 and 5 is not unique. On the other hand $z = 3 + 4i$ has a real part 3 and an imaginary part 4 and we cannot move things back and forth between the real and imaginary parts keeping the number fixed. *Thus if two complex numbers are equal, their real parts and imaginary parts are separately equal:*

$$z_1 = x_1 + iy_1 = z_2 = x_2 + iy_2 \text{ implies} \tag{5.2.8}$$

$$x_1 = x_2 \tag{5.2.9}$$

$$y_1 = y_2. \tag{5.2.10}$$

Suppose this were not true. This would imply $x_1 - x_2 = i(y_2 - y_1)$, without both of them vanishing separately. Squaring both sides, we would find a positive definite left-hand side and a negative definite right-hand side. The only way to avoid a contradiction is for both sides to vanish, giving us $0 = -0$, which is something we can live with.

Now, given any real number x, we can associate with it a unique number $-x$, called its negative. We can do that with a complex number $z = x + iy$ too, by negating x and y. This number is called $-z$. But now we have an intermediate choice in which we negate just y: the result

$$z^* = x - iy \tag{5.2.11}$$

pronounced "z-star" is called the *complex conjugate* of z. Some people like to write it $\bar{z}$ and call it "z-bar".

Note that

$$zz^* = x^2 + y^2 \equiv |z|^2 \geq 0. \tag{5.2.12}$$

One refers to $|z| = \sqrt{x^2 + y^2}$ as the *modulus* or *absolute value* of the complex number z. It is useful to know that given z and its z^*, we can recover the real and imaginary parts of z as follows:

$$\text{Re } z \equiv x \;=\; \frac{z + z^*}{2} \tag{5.2.13}$$

$$\text{Im } z \equiv y \;=\; \frac{z - z^*}{2i} \tag{5.2.14}$$

Note that we did not ever explicitly define the rules for division of complex numbers. This is because we can carry out division as the inverse of multiplication. Thus if $z_1/z_2 = z_3$, we can multiply both sides by z_2 (which we know how to do) and solve for x_3 and y_3 by equating real and imaginary parts in

$$x_1 + iy_1 = (x_3 + iy_3)(x_2 + iy_2) \tag{5.2.15}$$

to obtain (upon solving a pair of simultaneous equations)

$$x_3 \;=\; \frac{x_1 x_2 + y_1 y_2}{x_2^2 + y_2^2} \tag{5.2.16}$$

$$y_3 \;=\; \frac{y_1 x_2 - y_2 x_1}{x_2^2 + y_2^2}. \tag{5.2.17}$$

This result is more easily obtained by using the notion of complex conjugates. First note that

$$\frac{1}{z} \;=\; \frac{z^*}{zz^*} \tag{5.2.18}$$

$$\;=\; \frac{x - iy}{x^2 + y^2}. \tag{5.2.19}$$

Applying this result to the $1/z_2$ in the ratio z_1/z_2, we can obtain Eqns. (5.2.16, 5.2.17) more easily than before. In other words, by invoking the complex conjugate we have reduced the problem to division by a *real* number, namely z^*z, which is a familiar concept.

Complex conjugation can be viewed as the process of replacing i by $-i$ within the complex number. Stated this way it is clear that the complex conjugate of a product is the product of the complex conjugates

$$(z_1 z_2)^* = z_1^* z_2^* \tag{5.2.20}$$

as you may check by explicit evaluation of both sides. The same obviously is true for the sum of two complex numbers. Now, if two complex numbers are equal, their real imaginary parts are separately equal. Consider an equation with complex numbers on both sides. *If we replace all the numbers by their conjugates, the resulting quantities must still be equal.* This is because the imaginary parts, originally equal on both sides, will continue to be equal after signs are changed on both sides. Thus every complex equation implies another one obtained by complex conjugation of both sides. The latter does not contain any more or less information: both tell us the real and imaginary parts are separately equal.

Let us consider an illustrative example. Let

$$z_1 = \frac{\sqrt{2} + i}{1 - i} \qquad\qquad z_2 = \frac{1 + i}{\sqrt{2}} \qquad (5.2.21)$$

Let us first simplify z_1 to the form $x + iy$ by multiplying the numerator and denominator by the complex conjugate of the latter:

$$z_1 = \frac{(\sqrt{2} + i)(1 + i)}{(1 - i)(1 + i)} = \frac{(\sqrt{2} - 1) + i(\sqrt{2} + 1)}{2} \qquad (5.2.22)$$

Since z_2 is already in this form, let us move on to compute

$$z_1 z_2 = \frac{(\sqrt{2} - 1) + i(\sqrt{2} + 1)}{2} \cdot \frac{1 + i}{\sqrt{2}} \qquad (5.2.23)$$

$$= \frac{(\sqrt{2} - 1) + i(\sqrt{2} + 1) + i(\sqrt{2} - 1) + i^2(\sqrt{2} + 1)}{2\sqrt{2}} \qquad (5.2.24)$$

$$= i - \frac{1}{\sqrt{2}} \qquad (5.2.25)$$

and

$$\frac{z_1}{z_2} = \frac{(\sqrt{2} - 1) + i(\sqrt{2} + 1)}{\sqrt{2}(1 + i)} \qquad (5.2.26)$$

$$= \frac{[(\sqrt{2} - 1) + i(\sqrt{2} + 1)][1 - i]}{2\sqrt{2}} \qquad (5.2.27)$$

$$= 1 + \frac{i}{\sqrt{2}}. \qquad (5.2.28)$$

Problem 5.2.2. *Show that* $z_2^2 = i$.

The following exercises should give you some more practice with the manipulation of complex numbers.

Problem 5.2.3. *Solve for x and y given*

$$\frac{2 + 3i}{6 + 7i} + \frac{2}{x + iy} = 2 + 9i.$$

Problem 5.2.4. *Find the real part, imaginary part, modulus, complex conjugate, and inverse of the following numbers:* (i) $\frac{2}{3+4i}$, (ii) $(3+4i)^2$, (iii) $\frac{3+4i}{3-4i}$, (iv) $\frac{1+\sqrt{2}i}{1-\sqrt{3}i}$, *and* (v) $\cos\theta + i\sin\theta$.

Problem 5.2.5. *Show that a polynomial with real coefficients has only real roots or complex roots that come in complex conjugate pairs.*

Problem 5.2.6. (Very important). *Prove algebraically that*

$$|\operatorname{Re}\ z| \ \leq \ |z| \tag{5.2.29}$$

$$|\operatorname{Im}\ z| \ \leq \ |z|. \tag{5.2.30}$$

$$|z_1 + z_2|^2 = |z_1|^2 + |z_2|^2 + 2\operatorname{Re}(z_1 z_2^*) \tag{5.2.31}$$

$$|z_1 + z_2| \leq |z_1| + |z_2| \tag{5.2.32}$$

$$|z_1 z_2| = |z_1||z_2| \tag{5.2.33}$$

Problem 5.2.7. (Important). *Verify that the numbers z_1, z_2 from Eqn. (5.2.21) respect Eqns. (5.2.31-5.2.33).*

Recall that all real numbers can be visualized as points on a line, called the x axis. To visualize all complex numbers we introduce the *complex plane* which is just the $x - y$ plane. The complex number $z = x + iy$ is labeled as shown in Fig. 5.1. The conjugate is z^*. The significance of r and θ will now be explained.

5.3.   Polar Form of Complex Numbers

We begin this section with a remarkable identity due to Euler:

$$e^{i\theta} = \cos\theta + i\sin\theta, \tag{5.3.1}$$

where we will choose θ to be real. To prove this identity, we must define what we mean by e raised to a complex power $i\theta$. We define $e^{anything}$ to be the infinite power series associated with the exponential function e^x with x replaced by $anything$. Thus

$$e^{elephant} = \sum_{n=0}^{\infty} \frac{(elephant)^n}{n!} \tag{5.3.2}$$

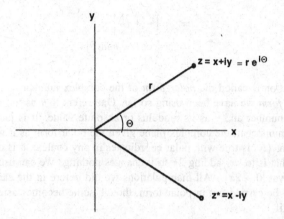

Figure 5.1. The complex plane.

which converges for any finite sized *elephant*.

Turning to our problem, we expand the infinite series for the exponential and collect the real and imaginary parts as follows:

$$e^{i\theta} \;=\; \sum_{n=0}^{\infty} \frac{(i\theta)^n}{n!} \tag{5.3.3}$$

$$=\; \sum_{n=0}^{\infty} \frac{(-1)^n(\theta)^{2n}}{(2n)!} + i \sum_{n=0}^{\infty} \frac{(-1)^n(\theta)^{2n+1}}{(2n+1)!} \tag{5.3.4}$$

$$=\; \cos\theta + i\sin\theta \tag{5.3.5}$$

where we have used the fact that $i^2 = -1$, $i^3 = -i$, $i^4 = 1$, and so on, as well as the infinite series that define the sine and cosine functions. (These expansions converge for all finite θ, as shown before. The presence of i does not in any way complicate the question of convergence since it either turns into a ± 1 or into $\pm i$.) Setting $\theta = \pi$ we obtain one of the most remarkable formulae in mathematics:

$$e^{i\pi} + 1 = 0. \tag{5.3.6}$$

Who would have thought that π which enters as the ratio of circumference to diameter, e, as the natural base for logarithms, i, as the fundamental imaginary unit and 0 and 1 (which we know all about from infancy) would all be tied together in any way, not to mention such a simple and compact way? I hope I never stumble into anything like this formula, for nothing I do after that in life would have any significance.

Look at Fig. 5.1 of the complex plane and note that

$$z \;=\; x + iy \tag{5.3.7}$$

$$= r\left[\frac{x}{r} + i\frac{y}{r}\right] \tag{5.3.8}$$

$$= r\left[\cos\theta + i\sin\theta\right] \tag{5.3.9}$$

$$= re^{i\theta}. \tag{5.3.10}$$

The last equation is called the *polar form* of the complex number as compared to the *cartesian form* we have been using so far. One refers to θ as the *argument* or *phase* of the number and r as its modulus or absolute value. It is just as easy to visualize the number in the complex plane given the polar form as it was with the cartesian form. (As is true with polar coordinates in any context, θ is defined only modulo 2π, that is to say adding 2π to it changes nothing. We can usually restrict it to the interval $[0 - 2\pi]$.) All manipulations we did before in the cartesian form can of course be carried out in polar form, though some become easier and some harder. Thus if

$$z = re^{i\theta} \text{ then} \tag{5.3.11}$$

$$z^* = re^{-i\theta} \tag{5.3.12}$$

$$zz^* = r^2 \quad (|z| = r) \tag{5.3.13}$$

$$\frac{1}{z} = \frac{1}{r}e^{-i\theta} \tag{5.3.14}$$

$$z_1 z_2 = r_1 r_2 e^{i(\theta_1 + \theta_2)}. \tag{5.3.15}$$

(We are using the fact that the law of composition of exponents under a product works for complex exponents as well. Indeed this is built into the exponential function defined by the infinite series. You may check that this works to any given order even for imaginary arguments. In Chapter 6, this will be proven more directly.) The last formula tells us how easy it is to multiply or divide two complex numbers in polar form:

To multiply two complex numbers, multiply their moduli and add their phases. To divide, divide by the modulus and subtract the phase of the denominator.

On the other hand to add two complex numbers we have to go back to the cartesian form, add the components and revert to the polar form.

Let us return to Eqn. (5.2.21) and manipulate the numbers in polar form. First

$$z_1 = \frac{(\sqrt{2} - 1) + i(\sqrt{2} + 1)}{2} \tag{5.3.16}$$

$$= \sqrt{\frac{(\sqrt{2} - 1)^2 + (\sqrt{2} + 1)^2}{4}} \exp\left[i \arctan\left[\frac{1 + \sqrt{2}}{\sqrt{2} - 1}\right]\right] \tag{5.3.17}$$

$$= \sqrt{\frac{3}{2}} e^{1.400i}. \tag{5.3.18}$$

As for z_2,

$$z_2 = \sqrt{\frac{1}{2} + \frac{1}{2}} e^{i \arctan 1} = e^{i\pi/4} = e^{.785i}. \tag{5.3.19}$$

Now it is easy to form the product and quotient

$$z_1 z_2 = \sqrt{\frac{3}{2}} \cdot 1 e^{(1.400 + .785)i} \tag{5.3.20}$$

$$= \sqrt{\frac{3}{2}} e^{2.185i} = 1.224 \left(\cos 2.185 + i \sin 2.185 \right) \tag{5.3.21}$$

$$= -\frac{1}{\sqrt{2}} + i \tag{5.3.22}$$

$$\frac{z_1}{z_2} = \sqrt{\frac{3}{2}} e^{(1.400 - .785)i} = 1.224 \left(\cos .615 + i \sin .615 \right)$$

$$= 1 + \frac{i}{\sqrt{2}} \tag{5.3.23}$$

in agreement with the calculation done earlier in cartesian form.

Complex numbers $z = re^{i\theta}$ with $r = 1$ have $|z| = 1$ and are called *unimodular*. We may imagine them as lying on a circle of unit radius in the complex plane. Special points on this circle are

$$\theta = 0 \quad (1) \tag{5.3.24}$$

$$= \pi/2 \quad (i) \tag{5.3.25}$$

$$= \pi \quad (-1) \tag{5.3.26}$$

$$= -\pi/2 \quad (-i). \tag{5.3.27}$$

You are expected to know these points at all times.

Problem 5.3.1. *Verify the correctness of the above using Euler's formula.*

When we work with real numbers, we know that multiplication by a number, say 4, rescales the given number by 4. *Multiplying a number in the complex plane by $re^{i\theta}$, rescales its length (or modulus) by r and also rotates it counterclockwise by θ.* Multiplying by a unimodular number simply rotates without any rescaling.

Problem 5.3.2. *For the following pairs of numbers, give their polar form, their complex conjugates, their moduli, product, the quotient z_1/z_2, and the complex conjugate of the quotient:*

$$z_1 = \frac{1+i}{\sqrt{2}} \quad z_2 = \sqrt{3} - i$$

$$z_1 = \frac{3+4i}{3-4i} \quad z_2 = \left[\frac{1+2i}{1-3i} \right]^2.$$

Problem 5.3.3. *Express the sum of the following in* polar form:

$$z_1 = 2e^{i\pi/4} \quad z_2 = 6e^{i\pi/3}.$$

Recall from Euler's formula that for real θ

$$\cos\theta \;=\; \mathrm{Re}\; e^{i\theta} = \frac{e^{i\theta} + e^{-i\theta}}{2} \qquad (5.3.28)$$

$$\sin\theta \;=\; \mathrm{Im}\; e^{i\theta} = \frac{e^{i\theta} - e^{-i\theta}}{2i}. \qquad (5.3.29)$$

You should remember the above results at all times.

Problem 5.3.4. *Check the following familiar trigonometric identities by expressing all functions in terms of exponentials:* $\sin^2 x + \cos^2 x = 1$, $\sin 2x = 2\sin x \cos x$, $\cos 2x = \cos^2 x - \sin^2 x$.

Problem 5.3.5. *Consider the series*

$$e^{i\theta} + e^{3i\theta} + \cdots + e^{(2n-1)i\theta}. \qquad (5.3.30)$$

Sum this geometric series, take the real and imaginary parts of both sides and show that

$$\cos\theta + \cos 3\theta + \cdots + \cos(2n-1)\theta = \frac{\sin 2n\theta}{2\sin\theta}$$

and that a similar sum with sines adds up to $\sin^2 n\theta / \sin\theta$.

Problem 5.3.6. *Consider* De Moivre's Theorem, *which states that* $(\cos\theta + i\sin\theta)^n = \cos n\theta + i\sin n\theta$. *This follows from taking the n-th power of both sides of Euler's theorem. Find the formula for* $\cos 4\theta$ *and* $\sin 4\theta$ *is terms of* $\cos\theta$ *and* $\sin\theta$. *Given* $e^{iA}e^{iB} = e^{i(A+B)}$ *deduce* $\cos(A+B)$ *and* $\sin(A+B)$.

5.4. An Application

We will now consider a situation where complex numbers not only help in solving the problem, but also appear as the *natural* things to use. Consider the simple LCR electrical circuit shown in the left half of Fig. 5.2.

Our goal is to calculate the current $I(t)$ in the circuit at time t, given the applied voltage $V(t)$. Our strategy will be to first calculate the charge on the capacitor $Q(t)$, and then use

$$I(t) = \frac{dQ}{dt}. \qquad (5.4.1)$$

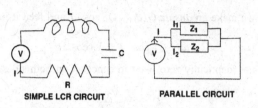

SIMPLE LCR CIRCUIT   **PARALLEL CIRCUIT**

Figure 5.2. The LCR circuit.

At any time, we may relate the charge $Q(t)$ to the voltage $V(t)$ in terms of the resistance R, inductance L and capacitance C as follows:

$$L\frac{d^2Q}{dt^2} + R\frac{dQ}{dt} + \frac{Q}{C} = V(t). \tag{5.4.2}$$

This is called a *differential equation*: it relates Q and its derivatives to the applied voltage and we are to determine Q using this information. Since we are not familiar with differential equations, we cannot take the problem head on. So will try to finesse the problem in stages.

One problem we can solve at this point is when only R is present. In that case the equation reduces to Ohm's Law for the current and gives

$$I(t) = \frac{V(t)}{R}. \tag{5.4.3}$$

In other words we have just an *algebraic equation* for the current. Finding the response is just as easy as dividing V by R.

To proceed further, let us specialize to the case of considerable practical importance, where

$$V(t) = V_0 \cos \omega t. \tag{5.4.4}$$

This form of voltage is interesting because many AC generators produce an output of essentially this form and because, as we shall see later, any periodic voltage can be written as a sum of periodic cosines and given the response to each term in the sum, the response to the total is the sum of the individual responses.

As stated above, in the absence of L and C we know the answer:

$$I(t) = \frac{V_0}{R} \cos \omega t. \tag{5.4.5}$$

Thus the process of finding the current for a given applied voltage reduces to dividing by a number R, in particular the cosine voltage leads to a cosine current.

Suppose only L is present. Then the equation tells to us find a function $Q(t)$ such that its second derivative is proportional to $\cos \omega t$. We can readily guess the answer, $Q(t)$ itself must be proportional to $\cos \omega t$. To find the constant of proportionality, we make an *ansatz* $Q(t) = Q_0 \cos \omega t$ and feed it into Eqn. (5.4.2). This gives

$$- \omega^2 L Q_0 \cos \omega t = V_0 \cos \omega t. \tag{5.4.6}$$

Since $\cos \omega t$ is not identically zero, we can cancel it on both sides to find that

$$Q_0 = \frac{V_0}{-\omega^2 L} \tag{5.4.7}$$

which solves the problem. Notice how we converted the differential equation (5.4.2) to an algebraic equation (5.4.7) by making the right *ansatz*. This will be our general strategy. (This idea of making an educated guess or *ansatz* with some parameters in it, feeding it into the equation and determining the parameters is a very useful one. We shall use it here and in later chapters.)

If we have just the capacitor, the equation for Q is again algebraic and we can solve for Q_0.

We can go a step further and solve the problem with both L and C since the *ansatz* of a cosine for the charge would work, with an amplitude determined by the equation to be

$$Q_0 = \frac{V_0}{-\omega^2 L + 1/C}. \tag{5.4.8}$$

Problem 5.4.1. *Verify the preceding equation.*

The problem when all three elements are present is that neither a cosine (nor sine) *ansatz* will work. That is to say, neither function gives a cosine when differentiated once, twice, or zero times.

Suppose however that the applied voltage is an exponential. We can then find an *ansatz* that is also exponential and tune its coefficient as dictated by the equation. *This is because the exponential remains an exponential when differentiated any number of times.* But the exponential will either grow or decay indefinitely with time and such problems are not of much interest. Consider now a *complex* exponential $e^{i\omega t}$. From Euler's formula we know that its real and imaginary parts execute bounded oscillations. Indeed its real part is just the cosine we want. But what shall we do with the imaginary part which inevitably comes along? We deal with this problem as follows.

If we examine Eqn. (5.4.2) we see that it is *linear*. This means that if Q_1 is produced by V_1 and Q_2 by V_2, then the linear combination $aV_1 + bV_2$ will produce a response which is given by the same linear combination of charges, $aQ_1 + bQ_2$. This is called the *superposition principle*.

Problem 5.4.2. (Important). *Verify this by writing down two equations, one for Q_1, V_1 and the other for Q_2, V_2 and forming the appropriate linear combinations.*

Let us now apply this logic to $V_1 = V_0 \cos \omega t$, $V_2 = V_0 \sin \omega t$, $a = 1$ and $b = i$. Thus the total applied voltage is $V = V_1 + iV_2 = V_0 e^{i\omega t}$. If we find the total

$Q = Q_1 + iQ_2$ due to this, the real part of the answer, Q_1 is attributable to the real part of the voltage since we know a real voltage will produce a real current and a purely imaginary voltage will produce a purely imaginary current. But the real part of the voltage is precisely the one we began with, $V_0 \cos \omega t$. Thus what we must do is

- Apply the voltage $V_0 e^{i\omega t}$.

- Find the response.

- Take the real part of the answer.

Let us now feed the *ansatz*

$$Q(t) = Q_0 e^{i\omega t} \qquad (5.4.9)$$

into Eqn. (5.4.2) with $V(t) = V_0 e^{i\omega t}$ to obtain (upon canceling the exponential function on both sides),

$$(-\omega^2 L + i\omega R + 1/C)Q_0 = V_0 \qquad (5.4.10)$$

the solution to which is

$$Q_0 = \frac{V_0}{(-\omega^2 L + i\omega R + 1/C)}, \qquad (5.4.11)$$

$$Q(t) = \frac{V_0 e^{i\omega t}}{(-\omega^2 L + i\omega R + 1/C)}. \qquad (5.4.12)$$

The current is now given by differentiation:

$$I(t) = I_0 e^{i\omega t} = \frac{V_0 e^{i\omega t}}{(i\omega L + R + 1/(i\omega C))} \qquad (5.4.13)$$

$$\equiv \frac{V_0 e^{i\omega t}}{Z} \qquad (5.4.14)$$

which defines the *impedance* Z. The current in the original problem is then

$$I(t) = \mathrm{Re}\left[\frac{V_0}{Z} e^{i\omega t}\right]. \qquad (5.4.15)$$

Let us analyze this result a bit. Let us assume without loss of generality that V_0 is real.[1] As for Z, it is generally complex:

$$Z = (i\omega L + R + 1/(i\omega C)) \qquad (5.4.16)$$

---

[1] If V_0 has a phase ψ, it will simply change the real part of $V(t)$ from $V_0 \cos \omega t$ to $V_0 \cos(\omega t + \psi)$. We can reset the zero of our clock so as to eliminate this phase. What we are looking for is the phase *difference* between the applied voltage and current, which cannot be altered by any such resetting of clocks.

$$= |Z|e^{i\phi} \quad \text{where} \tag{5.4.17}$$

$$|Z| = \sqrt{R^2 + (\omega L - 1/\omega C)^2} \tag{5.4.18}$$

$$\tan\phi = \frac{(\omega L - 1/\omega C)}{R}. \tag{5.4.19}$$

Thus

$$I_0 = \frac{V_0}{Z} \tag{5.4.20}$$

$$= \frac{V_0}{|Z|}e^{-i\phi} \quad \text{so that} \tag{5.4.21}$$

$$I(t) = \text{Re}\left[\frac{V_0}{|Z|}e^{i(\omega t - \phi)}\right] \tag{5.4.22}$$

$$= \frac{V_0}{|Z|}\cos(\omega t - \phi). \tag{5.4.23}$$

Let us observe several things. First, we have managed to solve a differential equation by reducing it to just an algebraic equation (5.4.20) by the use of complex numbers: the current (in amplitude) is given by the voltage (in amplitude) divided by the impedance. This is just like Ohm's law for a purely resistive circuit.

Next we note that the final current is neither $\cos\omega t$ nor $\sin\omega t$, it is $\cos(\omega t - \phi)$, which is a linear combination of cosine and sine. We say that the current here lags the voltage in phase by ϕ with respect to the applied voltage. (This is assuming ϕ is positive; if not, we say it leads by ϕ.) This is why we could not get away with any simple *ansatz*. Now it is clear in hindsight that an *ansatz* with a lagging or leading cosine would do the trick. So who needs complex numbers? There are several answers to this.

First note that unlike in a problem with just a resistor, where the current is in phase with the voltage and related to it by a single proportionality factor R, in the general case, it is shifted in phase as well. Thus we need two pieces of information to go from the applied voltage to the current it produces. Why not store them as a pair of real numbers? *We can, but note that in the present treatment, the information is very naturally stored in one complex number Z, which relates the amplitude of the current to that of the voltage by a natural generalization of Ohm's law, Eqn. (5.4.20).* We have seen earlier that multiplying (or dividing) a given number by another complex number has the twin effect of changing the magnitude of the former as well as rotating it in the complex plane. *But this is exactly what we want to do in going from the voltage to the current.* Again one can ask why we do not store the two pieces of information as a vector in two dimensions. The problem is that there is no natural way to define division by a vector. Thus even though complex numbers add like vectors in two dimensions, they behave more like real numbers in that they can be divided by each other.

For the diehards who will not give up real numbers, here is an example of a more complicated circuit, shown in the right half of Fig. 5.2. Try guessing the current in this problem using just real numbers! But with complex numbers, where the generalized Ohm's law works, we have the familiar result for adding resistances in parallel generalized as follows: the total impedance for this circuit is

$$\frac{1}{Z} = \frac{1}{Z_1} + \frac{1}{Z_2} \tag{5.4.24}$$

Problem 5.4.3. *Provide the arguments to validate the above equation. As a warm up, do the problem with two impedances in series.*

Problem 5.4.4. *Consider an LCR circuit with $R = 1$, $L = .02$ and $C = .001$ with an applied voltage $V(t) = 100\cos 200t$. What is the impedance of this circuit in cartesian and polar form? What is the amplitude of the current? By how much does the current lag or lead the voltage? If we can vary the frequency, at what value will the current have the largest amplitude? This is called* resonance.

Problem 5.4.5. *If $Z_1 = 3 + 4i$ and $Z_2 = 1 + i$, in the parallel circuit, what is the net impedance?*

Problem 5.4.6. *Consider a circuit in which L, C and R are connected in parallel to a voltage source at frequency ω. What is the impedance?*

Transients
Now for a technical point. The solutions $Q(t)$ and $I(t)$ that we found are not the most general solutions; it is possible to add to them extra functions of time, called *transients*, and still satisfy the circuit equation. This will now be explained.

Suppose you are asked to solve the equation

$$\frac{d^2y}{dx^2} = 2 \tag{5.4.25}$$

and come up with the solution $y = x^2$. You know that while this is a solution, it is not unique: we can add to it the function $y_c(x) = Ax + B$ where A and B are arbitrary. This is because the derivatives on the left-hand-side of the above equation kill this extra piece, called the *complementary function*.

In the same way, we can add to our solution for $Q(t)$ any function $Q_c(t)$ that obeys

$$L\frac{d^2Q_c}{dt^2} + R\frac{dQ_c}{dt} + \frac{Q_c}{C} = 0 \tag{5.4.26}$$

and still have a solution. Equivalently we can say that the applied voltage is $V(t) = V_0 e^{i\omega t} = V_0 e^{i\omega t} + 0$, and by the superposition principle, Q_c, the response to 0, must be added to the Q we got earlier. We shall refer to the latter as the *particular solution* $Q_p(t)$. Thus the full solution is

$$Q(t) = Q_p(t) + Q_c(t). \tag{5.4.27}$$

The physical origin of Q_c, the response to zero applied voltage, is easy to understand. Suppose we remove the external voltage, charge up the capacitor, and close the circuit. The charge will rush from one plate to the other to render them both at the same potential. But when this is done, the inductance, which does not like changes in current, will not let the current die down, it will keep it flowing until the capacitor gets charged the other way. The charge and current continue to slosh back and forth. They will however decay exponentially in amplitude since the resistor is dissipating the energy and we started out with a fixed amount of stored energy in the capacitor. The precise form of decay will be discussed in Chapter 10.

In circuit terminology, the complementary and particular solutions are called the *transient and steady state* responses since in the long run, only the response to the applied voltage will survive, the transients having died down to exponentially small values.

All of this has a mechanical counterpart. Suppose we have a mass m coupled to a spring of force constant k, subject to a frictional force $-\gamma \frac{dx}{dt}$ and an external force $F(t)$. The equation of motion

$$m\frac{d^2x}{dt^2} + \gamma\frac{dx}{dt} + kx = F(t) \tag{5.4.28}$$

is identical in form to the one for $Q(t)$ and differs only in the change of labels which do not affect the mathematics in any way:

$$x \;\leftrightarrow\; Q \tag{5.4.29}$$
$$F \;\leftrightarrow\; V \tag{5.4.30}$$
$$m \;\leftrightarrow\; L \tag{5.4.31}$$
$$\gamma \;\leftrightarrow\; R \tag{5.4.32}$$
$$k \;\leftrightarrow\; 1/C. \tag{5.4.33}$$

The transient response here corresponds to free oscillations of the system in the absence of any applied force, say the response we get when we pull the spring and release it.

In both the mechanical and electrical problems, the transient will contain two free parameters. They will be determined by the conditions at some time, say $t = 0$. In the mechanical example these conditions will describe the position and velocity of the mass. Since Newton's Law only fixes the acceleration in terms of the applied forces, these parameters are not restricted.

Problem 5.4.7. *Find the charge $Q(t)$ when $L = .5$, $R = 6$ and $C = .02$ are connected in series with a voltage $V(t) = 240 \sin 10t$. You are given both I and Q vanish at $t = 0$.*

5.5. Summary

Here are the highlights.

- If
$$z = x + iy,$$
  then x is the real part or Re z, y is the imaginary part or Im z. This is the cartesian form of z.

- The polar form is
$$z = x + iy = re^{i\theta} \quad r = \sqrt{x^2 + y^2}, \ \tan\theta = y/x,$$

where we have used Euler's identity:

$$e^{i\theta} = \cos\theta + i\sin\theta.$$

One calls r the modulus and θ the phase or argument of z.

- From Euler's theorem, by taking real and imaginary parts, we find

$$\cos\theta = \frac{e^{i\theta} + e^{-i\theta}}{2} \qquad \sin\theta = \frac{e^{i\theta} - e^{-i\theta}}{2i}$$

Forget these and you are doomed.

- The complex conjugate is

$$z^* = x - iy = re^{-i\theta}.$$

- The modulus squared is given by

$$|z|^2 = x^2 + y^2 = zz^*$$

- Some very useful relations and inequalities to remember:

$$|\operatorname{Re}\ z| \le |z| \qquad\qquad |\operatorname{Im}\ z| \le |z|.$$

$$|z_1 + z_2|^2 = |z_1|^2 + |z_2|^2 + 2\operatorname{Re}(z_1 z_2^*)$$

$$|z_1 + z_2| \le |z_1| + |z_2| \qquad\qquad |z_1 z_2| = |z_1||z_2|$$

$$(z_1 z_2)^* = z_1^* z_2^*$$

$$\frac{z_1}{z_2} = \frac{r_1}{r_2} e^{i(\theta_1 - \theta_2)}.$$

6

FUNCTIONS OF A COMPLEX VARIABLE

6.1. Analytic Functions

Now that we understand complex numbers, we ask about a function of a complex variable: a machine which takes in a complex number and spits out a complex number. Just as we can associate with points on the x-axis a function $f(x)$, we may associate with each point (x, y) in the complex plane a function

$$f(x, y) = u(x, y) + iv(x, y) \tag{6.1.1}$$

where u and v are real and represent the real and imaginary parts of the function f. Thus we have manufactured a complex function by patching together any two real functions of two real variables. Let us assume u and v have first order partial derivatives that are continuous. Continuity in two variables is defined as it was in one variable:

Definition 6.1. *A function of two variables is continuous at a point if*

- *The function approaches a definite limit as we approach the point from any direction.*

- *The limit coincides with the value ascribed to that point in the definition of the function.*

Thus for example

$$f(x, y) = \frac{x^2}{x^2 + y^2}$$

is not continuous at the origin because there is no unique value that we can assign to the function at the origin. If we simply set $x = y = 0$ we get the indeterminate value $0/0$. On the other hand if try to define the value at the origin by the limit of what we encounter as we approach the origin, the answer depends on the direction of approach.

Problem 6.1.1. *Explore the above claim by approaching the origin along the x and y axes. Then look at all other approaches by writing it in polar coordinates. What happens at some other point, say $(x = 1, y = 1)$?*

The function in Eqn. (6.1.1) is generally a function of *two complex variables* z *and* z^* related to (x, y) as follows:

$$z = x + iy \tag{6.1.2}$$
$$z^* = x - iy \quad \text{or conversely} \tag{6.1.3}$$
$$x = \frac{z + z^*}{2} \tag{6.1.4}$$
$$y = \frac{z - z^*}{2i}. \tag{6.1.5}$$

That f is generally a function of z and z^* follows from the fact that you can only trade a pair (x, y) for another pair. For example the other pair could have been

$$x_\pm = x \pm y \quad \text{or conversely} \tag{6.1.6}$$
$$x = \frac{x_+ + x_-}{2} \tag{6.1.7}$$
$$y = \frac{x_+ - x_-}{2}. \tag{6.1.8}$$

Expressing x and y in terms of the new variables will produce in general a function of both new variables. Take for example the function

$$f = x^2 - y^2. \tag{6.1.9}$$

We can write it in terms of $x_\pm$ as

$$f = x_+ \cdot x_- \tag{6.1.10}$$

or in terms of (z, z^*) as

$$f = \left(\frac{z + z^*}{2}\right)^2 - \left(\frac{z - z^*}{2i}\right)^2 \tag{6.1.11}$$
$$= \frac{z^2 + (z^*)^2}{2}. \tag{6.1.12}$$

In either case we see that f depends on both the new variables. It is all right to study such functions of two variables, take their partial derivatives and so on. *But a special place is reserved for functions of* (x, y) *that depend on just one of the two complex variables* z *or* z^*. This is no different from $f = (x - y)^2$ depending on just one of the two possible new variables, x_+ and x_-, namely the latter.

Definition 6.2. *We say f is an* analytic *function of z if it does not depend on z^*:*

$$f(z, z^*) = f(z) \qquad (6.1.13)$$

In other words

$$f(x, y) = f(x + iy) \qquad (6.1.14)$$

i.e., x and y enter f only in the combination $x + iy$.

Suppose we are given $f(x, y)$. We check if it is analytic, i.e., that x and y enter f only via $z = x + iy$, as follows. Suppose we change x by dx and keep y fixed. What happens to f? It only responds to the change in z, and its response is

$$df = \frac{df}{dz} dz = \frac{df}{dz} \left(\frac{\partial z}{\partial x} \right)_y dx = \frac{df}{dz} dx \qquad (6.1.15)$$

so that

$$\frac{\partial f}{\partial x} = \frac{df}{dz}. \qquad (6.1.16)$$

Suppose we now change just y. Now $dz = idy$ and we get by similar reasoning

$$\frac{\partial f}{\partial y} = \frac{df}{dz} \cdot i. \qquad (6.1.17)$$

In other words, if f is a function of only $z = x + iy$, the partial derivatives with respect to x and y are related as follows:

$$f_y = i f_x \qquad (6.1.18)$$

If we write $f = u + iv$ we find

$$u_y + i v_y = i(u_x + i v_x) \qquad (6.1.19)$$

upon comparing the real and imaginary parts of which we find we obtain the *Cauchy–Riemann Equations (CRE)* that define an analytic function of z:

$$u_x = v_y \qquad (6.1.20)$$
$$u_y = -v_x. \qquad (6.1.21)$$

(We could just as easily find the diagnostic for an *anti-analytic function*, a function of just z^*. But we will not, since the mathematical content of analytic and anti-analytic functions is identical.)

As an example let us consider

$$f = x^2 + y^2 \qquad (6.1.22)$$

and

$$f = x^2 - y^2 + 2ixy. \qquad (6.1.23)$$

In the first case we find $u_x = 2x, u_y = 2y$, and $v \equiv 0$. The function is clearly not analytic. In the second case we find $u_x = 2x = v_y$ and $u_y = -2y = -v_x$. So the function is analytic.

Problem 6.1.2. *Find the equations obeyed by an anti-analytic function.*

Problem 6.1.3. *For the real function $f(x, y)$ find the condition which ensures that it depends only on x_+.*

Do not lose sight of the fact that the CRE merely ensure that f is a function of just $z = x + iy$. Thus for example $f = x^2 - y^2 + 2ixy$ is guaranteed to obey the CRE because it is just $(x + iy)^2 = z^2$ whereas $f = x^2 - y^2$, which we saw depended on both z and z^*, is guaranteed not to satisfy the CRE. Thus if you can tell by inspection that the given $f(x, y)$ can or cannot be written as some $f(z)$, you need not bother with the CRE.

Now it is possible that the CRE are obeyed in some parts of the complex plane and not others.

Definition 6.3. *We say a function is* analytic *in a domain $\mathcal{D}$ if the first partial derivatives of u and v exist, are continuous and obey the CRE equations everywhere inside it.*

(Some of the conditions on u and v can be relaxed without changing many of the results derived here. But the corresponding proofs are more complicated. We will not follow that path in this pedestrian course.)

Consider in particular the domain called the *ε-neighborhood of any point*, (or simply the neighborhood of a point) which is a collection of all points within a disc of radius $\varepsilon > 0$ centered at the given point.

Definition 6.4. *We say $f = u + iv$ is* analytic *at z_0 if u and v have continuous partial derivatives that obey the CRE in its ε-neighborhood.*

No importance is given to functions which obey the CRE only at isolated points or on lines.

The following is a sample of things that can happen when we test a function for analyticity.

- Consider $f = x^2 - y^2 + 2ixy$. The CRE are obeyed and the partial derivatives of u and v are continuous everywhere. The function is analytic at all finite points in the plane. This is expected since $f = z^2$.

- Consider $f = \cos y - i \sin y$. The CRE will require that $\cos y = 0 = \sin y$ which is impossible. This result is expected from the fact that f is a function of just $y = (z - z^*)/2i$ and therefore clearly a function of z *and* z^*. The CRE are not satisfied anywhere. The function is not analytic in any domain.

- Consider $f = x^2 + y^2$. The CRE are satisfied only at the origin where all partial derivatives vanish. This is to be expected since $f = zz^*$ depends on z^*. This dependence is merely hidden at the origin since $\partial f / \partial z^* = 0$ there. At any point off the origin, the CRE fail, i.e., there is no neighborhood where the CRE are obeyed. We are not interested in this function, we cannot get any mileage out of its special properties at one point, the origin.

- Consider $f = x^2 - iy^2$ which satisfies the CRE on the line $x = -y$. We can easily see that this f also depends on both z and z^*. This will be apparent in the CRE if we move off the line. We are not interested in this function. Useful results can be derived only if the function has continuous partial derivative obeying the CRE in a two-dimensional domain.

Problem 6.1.4. *Verify that $f = e^x \cos y + ie^x \sin y$ obeys the CRE.*

Problem 6.1.5. *Show that $f = x^3 - 3xy^2 + i(3x^2 y - y^3)$ is analytic by any means.*

Problem 6.1.6. *Show by any means that $f = x^2 + y^2$ is not analytic.*

6.1.1. Singularities of analytic functions

Assume now that we have verified that a function obeys the CRE in some domain. Will it obey them in the entire complex plane? In principle anything can happen since nothing has been stated about the function outside the domain. But typically what happens is this. There will be *isolated points* where the equations fail. These points are called *singularities*. Since we have equated the validity of the CRE to the absence of any z^*-dependence the last remark may seem paradoxical. If the CRE are valid in some domain so that the function is known to have no z^*-dependence, how can they be invalid at some other places? Does z^* suddenly creep in at these points? No, what happens is that the conditions of continuity of the partial derivatives breaks down. Let us consider the example $f = 1/z$ which will clarify these points. Now this f is manifestly a function of just z. It is easily shown that

$$\frac{1}{z} = \frac{z^*}{zz^*} = \frac{x - iy}{x^2 + y^2} \tag{6.1.24}$$

$$u_x = v_y = \frac{y^2 - x^2}{(x^2 + y^2)^2} \tag{6.1.25}$$

$$u_y = -v_x = \frac{-2xy}{(x^2 + y^2)^2} \tag{6.1.26}$$

We see that the CRE hold at any point where the various derivatives are finite and well-defined. In particular they will hold in any neighborhood that excludes the origin. As we approach the origin along different directions, we find that the derivative approaches a different limit. For example along the x axis, u_x blows up as $-1/x^2$ whereas along the y axis it blows up as $1/y^2$. While it is true that v_y is sick in exactly the same fashion (because of the CRE), neither has a definite value at the origin. Indeed even before we consider these derivatives, we see that $f = 1/z$ itself has no unique limit as we approach the origin, blowing up to $\pm\infty$ as we approach the origin from the right/left and to $\pm i\infty$ as we approach it from below/above.

To conclude, the CRE breakdown at a singularity because the function or its derivatives (which enter the CRE's) become ill-defined as we approach that point.

Here is a list of important singularities you will encounter.

- *Simple pole* The case we just saw, $f(z) = 1/z$ is an example of a simple pole. Clearly the function c/z, where c is a constant, also has a simple pole. We say the *residue* of the pole at the origin is c.

Consider next

$$f(z) = \frac{1}{(z + i)} \frac{1}{(z - i)} \tag{6.1.27}$$

This function has poles at two points $z = \pm i$. As we approach $z = i$, the factor $1/(z + i)$ approaches $1/(2i)$ and may be replaced by it. On the other hand, we cannot replace $1/(z - i)$ by anything simple since the function is very singular and varies wildly. Thus in the immediate vicinity of $z = i$ we may trade the given function for $\frac{(1/2i)}{(z-i)}$. One says the function has a pole at $z = i$ with a residue $1/2i$. Equivalently, the residue $R(z = i)$ is given by

$$R(z = i) = \lim_{z \to i}(z - i)f(z) = \frac{1}{2i}. \tag{6.1.28}$$

The function also has a pole at $z = -i$ with a residue

$$R(z = -i) = \lim_{z \to i}(z + i)f(z) = \frac{1}{-2i} \tag{6.1.29}$$

In the present problem we can obtain the above results simply by using partial fractions in Eqn. (6.1.27):

$$f(z) = \frac{1}{(z+i)(z-i)} = \frac{1}{2i}\left[\frac{1}{z-i} - \frac{1}{z+i}\right]. \tag{6.1.30}$$

More generally, if a function has a simple pole at z_0, its residue there is

$$R(z_0) = \lim_{z \to z_0} (z - z_0)f(z). \tag{6.1.31}$$

- *n-th order pole* We say that f has an n-th order pole at z_0 if

$$f(z) \to \frac{R(z_0)}{(z - z_0)^n}(1 + \mathcal{O}(z - z_0)). \tag{6.1.32}$$

as $z \to z_0$, where $\mathcal{O}(z - z_0)$ denotes terms that vanish as $z \to z_0$.

- *Essential singularity* A function has an essential singularity at a point z_0 if it has poles of arbitrarily high order which cannot be eliminated by multiplication by $(z - z_0)^n$ for any finite choice of n. An example is the function

$$f(z) = \sum_0^\infty \frac{1}{z^n n!} \tag{6.1.33}$$

which has poles of arbitrarily high order at the origin.

- *Branch point* A function has a branch point at z_0 if, upon encircling z_0 and returning to the starting point, the function does not return to the starting value. Thus the function is multiple valued. An example is

$$f(z) = z^{\frac{1}{2}} = r^{\frac{1}{2}}e^{i\theta/2}. \tag{6.1.34}$$

Notice that as we increase θ from 0 to 2π, at some fixed r, the function goes from $r^{\frac{1}{2}}$ to $r^{\frac{1}{2}}e^{i\pi} = -r^{\frac{1}{2}}$. We say f has a branch point at the origin.

You might think that among analytic functions, those with no singularities anywhere are specially interesting. This is not so, because the only example of such a function is a constant. (This can be proven, though we will not prove it here.) Even polynomials misbehave as $z \to \infty$; they blow up.

In this chapter we will mainly emphasize *meromorphic* functions:

Definition 6.5. *A function $f(z)$ is meromorphic if its only singularities for finite z are poles.*

Problem 6.1.7. *Locate and name the singularities of $f = 1/(1 + z^4)$ and $f = 1/(z^4 + 2z^2 + 1)$.*

Problem 6.1.8. *Show that $dz = (dr + ird\theta)e^{i\theta}$ given $z = re^{i\theta}$. Interpret the two factors in dz, in particular the role of the exponential factor. Equating the derivatives in the radial and angular directions, find the CRE in terms of $u_r, u_\theta, v_r, v_\theta$. Begin now with $f(r, \theta) = f(re^{i\theta})$ as the definition of analyticity, relate r and θ derivatives of f, and regain the CRE in polar form.*

Problem 6.1.9. *Confirm the analyticity of:*

$$\frac{x - iy}{x^2 + y^2},$$

$$\sin x \cosh y + i \cos x \sinh y.$$

You can either use the CRE or find $f(z)$.

6.1.2. Derivatives of analytic functions

When u and v possess continuous partial derivatives that obey the CRE we may define the z-derivative of $f = u + iv$ as follows. If the function is known explicitly in terms of z, say $f = z^2$, we take derivatives as we did with a real variable x:

$$\Delta f = (z + \Delta z)^2 - z^2 = 2z\,\Delta z \qquad (6.1.35)$$

$$\frac{df}{dz} = 2z. \qquad (6.1.36)$$

Suppose we have the function in terms of x and y and only know that it obeys the CRE, so that in principle it can be written in terms of just z. Let us vary both x and y to obtain to *first order*

$$df = \frac{\partial f}{\partial x}dx + \frac{\partial f}{\partial y}dy \qquad (6.1.37)$$

$$= (u_x + iv_x)dx + (u_y + iv_y)dy \qquad (6.1.38)$$

$$= (u_x + iv_x)dx + (-iu_y + v_y)idy. \qquad (6.1.39)$$

If we now invoke the CRE we see that both brackets are equal and the change in f is proportional to $dx + idy = dz$. It follows that the derivative is

$$\frac{df}{dz} = \underbrace{(u_x + iv_x)}_{\frac{\partial f}{\partial x}} = \underbrace{(-iu_y + v_y)}_{\frac{\partial f}{i\partial y}} \qquad (6.1.40)$$

in accordance with Eqn. (6.1.18).

Thus for example the derivative of

$$f = x^2 - y^2 + 2ixy.$$ (6.1.41)

is

$$\frac{df}{dz} = \frac{\partial f}{\partial x} = 2x + 2iy = 2z.$$ (6.1.42)

Notice that the z-derivative of an analytic function is independent of the direction in which we make the infinitesimal displacement that enters the definition of the derivative. You can choose $dz = dx$ or $dz = idy$ or any general case $dz = dx + idy$ and get the same answer. The following exercise looks at this in more detail.

Problem 6.1.10. (Important). *Consider df, the first order change in response to a change dx and dy in the coordinates, of a function $f = u + iv$ where u and v have continuous derivatives which do not however obey the CRE. The shift in x and y corresponds to changing z by $dz = dx + idy$ and z^* by $dz^* = dx - idy$. In other words, as we move in the complex plane labeled by x, y, we change both z and z^*. Show that df generally has parts proportional to both dz and dz^* by reexpressing dx and dy in terms of the former. Show that as a result the symbol $\frac{df}{dz}$ makes no sense in general: it is like trying to define $df(x, y)/dx$ for a function of two variables, when all one can define is the partial derivative. If however, the function of x and y happened to have no dependence on y, we could define the total derivative with respect to x. That is what is happening with analytic functions f which depend only on z.*

It follows from the preceding discussion that if f is analytic at a point, $\frac{df}{dz}$ exists in a neighborhood of that point.

From the CRE, we can easily show, by taking derivatives that

$$u_{xx} + u_{yy} = 0 \quad v_{xx} + v_{yy} = 0.$$ (6.1.43)

Note that so far we have only made assumptions about the continuity of first derivatives of u and v. The above equation assumes the second derivatives are also well-defined. We shall see later that this is not an additional assumption: it will be shown towards the end of the chapter, (see discussion following Eqn. (6.4.22)), that if $f(z)$ has a first derivative in a domain, it has all higher derivatives in that domain.

The equation obeyed by u

$$\frac{\partial^2 u}{\partial x^2} + \frac{\partial^2 u}{\partial y^2} = 0$$ (6.1.44)

(and similarly v) is called *Laplace's equation* and u and v are said to be *harmonic functions*.

Problem 6.1.11. *Prove that u and v are harmonic given the CRE.*

Problem 6.1.12. *You are given that $u = x^3 - 3xy^2$ is the real part of an analytic function. (i) Find v. Reconstruct $f(z)$. Verify that u and v are harmonic. (ii) Repeat for the case $v = e^{-y} \sin x$. (iii) You are given $u = x^3 - 3x^2 y$ and asked to find v. You run into real problems. Why?*

Problem 6.1.13. *Prove that $(\frac{\partial^2}{\partial x^2} + \frac{\partial^2}{\partial y^2})|f(z)|^2 = 4|f'(z)|^2$.*

Problem 6.1.14. *Given that (u, v) form a harmonic pair, show without brute force that $2uv$ and $u^2 - v^2$ are also such a pair.*

6.2. Analytic Functions Defined by Power Series

We can build any number of analytic functions by simply writing down any expression that involves just z. Take for example the function

$$f(z) = z^2 + z \tag{6.2.1}$$

which assigns definite values to each complex number z. For example at the point $z = 3 + 4i$, $f(z) = -4 + 28i$.

This is a special case of n-th order polynomials $P_n(z)$

$$P_n(z) = a_0 + a_1 z + \ldots + a_n z^n \tag{6.2.2}$$

which will obey the CRE for any finite z. Equivalently they will have well-defined derivatives for all finite z. They will however misbehave as we approach infinity: the n-th order polynomial will (mis)behave as $z^n = r^n e^{in\theta}$ blowing up in different ways in different directions.

Now we turn to a very important notion: analytic functions defined by *infinite series*. As a prelude we must discuss the notion of convergence of complex series.

Definition 6.6. *The infinite series of complex terms a_n,*

$$S = \sum_{0}^{\infty} a_n, \tag{6.2.3}$$

is said to converge if its real and imaginary parts, i.e., the series that sum the real and imaginary parts of a_n, converge.

Since we already know how to test real series for convergence, no new tools are needed. Consider for example

$$S = \sum_0^\infty (2^{-n} + ie^{-n}) = \sum_0^\infty 2^{-n} + i \sum_0^\infty e^{-n}. \tag{6.2.4}$$

The ratio rest on the real and imaginary parts show that both converge (being geometric series with $r < 1$) and the full series itself converges to

$$S = \frac{1}{1 - 1/2} + \frac{i}{1 - e^{-1}}. \tag{6.2.5}$$

Recall that in the case of real series, we introduced a more stringent notion of *absolute convergence*: the series converged absolutely if it converged upon replacing each term by its absolute value. A similar notion exists here as well:

Definition 6.7. *The series*

$$S = \sum_0^\infty a_n$$

is said to converge absolutely if

$$S = \sum_0^\infty |a_n| \tag{6.2.6}$$

does, i.e., if

$$r = \lim_{n \to \infty} \frac{|a_{n+1}|}{|a_n|} < 1. \tag{6.2.7}$$

Since the real or imaginary part of a complex number is bounded in magnitude by its absolute value, both the real and imaginary sums are dominated by the series with absolute values. Thus a series which converges absolutely also converges.

Problem 6.2.1. *Show why the absolute value of a sum of complex numbers is bounded by the sum of their absolute values. You may do this algebraically (in which case it is better to square both sides of the inequality) or by graphical arguments, i.e., by viewing the sum as addition of arrows in the complex plane. In the first approach you may want to start with a sum with just two numbers if you are having trouble.*

In the real case, absolute convergence meant that the series converged even without relying on cancellations between positive and negative terms in the series. Here it means that even if all the complex numbers in the sum (which can be viewed as vectors in the plane pointing in different directions) were lined up along the real axis, the sum would still converge.

Let us now turn to the power series

$$S = \sum_0^\infty a_n z^n. \tag{6.2.8}$$

This clearly defines an analytic function as long as the series converges. *We shall be only interested in absolute convergence.* This in turn means that

$$|z| < R = \lim_{n \to \infty} \frac{|a_n|}{|a_{n+1}|} \tag{6.2.9}$$

where R is now called the *radius of convergence* since the series converges for all z lying strictly inside a circle of radius R centered at the origin. (This is to be compared to real series which converge in an interval $|x| < R$.) We could similarly consider series

$$S = \sum_0^\infty a_n (z - z_0)^n \tag{6.2.10}$$

which would converge within a circle of radius R centered at z_0. In our discussions we will typically choose $z_0 = 0$.

As a concrete example, consider the series with all $a_n = 1$, which clearly has $R = 1$. Within this radius we can sum the series. That is

$$f(z) = 1 + z + z^2 + \cdots \quad (|z| < 1). \tag{6.2.11}$$

is a well-defined number for each z within the unit disk, the number being the value of the convergent infinite sum at each value of z. In the present case we have the luxury of knowing this limiting value in closed form. Using the trick for summing a geometric series from Chapter 4, we see that the partial sum with N terms is

$$\frac{1 - z^{N+1}}{1 - z}$$

so that if $|z| < 1$, we may drop z^{N+1} as $N \to \infty$ and obtain

$$f(z) = \frac{1}{1 - z}. \tag{6.2.12}$$

Thus for example, the series will sum to $3/2$ at the point $z = 1/3$ or to $\frac{1}{1-(1/2+i/2)} = 1 + i$ at $z = (1 + i)/2$.

Notice that Eqn. (6.2.8) is just a Taylor series at the origin of a function with $f^n(0) = a_n n!$. (Verify this by taking the n-th derivative and then setting $z = 0$.) Every choice we make for the infinitely many variables a_n, defines a new function with its own set of derivatives at the origin. Of course this defines the function only within the radius of convergence. Later we shall see how we can go beyond this initial circle. For the present let us take series for other known functions of real variables and define complex functions, being content to work within this circle of convergence. As a concrete example consider

$$e^x = \sum_0^\infty \frac{x^n}{n!}$$

(6.2.13)

which corresponds to the choice $a_n = 1/n!$. We now replace x by z and obtain a new function which we call e^z:

$$e^z = \sum_0^\infty \frac{z^n}{n!}.$$

(6.2.14)

The radius of convergence of the series for e^z is seen to be infinite since the coefficients are just those of e^x. *Thus the exponential function is defined by this series for all finite z.* For example its value at the point $z = 3 + 4i$ is $e^{3+4i} = e^3 e^{4i} = e^3 (\cos 4 + i \sin 4)$.

What can we say about this function other than the fact that on the real axis it reduces to e^x?

It obeys

$$\frac{de^z}{dz} = e^z$$

(6.2.15)

(as is clear from differentiating the series and noting that the derivative also has infinite radius of convergence).

Next it satisfies

$$e^{z_1} \cdot e^{z_2} = e^{z_1 + z_2}$$

(6.2.16)

as can be seen by comparing the product of the Taylor series of the factors on the left-hand side to the series for the term on the right, to any order. Equivalently consider the function $e^z e^{a-z}$ which has zero derivative as per the product and chain rules. It is therefore a constant, the constant value is e^a as can be seen at $z = 0$. Thus $e^z e^{a-z} = e^a$. Calling z as z_1, $a - z$ as z_2, we get the advertised result.

We now go on and define trigonometric and hyperbolic functions in the same way:

$$\sin z = \sum_0^\infty (-1)^n \frac{z^{2n+1}}{(2n+1)!}$$

(6.2.17)

$$\cos z = \sum_0^\infty (-1)^n \frac{z^{2n}}{(2n)!}$$

(6.2.18)

$$\sinh z = \sum_0^\infty \frac{z^{2n+1}}{(2n+1)!}$$

(6.2.19)

$$\cosh z = \sum_0^\infty \frac{z^{2n}}{(2n)!}$$

(6.2.20)

Let us take a look at these series. We notice that these functions are related to the exponential function in the same way as they were in the real case. For example

$$\cos z = \frac{e^{iz} + e^{-iz}}{2}, \tag{6.2.21}$$

$$\sin z = \frac{e^{iz} - e^{-iz}}{2i} \tag{6.2.22}$$

and likewise for the hyperbolic functions. We shall see towards the end of this chapter that this fact has a very general explanation.

The series above also have $R = \infty$, as can be seen by the ratio test or by appealing to their relation to the exponential series which is defined for all finite z.

What are the properties of the trigonometric and hyperbolic function in the complex plane? Is it true for example that

$$\cos^2 z + \sin^2 z = 1 \tag{6.2.23}$$

even for complex z? In the case of real z, we could relate these functions to sides of a right triangle, use the Pythagoras theorem, and prove the result. Clearly this does not work when the angle z is complex; we have to rely on the series that define them. While squaring and adding the two series (and keeping as many terms as we want), will provide circumstantial evidence, it is much easier to relate them to exponentials and argue as follows:

$$\cos^2 z + \sin^2 z = \left[\frac{e^{iz} + e^{-iz}}{2}\right]^2 + \left[\frac{e^{iz} - e^{-iz}}{2i}\right]^2$$

$$= \frac{e^{2iz} + e^{-2iz} + 2 - e^{2iz} - e^{-2iz} + 2}{4} = 1. \tag{6.2.24}$$

All the usual properties of these functions are likewise preserved under extension to the complex plane. A few are taken up in the exercises below.

Problem 6.2.2. *Show that the hyperbolic functions obey* $\cosh^2 z - \sinh^2 z = 1$ *by writing the functions in terms of exponentials.*

Problem 6.2.3. *Argue that the old formula for* $\sin(z_1 + z_2)$ *must be valid for complex arguments by writing* $\sin z$ *in terms of exponentials. Find the real part, imaginary part, and modulus of* $\sin(x + iy)$. *Where all does the function vanish in the complex plane?*

Problem 6.2.4. *Locate the zeros in the complex plane of* $\sin z$, $\cos z$, $\sinh z$, *and* $\cosh z$. *A suggestion: work with the exponential function as far as possible.*

Problem 6.2.5. *Show that* $|\sin z|^2 = \sin^2 x + \sinh^2 y$ *and* $|\cos z|^2 = \cos^2 x + \sinh^2 y$. *What does this tell you about the zero's of* $\sin z$?

Problem 6.2.6. *Show that the solutions to* $\sin z = 5$ *are* $z = \frac{\pi}{2} + 2n\pi + i \cosh^{-1} 5$

There are also new insights we get in the complex plane. For example we know that for $z = x$, sines and cosines are just shifted versions of each other: $\sin x = \cos(x - \pi/2)$. What one function does here, the other does at a related point. Now we learn that the hyperbolic and trigonometric functions are similarly related:

$$\cos iz = \cosh z \tag{6.2.25}$$

$$\sin iz = i \sinh z, \tag{6.2.26}$$

i.e., what the trigonometric functions do on the real axis, the hyperbolic functions do on the imaginary axis and vice versa.

So far we have focused on functions (related to e^z) whose Taylor series at the origin converged for all finite z. Thus our detailed knowledge at the origin sufficed to nail them down in the entire complex plane. But this is not generic. Consider the function $f(z) = 1/(1 - z)$ whose series $1 + z + z^2 + \ldots$ converges only for $|z| < 1$. How are we to give a meaning to the function outside the unit disc? The closed form of the function tells us the function certainly can be defined outside the unit disc and that its only genuine problem is at $z = 1$ where it has a pole. Suppose we were not privy to this closed form. Could we, armed with just the Taylor coefficients at the origin, reconstruct the function, say at $z = 2$ where it has the value -1? The answer is affirmative and relies on *Analytic Continuation*, which will be discussed towards the end of this chapter.

The logarithm

Now we turn to another function in the complex plane, the logarithm. Recall that when we first discussed logarithms, it was pointed out that we could not define the log of a negative number since there was no real number whose exponential was negative. Clearly all this is changed if we admit complex numbers, since we have for example $e^{i\pi} = -1$. Let us consider the ln function defined by its series

$$\ln(1 + z) = \sum_1^\infty \frac{(-1)^{n+1}}{n} z^n \tag{6.2.27}$$

which defines a function within a circle of radius 1. (At the circumference is a branch point, to be discussed later.) Within this circle, the ln and exponential functions are inverses, that is to say, $e^{\ln(1+z)} = 1 + z$.

Problem 6.2.7. *Verify* $e^{\ln(1+z)} = 1 + z$ *to order* z^3 *by using the respective series.*

The equality of these two functions ($e^{\ln(1+z)} = 1 + z$) within the circle tells us they are forever equal, a point that will be elaborated on towards the end of this chapter. Replacing $1 + z$ by z everywhere we have

$$e^{\ln z} = z. \tag{6.2.28}$$

We originally introduced the ln function in our quest for arbitrary real powers of any given number. We found

$$a^x = e^{x \ln a}. \tag{6.2.29}$$

Since we could evaluate both the ln and exponential functions to any desired accuracy via their series, we could give an operational meaning to a^x.

Let us now try to see how all this goes in the complex plane, first without invoking the logarithm. Let us begin with the square root of $z = re^{i\theta}$. It is clearly that number, which on multiplying by itself, gives us z. Since this process squares the modulus and doubles the phase,

$$z^{1/2} = r^{1/2}e^{i\theta/2} \tag{6.2.30}$$

does the job. Now let $z = 4$. The above answer gives us $z^{1/2} = 2$, which is fine. But we also know that there is one more solution, namely -2. How does that emerge from this approach? Perhaps, with the advent of complex numbers there are even more solutions for the square root?

Starting with the former question, we first observe that we can write z in two equivalent ways:

$$z = re^{i\theta} = re^{i\theta+2\pi i} \tag{6.2.31}$$

since adding $2\pi i$ to the phase does not affect anything. (Recall Euler's formula and the period of the sines and cosines.) But the second version gives us the other square root, for when we halve the phase there, we get an extra π, leading to

$$z^{1/2} = r^{1/2}e^{i\theta/2+i\pi} = -r^{1/2}e^{i\theta/2} \tag{6.2.32}$$

since $e^{i\pi} = -1$. What if we continue and start with $z = r\exp(i\theta + 4\pi i)$? We do not get any new answers since halving this phase adds an extra $2\pi i$ to the phase in the square root and that makes no difference even to the square root. Thus a complex number has only two square roots. Let us ponder on this, focusing on the case $r = 1$, i.e., the unimodular numbers, since $r \to r^{1/2}$ in all cases. Our starting number is $\exp(i\theta)$, which makes an angle θ with the x axis. One of its roots makes half the angle, so that when squared, it gives us the original number. The other has half the angle plus π, which puts it at the diametrically opposite point. When squared, doubling the π becomes unobservable and we end up where we did when we squared the other root.

It is clear what happens when we consider the cube root. There will be three of them. The first will have one-third the phase of z. The other two will be rotated

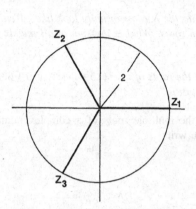

Figure 6.1. The three cube roots of $z = 8$. Note that all have length 2 and phases that equal 0 modulo 2π when tripled.

by 120 degrees ($2\pi/3$) and 240 degrees ($4\pi/3$) relative to the first. These extra phase factors, which distinguish the roots, will all become invisible when the cube is taken.

Let us consider as an example the cube roots of 8. By the preceding discussion, they are

$$z_1 = 2 \qquad z_2 = 2e^{2\pi i/3} = -1 + i\sqrt{3} \qquad z_3 = 2e^{-2\pi i/3} = -1 - i\sqrt{3}.$$
$$(6.2.33)$$

The roots are plotted in Fig. 6.1.

Problem 6.2.8. (Important). *Verify that the cubes of the roots of 8 (in cartesian form) in the preceding equation indeed give 8.*

Consider the function $f(z) = z^{1/2}$. Let us visualize at each point z the associated value of the function. Starting with the real branch $f = \sqrt{r}$ on the positive real axis, let us go around the origin along the path $z = re^{i\theta}$. As we traverse this path, f is given by $\sqrt{r}e^{i\theta/2}$. When we describe a full circle and return to the starting point, f returns to the other root, $\sqrt{r}e^{i\pi}$. Thus the function is multiple valued: as we go around the branch point at the origin, f changes from one branch to the other. Likewise, the function $\sqrt{1-z}$ has a branch point at $z = 1$. Its Taylor series about the origin, which goes as

$$f(z) = 1 - \frac{1}{2}z - \frac{1}{8}z^2 + \cdots \qquad (6.2.34)$$

breaks down on the unit circle. This is because this power series of single valued terms cannot possibly reproduce the double-valuedness around the branch point.

Problem 6.2.9. *What do the N roots of unity look like? Write down explicit carte-sian forms for the sixth roots. What is the sum of all the six roots?*

Problem 6.2.10. *Find the roots of* $z^2 - (15 + 9i)z + (16 + 63i)$. *(Hint: the radical contains a perfect square.)*

Let us see how the multiple roots of a complex number emerge from the logarithm. Suppose we write

$$z = e^{\ln z} \qquad (6.2.35)$$

and say

$$z^{1/N} = e^{\frac{1}{N} \ln z}; \qquad (6.2.36)$$

where does the multiplicity of roots arise? It must come from the logarithm. What is $\ln z$? In polar form,

$$\ln z = \ln r + i\theta. \qquad (6.2.37)$$

But we can add any integral multiple of 2π to the phase of z without affecting it. But each such choice gives a different value for the logarithm:

$$\ln z = \ln r + i(\theta + 2\pi n) \qquad (6.2.38)$$

One says that there are infinitely many *branches* of the logarithm. Thus there is no unique value associated with the logarithm of a given number—there is a lot of latitude! Why were we denied this freedom (along with many others) in high school? Because back then we insisted that the log be real, and there is just one branch. Stated differently, we asked for example: to what power must we raise e to get say, 10? The answer was: roughly 2.303. But there are other answers such as $2.303 + 2\pi i$.

The function $\ln z$ clearly has a problem at the origin since $\ln z \to -\infty$ as we approach the origin from the right. But the problem there is not a pole but a branch point. By this one means the following. Start at $z = 2$. Choose the branch which is real, i.e., has $\ln 2 = .693$. Move along a circle of radius 2 and follow what is happening to the $\ln z$ function. Its real part will be fixed at .693 while its imaginary part will grow from 0 to 2π as we return to our starting point. Thus the function does not have a unique value at every point in the plane: each time we go around the origin, the imaginary part of the logarithm goes up by 2π. Note that a circuit that does not enclose the origin does not produce this effect. We say the function is multiple valued and that the origin is a branch point. To keep the function single valued, we must draw a line from the origin to infinity and agree never to cross it.

Let us now see how the multiplicity of the logarithm leads to N distinct N-th roots of z:

$$z^{1/N} = \exp\left[\frac{1}{N}\ln z\right] \qquad (6.2.39)$$

$$= \exp\left[\frac{1}{N}[\ln r + i(\theta + 2\pi n)]\right] \qquad (6.2.40)$$

$$= r^{1/N}\exp\left[\frac{i\theta}{N}\right]\exp\left[\frac{2\pi i n}{N}\right] \qquad [n = 0, 1, \ldots N - 1]. \; (6.2.41)$$

We limit n to take only N values since after that we start repeating ourselves. This is a special property of the N-th root. We do not find this in the general case we now turn to: complex powers of complex numbers. We define these by

$$z^a = e^{a\ln z}. \qquad (6.2.42)$$

Since the exponential and ln function are defined by their series, the above expression has an operationally well-defined meaning. This power z^a can have infinitely many values because adding $2n\pi i$ to $\ln z$ upstairs can affect the left-hand side since $2n\pi a$ need not be a multiple of 2π for any n.[1] That is the bad news. The good news is that it is still a complex number! By this I mean the following. We did not go looking for complex numbers. They were thrust upon us when we attempted to solve equations such as $x^2 + 1 = 0$, written entirely in terms of real numbers. Thus it is entirely reasonable to ask if equations with complex coefficients will force us to introduce something more bizarre. How about complex roots or powers of complex numbers? *The answer to all this is that we will never go outside the realm of complex numbers by performing any of the algebraic operations of adding, multiplying, and taking roots.*

There are however other ways to generalize complex numbers. We can look for some hyper-complex numbers h which have several distinct separable parts (like the real and imaginary parts), absolute values which are real and nonnegative and vanish only if all components do, and multiply when the numbers are multiplied; and for which division by nonzero h makes sense. There are two extensions possible: quaternions (which can store four bits of information but do not commute in general under multiplication, $h_1 h_2 \neq h_2 h_1$) and octonions (which store eight pieces of information and are not even associative, i.e., $(h_1 h_2)h_3 \neq h_1(h_2 h_3)$).

Problem 6.2.11. *Write down the two square roots of (i) i and (ii) $3+4i$ in cartesian form.*

Problem 6.2.12. *Find all the roots of $e^z + 2 = 0$.*

---

[1]Consider for example $(3 + 4i)^i = \exp[i(\log 5 + i(2\pi n + \arctan(4/3))] = \exp[-(2\pi n + \arctan(4/3))]\exp[i\log 5]$ which has infinitely many values, one for each integer n.

Problem 6.2.13. *Express the following in cartesian form: (i)* 3^i, *(ii)* $\ln \frac{1+i}{1-i}$. *First express in polar form and take the square roots of: (iii)* $(3+4i)(4+3i)$, *(iv)* $\frac{1+i\sqrt{3}}{1-i\sqrt{3}}$.

Problem 6.2.14. *(i) Find the roots of* $z^4 + 2z^2 + 1 = 0$. *How many roots do you expect? (ii) Repeat with* $z^6 + 2z^3 + 1$.

Problem 6.2.15. *Write* $1 + i$ *in polar form and find its cube roots in cartesian form. Check that their cubes do give the original number.*

Problem 6.2.16. *Give all the values of* i^i.

6.3. Calculus of Analytic Functions

Consider first the derivative of functions represented by infinite series. As in the real case, the function and the series are fully equivalent within the circle of convergence. If we multiply the series for two function we will get the series for the product function assuming we are within the common domain of convergence for all three.

We will now argue that a *power series may be differentiated term by term to obtain another series with the same radius of convergence.* The new series will converge to the derivative of the original function. Why is this true? Suppose

$$f(z) = \sum_0^\infty a_n z^n \text{ and} \tag{6.3.1}$$

$$\lim_{n\to\infty} \left| \frac{a_n}{a_{n+1}} \right| = R. \tag{6.3.2}$$

Consider now the term-by-term differentiation of the series. This sends $a_n z^n$ to $n a_n z^{n-1}$ and $R \to \lim_{n\to\infty} R \frac{n}{n+1} = R$. What goes for the first derivative goes for all higher ones, since convergence properties of f are clearly inherited by its derivative, which passes it on to the next, and so on.

Problem 6.3.1. *Verify that the derivative of the series for* $\ln(1+z)$ *gives the series for* $1/(1+z)$. *Verify that the derivative of the sine is the cosine by looking at the series. Make sure R is the same for the function and the derivative.*

Consider now Fig. 6.2A and the integration of a function $f(z)$.

We now define the integral of analytic functions in analogy with the real integral. First we pick two end points z_1 and z_2 in the complex plane. *These two*

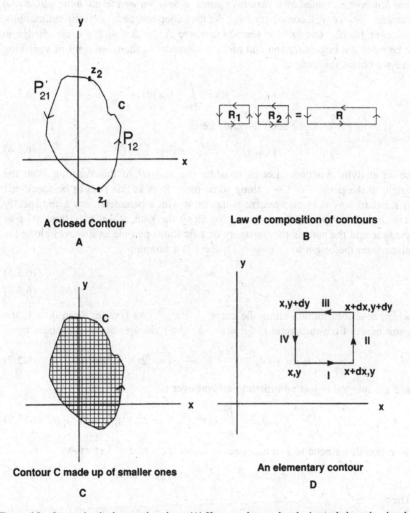

A Closed Contour

A

Law of composition of contours

B

Contour C made up of smaller ones

C

An elementary contour

D

Figure 6.2. Integration in the complex plane. (A) Here we observe that the integral along the closed path from z_1 to z_2 and back, is the difference of two integrals, in which we go from z_1 to z_2 along the two paths joining these end points. Thus if the integral on the loop vanishes, the integral along the two paths are equal. (B) Here we see that the sum of the integrals around two adjacent rectangles is the integral along a large one formed by gluing them and deleting the common edge. (C) We see how the (counterclockwise) integral around some closed contour is equal to the integral (counterclockwise) around the tiny rectangles that tile it. (D) The infinitesimal rectangle around which the analytic function is to be integrated. Since this vanishes by CRE, so does the integral around any bigger loop in the domain of analyticity.

can however be joined by a variety of paths, a freedom we did not have with a real variable. So we pick some path P_{12}. We then chop this path into infinitesimal bits. We sum $f(z)\Delta z$ over all the segments, where $\Delta z = \Delta x + i\Delta y$ is the change in z between the beginning and end of each segment. In the usual limit of vanishing Δz we obtain the integral

$$F(z_1, z_2, P) = \int_{z_1}^{z_2} f(z)dz. \tag{6.3.3}$$

Let us perform one such integral. Let

$$f(x, y) = u(x, y) + iv(x, y) \tag{6.3.4}$$

be an analytic function. Let us consider the integral of this function from the origin to the point $z = 1 + i$ along some path. How is this path to be specified? A standard way is to parametrize points on it with a parameter, say t and specify how x and y depend on t as we move along the path. (You may think of t as the time and the path as the trajectory of a fictitious particle as it moves along the plane from the origin to $(x = y = 1)$. Here is a sample:

$$x = t^2 \tag{6.3.5}$$
$$y = t^3 \tag{6.3.6}$$

which describes motion along the curve $y = x^{3/2}$. As t varies from 0 to 1, the point moves from the origin to $(x = y = 1)$. As t changes by dt, z changes by

$$dz = dx + idy = (\frac{dx}{dt} + i\frac{dy}{dt})dt = [2t + 3t^2i]dt \tag{6.3.7}$$

and the integral is just an ordinary integral over t:

$$F(z_1, z_2, P) = \int_0^{1+i} f(z)dz = \int_0^1 [u(t^2, t^3) + iv(t^2, t^3)][2t + 3t^2i]dt. \tag{6.3.8}$$

To proceed, we need to assume specific forms for u and v. Let us say

$$u(x, y) = x^2 - y^2 \qquad v(x, y) = 2xy. \tag{6.3.9}$$

Then

$$\int_0^{1+i} f(z)dz = \int_0^1 [t^4 - t^6 + 2it^5][2t + 3t^2i]dt = -\frac{2}{3} + \frac{2}{3}i. \tag{6.3.10}$$

In some cases we can use x or y itself as a parameter instead of t. For example in the path $x = t$, $y = t^2$, we can get rid of t and use just x: thus the path is given by $y = x^2$, on this path $dz = dx + i2xdx$ and $f(x, y) = f(x, x^2)$.[2]

---

[2]Note that the integral is independent of the parametrization. Thus $x = 4t^2, y = 8t^3$ with $[0 \le t \le 1/2]$ is also path $y = x^{3/2}$ joining the origin to $(1, 1)$ and will give the same answer as Eqns. (6.3.5 -6.3.6).

Consider now a function which is analytic inside a *simply connected domain* $\mathcal{D}$. We will now see that

Theorem 6.1. *The integral of $f(z)$ within a simply-connected domain $\mathcal{D}$ depends only on the end points and is path-independent.*

Proof: Let us first understand the condition "simply connected": it means that any closed curve in $\mathcal{D}$ must be contractible to a point without leaving $\mathcal{D}$. Thus the interior of a circle (or any smooth deformation of it) is simply connected whereas an annulus is not: a loop going all the way around cannot be collapsed to a point without leaving the domain.

The proof will be provided in two stages.

First it will be shown that

$$\oint_{C \in D} f(z)dz = 0 \qquad (6.3.11)$$

which means the integral of f around any closed path C entirely in $\mathcal{D}$ vanishes. Given this fact consider part A of Fig. 6.2, which shows a closed contour with two points on it called z_1 and z_2. Clearly

$$\oint_C = \int_{P_{12}} + \int_{P'_{21}} = 0 \text{ (given)} \qquad (6.3.12)$$

which states that the closed contour equals the path from 1 to 2 along P and back from 2 to 1 along P'. Using the fact that the integral changes sign when we reverse the path (since every dz goes to $-dz$), we may rewrite the above equation as

$$\int_{P_{12}} = \int_{P'_{12}} \qquad (6.3.13)$$

by reversing the path P'. This is the advertised result: the integral depends on only the end points and not the path if the integral on any closed loop vanishes. The converse is also true: if the integral is the same on all paths with the same end points, the integral around any closed path is zero. The proof is obtained by running the above proof backwards.

We now have to prove Eqn. (6.3.11). We use the following general argument. Consider a rectangle R obtained by gluing together two smaller rectangles R_1 and R_2 so that they share a common edge, as shown in part B of Fig. 6.2, where the rectangles are slightly displaced for clarity. The integral of f around the perimeter of R equals the sum of the integrals around R_1 and R_2, all integrals being done counter-clockwise. This follows from the fact that when we add the separate contributions from the two rectangles, the common edge occurs twice, with opposite sign, since it's traversed in opposite directions.

By gluing together more and more rectangles we will continue to find that the shared edges cancel, leaving behind only the exposed edges, i.e., the edges that line the perimeter of the composite object they form. *Conversely, the integral around a closed curve is the sum of the contributions from smaller pieces that tile the interior of the given contour.* If we make these pieces arbitrarily small we can use them to tile the area enclosed by any contour C. This is schematically shown in part C of the figure. Thus to find the integral of f around C, we need to know the integral around the tiny tiles that make up the area. We will now show that for an analytic function the contribution from every tiny rectangle is zero, which establishes our claim.

Now to furnish this last link in our arguments, let us turn to part D of the figure wherein an infinitesimal rectangle is magnified for our benefit. The integral has four contributions from the four sides. Let us pair the contributions from segments I and III (which run in opposite directions) as follows:

$$I + III = \int_x^{x+dx} f(x,y)dx - \int_x^{x+dx} f(x, y + dy)dx \qquad (6.3.14)$$

$$= \int_x^{x+dx} \left(-\frac{\partial f(x,y)}{\partial y} dy\right)dx \qquad (6.3.15)$$

$$= -\frac{\partial f(x,y)}{\partial y} dx\; dy + O(\text{cubic}) \qquad (6.3.16)$$

In the penultimate equation we have used the definition of the partial derivative, while in the last we have used the fact that the integral of a function over an infinitesimal segment equals the integrand times the length of the segment. Where should the partial derivative be evaluated? The answer is: anywhere in the rectangle. The difference between any two choices will be proportional to the shift in location and make a contribution that is cubic in the differentials since we already have a factor $dxdy$. (Here is where we assume the continuity of the derivatives.) However in the sum over tiny areas (which will give the integral around the bounding contour C) only the contribution proportional to the area itself ($dx\; dy$) will survive, (becoming the double integral over the enclosed area in the limit of infinitesimal areas) while higher powers will not. Recall the integral over a function of one variable where we divide the interval into N slices of width dx and form the product $f(x)dx$ in each slice and sum over slices. Assume each slice is of the same width for this argument. One can ask where within the slice the x in $f(x)$ lies. We saw that it does not matter in the limit $N \to \infty$ and $dx \to 0$. If we expanded f in a Taylor series at the left end of the interval $x = x_L$, $f(x) = f(x_L) + f'(x_L)(x - x_L) + \ldots$, and use $f(x_L)$ the additional terms shown are seen to be at least of order $f'(x_L)dx^2$ in each interval. In the limit $dx \to 0$, the sum over fdx will approach the area and terms like $f'dx^2$ will be of order $dx \simeq 1/N$ times the area and vanish in the limit. In other words, although dx and dx^2 are both going to zero, the former is going to zero as $1/N$ and is precisely compensated by the corresponding increase in N, the number of intervals or the number of terms in the sum, while the latter is going as $1/N^2$ and cannot be compensated by the growth in N. Likewise in our problem, $dx\,dy$ decreases like $1/N$, N being the number of tiny squares, while $dx^2\,dy$ falls faster and make no contribution in the limit we are taking.

Adding now the similar contribution from sides II and IV, we find

$$\oint \;=\; I + III + II + IV \tag{6.3.17}$$

$$= \; (-f_y + if_x)dxdy \tag{6.3.18}$$

$$= \; [(-u_y - iv_y) + i(u_x + iv_x)]\,dxdy \tag{6.3.19}$$

$$= \; 0 \text{ by Cauchy–Riemann.} \; \blacksquare \tag{6.3.20}$$

Problem 6.3.2. *Calculate the integral of $f = (x + iy)^2$ around a unit square with its lower left corner at the origin and obtain zero as you must.*

Let us explore the fact that the integral of $f(z)$ is path independent within the domain of analyticity. Consider the simple case of integrating $f(z) = z^2$ from the origin to the point $1 + i$. Since f is analytic for all finite z, we know the answer is the same for any path in the finite plane connecting these points. Let us try three paths. First let us go along the x-axis to $1 + 0i$, then up to $1 + i$. In the first part $f = x^2 - y^2 + 2ixy = x^2$ and $dz = dx + idy = dx$. Doing the x integral gives

$$\int_0^1 x^2 dx = 1/3. \tag{6.3.21}$$

In the second part $x = 1$ and y varies from 0 to 1 and $dz = dx + idy = idy$. Thus it contributes

$$\int_0^1 (1^2 - y^2 + 2i \cdot 1 \cdot y)idy = i(\frac{2}{3} + i) \tag{6.3.22}$$

giving a total of $\frac{2}{3}(i - 1)$. We have already seen that the same result obtains on the path $y = x^{3/2}$. The same result will obtain if we first go up the y-axis to i and then turn right or go along the line at 45 °, in which case we can write (since $x = y$), $f = x^2 - x^2 + 2ixx$, $dz = dx(1 + i)$. Having written everything in terms of x, we can do the x-integration from 0 to 1 and get the same result.

Problem 6.3.3. *Verify that the last two methods reproduce the answer derived above.*

There is a short cut for doing this. That is to treat z just like we treat x and write

$$\int_{z_1}^{z_2} z^2 dz \;=\; \left.\frac{z^3}{3}\right|_{z_1}^{z_2} \tag{6.3.23}$$

$$= \; \frac{(1+i)^3}{3} = \frac{2}{3}(i - 1) \tag{6.3.24}$$

which reproduces the earlier result. This is not an accident. More generally we can assert that

$$\int^z f(z')dz' = F(z) \tag{6.3.25}$$

where $F(z)$ is any function whose derivative is $f(z)$. The proof is as in the case of real integrals. We take the derivative of $F(z)$ by changing the upper limit by dz. The increment is then $f(z)dz$ and it follows $f = \frac{dF}{dz}$. Thus $F(z)$ gives the right dependence of the integral on the upper limit. (The answer does not depend on the direction in which we change the upper limit. Thus F has a nice direction-independent derivative at z as an analytic function should.) The above F is an indefinite integral; there is a free parameter in its definition corresponding to the free lower limit in the integral, just as in real integrals. *But the possible dependence on path is gone, given the path independence of such integrals. Thus any two indefinite integrals can only differ by a constant.* Notice however that all this is true only if we are operating within $\mathcal{D}$, the domain of analyticity.

Problem 6.3.4. *Evaluate the integral of z^2 on a rectangular path going from -1 to 1 to $1+i$ to $-1+i$ and back.*

Problem 6.3.5. *Integrate $z^2 - z^3$ from the origin to $1+i$ by first going to $z = 1$ along the real axis and then moving vertically up. Compare to what you get going at $45°$. Now compare this to the short cut in which you treat z just like x.*

Once we know how to integrate z^n, we can integrate any function given by a power series. For example the integral of $1/(1+z)$ is $\ln(1+z)$. As for the constant of integration, it is chosen so that the integral vanishes when $z = 0$, that is to say, the integral is from 0 to z. *Inside $\mathcal{D}$, the power series can be integrated term by term any number of times. The result will converge within the same $\mathcal{D}$ as can be seen by doing the ratio test.*

Problem 6.3.6. *Verify the above claim by carrying out the ratio test for the integral.*

6.4. The Residue Theorem

We close this chapter with a very important result on the integral of a function with a pole. Consider $f(z) = 1/z$ and a contour that goes around the origin along a circle of radius r so that $z = re^{i\theta}$ on it. At fixed r, we also have $dz = re^{i\theta}id\theta$. Thus

$$\oint \frac{1}{z}dz = \int_0^{2\pi} \frac{re^{i\theta}id\theta}{re^{i\theta}} \tag{6.4.1}$$

$$= i\int_0^{2\pi} d\theta = 2\pi i. \tag{6.4.2}$$

First note that we got a nonzero integral. Are we not supposed to get zero when we integrate an analytic function around a closed loop, and isn't being an explicit function of just z a criterion for analyticity? The point is that $f(z) = 1/z$ is indeed a function of only z, but has no definite value at the origin, blowing up as we approach it, in different ways for different paths. For the theorem to work, f must not only be a function of just z, it must have a unique and finite value. In other words, the CRE have no real meaning if u_x, v_y etc., are infinite as is the case in our problem at the origin. It also follows that if the contour does not enclose the singular point at the origin, the integral indeed will be zero.

Next note that the radius of the contour dropped out of the answer. Thus the integral is the same on circles of any radius surrounding the pole. In fact the answer is the same if we change the circle to any other shape that still encloses the origin. To understand this, start at some point z_1 on the circle, leave the circle in favor of another route and rejoin the circle at z_2, say by going along the chord joining these points. *The contribution must be the same from the two paths joining z_1 and z_2 since they lie entirely within the domain of analyticity of this f.* Thus the integral on the circle is equal to that on the contour that differs from it between z_1 and z_2. Of course if we can do this once and get away with it, we can do it any number of times and vary the contour at will. We cannot however deform it so as to exclude the origin, for in doing so we will necessarily be dealing with two paths that enclose a nonanalytic point, i.e., two paths that do not lie entirely in $\mathcal{D}$.

Consider a non-simply-connected domain $\mathcal{D}'$ that is shaped like an annulus centered at the origin, but such that the origin is excluded. The given f is analytic in $\mathcal{D}'$ and yet its integral around a closed loop that goes all the way around the annulus is not zero. The reason is that the integral over this loop cannot be written as a sum over integrals around small tiles because the tiling will stop at the inner perimeter of the annulus which will be another uncanceled boundary in addition to the loop we started with.

Returning to the main theme, is clear that if, for some constant A, we integrate A/z around any contour that encloses the origin, the result will be $2\pi i A$ and that *if we went around clockwise there would be a change of sign.*

Consider finally a function f which has several poles at points z_i and a contour that encloses a subset of them. Here is an example:

$$f(z) = \frac{1}{(z+i)(z-i)}. \tag{6.4.3}$$

It has a pole at $z = \pm i$, with residues $\pm \frac{1}{2i}$. Recall that in general

$$f(z) \simeq \frac{R(z_i)}{z - z_i} \quad \text{near} \quad z = z_i \tag{6.4.4}$$

where $R(z_i)$ is the residue at the pole z_i. We are now ready to prove

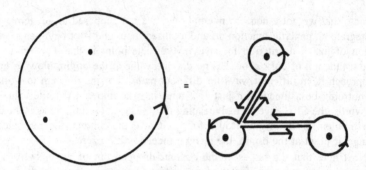

Figure 6.3. Two equivalent contours. We shrink the first one to the second one without crossing any poles.

Theorem 6.2. (Cauchy's Residue Theorem). *If* $f(z)$ *is meromorphic and has only simple poles,*

$$\oint_C f(z)dz = 2\pi i \sum_{z_i \in C} R(z_i) \tag{6.4.5}$$

where $z_i \in C$ *means the point* z_i *is inside the contour* C *and the latter is traversed counterclockwise.*

Proof: Consider Fig. 6.3.

At the left we see the contour C enclosing the poles marked by dots. Using the freedom to deform the contour *without crossing any poles* we can modify C to what we see in the top right of the figure. The circular paths around each pole can be made infinitesimal. Finally we can ignore the two way paths that connect the poles since they cancel going and coming. Thus we need to just go around each pole in an infinitesimally small circle. The contribution from each circle is deduced as follows. Given that f has a pole at some z_i we know that it must have the form

$$f(z) = \frac{R(z_i)}{z - z_i} \tag{6.4.6}$$

in the immediate vicinity of the pole. That is to say, the function factors into a part which has a pole $1/(z - z_i)$ and thus varies rapidly and a part (the residue) that is essentially constant over the infinitesimal circle around z_i. It is now clear that each pole makes a contribution that is $2\pi i$ times the residue there and Eqn. (6.4.5) follows. ∎

Once again consider $f(z) = 1/[(z - i)(z + i)]$. If C encloses the pole at i but not the one at $-i$, the integral is $2\pi i \cdot \frac{1}{2i}$, while if it encloses the lower pole, the answer is reversed in sign. What if C encloses both poles? Adding the two previous results, we get zero. This can also be obtained another way. Let us take this C and go on enlarging it. The answer cannot change since we are running

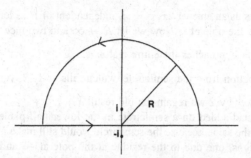

Figure 6.4. Application of the residue theorem.

away from the only singularities the function has. But as we go to infinitely large z the integrand approaches $1/z^2$ and on a circle of radius R,

$$\oint_R \simeq \int_0^{2\pi} \frac{Re^{i\theta}id\theta}{R^2 e^{2i\theta}} \qquad (6.4.7)$$

$$= 0. \qquad (6.4.8)$$

The integral vanishes for two reasons. First, the whole thing falls like $1/R$ which kills it as $R \to \infty$. Secondly, the angular integral is identically zero since the final integrand $e^{-i\theta}$ integrates to zero between the limits $0-2\pi$. Thus we do not need the argument based on counting powers of R here. It is provided in view of integrals where the range of θ is not the full circle and the integral in question vanishes only due to the inverse power(s) of R as in the following example. Consider

$$\int_0^\infty \frac{dx}{1+x^2}. \qquad (6.4.9)$$

We already know this equals $\pi/2$ by doing the integral with a trigonometric substitution. We will now rederive this result using the residue theorem. But this needs some ground work. First of all, we have a real integral and not a complex one in the complex plane. That is no problem since we can see the above as the integral of $f(z) = 1/(1+z^2)$ on the real axis where $z = x$ and $dz = dx$. The next problem is that this is not an integral over a closed contour. This we remedy in two stages. First we extend the limits to $-\infty \le z \le \infty$ and compensate with a factor of $\frac{1}{2}$ since the integrand is an even function. We now have a contour going from one end of the real axis to the other, but it is not yet a closed contour. To fix this up, we add the contribution from a large semicircle of radius R and argue that as $R \to \infty$, this will not affect the answer. To be precise we consider the contour shown in Fig. 6.4.

Now we have the good feature that the residue theorem applies, but the bad feature that this is not the contour given to us! Forging ahead, we find that the pole at $z = i$ gives us an answer $2\pi i \frac{1}{2i} = \pi$, independent of R as long as it is large enough to enclose the pole at i. Now we let $R \to \infty$ and two nice things happen:

- The diameter approaches the entire real axis.

- The contribution from the semicircle, which falls as $1/R$, vanishes.

Dividing the answer by 2 we regain the old result.

What if we had added on a semicircle in the lower half-plane? The answer would have been the same because the semicircle would still make no contribution, and two minus signs, one due to the residue at the pole at $-i$ and the other due to the fact that the residue theorem has an extra sign if we go around the pole *clockwise*, would cancel.

Here is another important thing to remember when you use the residue theorem: the residue is the coefficient of $1/(z - z_0)$. Thus the function $1/(2z - 4)$ has a pole at $z = 2$, the residue there is $1/2$. We must always recast the pole so that z has unit coefficient. In the example considered this means writing $1/(2z - 4) = \frac{1/2}{z-2}$.

Problem 6.4.1. *What is $\oint_C \frac{e^{iz} dz}{z^2+1}$ where C is a unit circle centered at $z = i$? Repeat when the center is at $z = -i$.*

Let us consider next an example that typifies a frequently encountered family:

$$\int_{-\infty}^{\infty} \frac{\cos x\, dx}{1+x^2}. \tag{6.4.10}$$

Let us view this as the integral of a complex function on the real axis. To use the residue theorem, we must be able to add on the semicircle at infinity. Now we have a problem. Without the cosine the integral vanishes like $1/R$. What does the cosine do? Let us examine the cosine in the complex plane:

$$\cos z = \frac{e^{iz} + e^{-iz}}{2} \tag{6.4.11}$$

$$= \frac{e^{ix-y} + e^{-ix+y}}{2}. \tag{6.4.12}$$

Notice that whereas the first exponential blows up in the lower half plane and converges in the upper half plane, the second does the reverse. Thus no matter which half plane we add the semicircle in, its contribution is not negligible. The way out is the following. We replace the cosine in the given x integral by e^{ix}, calculate the integral by closing the contour in the upper half plane and then take the real part of the answer. The pole at i has a residue

$$R(z = i) = \frac{e^{-1}}{2i}, \tag{6.4.13}$$

which means the integral is π/e.

Problem 6.4.2. *There was no need to take the real part of this answer since it was already real. Can you argue using symmetries of the original integral that changing* $\cos x$ *to* e^{ix} *does not change the integral? How would you deal with a function that had* $\sin x$ *in the numerator instead of cosine?*

Problem 6.4.3. (Excellent practice on closing contours). *Show that*

$$\int_0^\infty \frac{\cos ax + x \sin ax}{1 + x^2} dx = \pi e^{-a} \ \ if \ a > 0, \ \frac{\pi}{2} \ \ if \ a = 0, \ 0 \ if \ a < 0$$

We finally consider another class of integrals which can be done by contour integration. Here is a concrete example:

$$I = \int_0^{2\pi} \frac{d\theta}{a + \cos\theta} \quad (a > 1). \tag{6.4.14}$$

Where is the closed contour here? It appears if we interpret θ as the phase of a unimodular complex number $z = e^{i\theta}$. Consider an integral along the unit circle in the z-plane. We would have (*on this circle*)

$$z = e^{i\theta} \tag{6.4.15}$$

$$dz = id\theta e^{i\theta} = id\theta z \tag{6.4.16}$$

$$\cos\theta = \frac{e^{i\theta} + e^{-i\theta}}{2} = \frac{z + 1/z}{2} \tag{6.4.17}$$

$$I = \oint_{|z|=1} \frac{dz}{iz} \frac{1}{a + \frac{z+1/z}{2}} \tag{6.4.18}$$

$$= \oint_{|z|=1} \frac{-2idz}{2az + z^2 + 1}. \tag{6.4.19}$$

We must now look for the poles of the integrand that lie in the unit circle and evaluate the integral by residues. The roots of the quadratic function in the denominator are readily found and the result is

$$I = \frac{2\pi}{\sqrt{a^2 - 1}} \tag{6.4.20}$$

as detailed in the following exercise.

Problem 6.4.4. *Establish Eqn. (6.4.20) as follows. First show that when* $a > 1$ *the roots are real and their product is unity. Thus both roots must lie on the real axis, with one inside and one outside the unit circle. Using the contribution from the former obtain the above result. The case* $a < 1$ *is much more complicated. Show that the roots form complex conjugate pairs lying on the unit circle. Integrals with poles on the contour are not well-defined. You should consider a more advanced book if you wish to pursue this. Look for words like* **Principal Value** *in the index.*

Problem 6.4.5. *Evaluate by the residue theorem:*

(i) $\int_0^\infty \frac{dx}{1+x^4}$,

(ii) $\oint \frac{\sin z\, dz}{z-\pi}$ *on* $|z| = 1, 2$,

(iii) $\int_0^\infty \frac{x^2\, dx}{(x^2+25)(x^2+16)}$,

(iv) $\int_0^{2\pi} \frac{\cos\theta\, d\theta}{5+4\cos\theta}$,

(v) $\int_0^{2\pi} \frac{\sin\theta\, d\theta}{5+4\cos\theta}$, *and*

(vi) $\int_0^{2\pi} \frac{\sin^2\theta\, d\theta}{5+4\cos\theta}$.

Problem 6.4.6. *Show that*

$$(A) \qquad \int_{-\infty}^\infty \frac{x\sin x}{1+x^4}\, dx \;=\; \pi e^{-1/\sqrt{2}} \sin\frac{1}{\sqrt{2}},$$

$$(B) \qquad \int_0^\infty \frac{x\sin 3x}{x^2+9}\, dx \;=\; \frac{\pi}{2} e^{-9},$$

$$(C) \qquad \int_0^\infty \frac{x^2+1}{1+x^4}\, dx \;=\; \frac{\pi}{\sqrt{2}},$$

$$(D) \qquad \int_{-\infty}^\infty \frac{x\sin x}{x^2+2x+17}\, dx \;=\; \frac{(4\cos 1 + \sin 1)\pi}{4e^4}.$$

Hints: (A) and (C): Find the roots of $z^4 + 1 = 0$. For all problems except (C) recall $\sin x = \operatorname{Im} e^{ix}$. (B) and (D): The hard part here is to show that the big semicircle can be added for free. There are no inverse powers of R as in the other two problems. Analyze the exponential function and show that it vanishes as $R \to \infty$ except for an infinitesimal angular region which makes an infinitesimal contribution.

As a final application of the residue theorem, consider the equation

$$\frac{1}{2\pi i} \oint_{C \in D} \frac{f(z)\, dz}{z - z_0} = f(z_0). \qquad (6.4.21)$$

which states that if $f(z)/2\pi i(z - z_0)$ is integrated over a contour C entirely in the domain $\mathcal{D}$ of analyticity of f, the answer is $f(z_0)$. The result follows just from the residue theorem, since by construction, the integrand has only one pole, where we put it by hand, namely at z_0 with a residue $f(z_0)$.

This is a remarkable result: *given the values of an analytic function on a closed contour lying inside its domain of analyticity, its values inside the contour are fully determined.* But if we know the function inside, we also know in principle its derivatives. Indeed by taking the derivative of both sides with respect to z_0 we find

$$\frac{1}{2\pi i} \oint_{C \in D} \frac{f(z)\, dz}{(z - z_0)^2} = \frac{df(z_0)}{dz_0}. \qquad (6.4.22)$$

In other words, the integral of f around a double pole pulls out the derivative *of the residue.*

If we take higher derivatives we will find a similar result:

$$\frac{n!}{2\pi i} \oint_{C \in D} \frac{f(z)dz}{(z - z_0)^{n+1}} = \frac{d^n f(z_0)}{dz_0^n} \equiv f^{(n)}(z_0). \qquad (6.4.23)$$

Here is an illustration of Eqn. (6.4.23):

$$\oint_{|z|=1} \frac{\cos z\, dz}{z^3} = \frac{2\pi i}{2!} \left. \frac{d^2 \cos z}{dz^2} \right|_0 = -i\pi. \qquad (6.4.24)$$

Note that the n-th derivative defined by Eqn. (6.4.23) is well-behaved: the integrand is nonsingular since f is bounded on $C \in D$ as is the factor $\frac{1}{(z-z_0)^{n+1}}$ since z lies on the contour while z_0 is strictly in the interior. In the next subsection we will obtain a more precise bound on such integrals.

Problem 6.4.7. *Evaluate* $\oint \frac{dz}{(z-2)^2 z^3}$ *on* $|z - 3| = 2$ *and* $|z - 1| = 3$.

When we first began with analytic functions, we demanded that they have a first derivative. Now we see that having a first derivative in a neighborhood implies having all derivatives therein.

Problem 6.4.8. *Evaluate* $\int_0^\infty \frac{\cos x\, dx}{(x^2+9)^2}$.

6.5. Taylor Series for Analytic Functions

We have seen that within the radius of convergence R, a power series can be used to define an analytic function. We now consider the theorem which addresses the reverse question of whether every analytic function can be expanded in a Taylor series. We will first discuss the proof and then consider the importance of the result. Even if you are not planning to follow the proof in detail, you may wish to digest the discussion that follows it.

Theorem 6.3. (Taylor). *Let a be a point of analyticity of an analytic function $f(z)$ and R the distance to the nearest singularity. Then f can be expanded in a Taylor series centered at $z = a$ for $|z - a| < R$.*

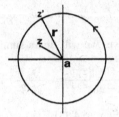

Figure 6.5. Taylor series for $f(z)$ about $z = a$.

Proof: Let us consider a circle of radius $r < R$ centered at a as shown in Fig. 6.5. We then have due to Eqn. (6.4.21)

$$f(z) - f(a) = \frac{1}{2\pi i} \int_{|z'-a|=r} f(z') \left[\frac{1}{z'-z} - \frac{1}{z'-a} \right] dz' \tag{6.5.1}$$

$$= \frac{z-a}{2\pi i} \int_{|z'-a|=r} f(z') \frac{1}{(z'-z)(z'-a)} dz'. \tag{6.5.2}$$

Now we use

$$\frac{1}{z'-z} = \frac{1}{(z'-a) - (z-a)} \tag{6.5.3}$$

$$= \frac{1}{z'-a} \left[1 + \frac{z-a}{z'-a} + \frac{(z-a)^2}{(z'-a)^2} + \cdots + \frac{(z-a)^{n-1}}{(z'-a)^{n-1}} + \frac{\left[\frac{z-a}{z'-a} \right]^n}{1 - \frac{z-a}{z'-a}} \right], \tag{6.5.4}$$

where we have invoked an old result about series:

$$1 + r + \cdots r^{n-1} = \frac{1}{1-r} - \frac{r^n}{1-r}. \tag{6.5.5}$$

Here we have chosen $r = (z-a)/(z'-a)$ and shifted the $r^n/(1-r)$ term to the left-hand side. If we now invoke Eqns. (6.4.23, 6.5.2) we find

$$f(z) - f(a) = (z-a) f^{(1)}(a) + \frac{1}{2!}(z-a)^2 f^{(2)}(a) + \cdots + \frac{1}{n!}(z-a)^n f^{(n)}(a) + \mathcal{R}, \tag{6.5.6}$$

where the remainder $\mathcal{R}$ is given by:

$$\mathcal{R} = \frac{1}{2\pi i} \int_{|z'-a|=r} \frac{f(z') dz'}{z'-z} \left(\frac{z-a}{z'-a} \right)^{n+1}. \tag{6.5.7}$$

On the circle of integration $|z'-a| = r$, let us write $dz' = re^{i\theta}$ and let M be the largest absolute value of f. Then

$$|\mathcal{R}| \leq \frac{M|z-a|^{n+1} r}{||z'-a| - |z-a|| r^{n+1}}. \tag{6.5.8}$$

The details are left to the following exercise.

Problem 6.5.1. *First recall that the sum of a set of complex numbers is bounded by the sum of the absolute values of the individual terms. Since the integral is a sum, this implies the integral is bounded by one in which we replace the integrand and dz by their absolute values. You can now bound every factor in the integrand, calculate the length of the integration contour, and argue that $1/(|z - z'|) \leq ||z' - a| - |z - a||^{-1}$ by looking at Fig. 6.5.*

Returning to Eqn. (6.5.8), since $|z - a|/r < 1$, and

$$\frac{Mr}{||z' - a| - |z - a||}$$

is bounded, we have a series expansion where the remainder after n terms can be made arbitrarily small by sending $n \to \infty$, i.e., a genuine Taylor series. ∎

Let us note the following points implied by this theorem:

- The Taylor series for a function fails to converge only when a singularity is encountered as we go outward from the point of expansion. There are no mysterious breakdowns of the series as in real variables: Why is the series for $f(x) = 1/(1+x^2)$ ill behaved at and beyond $|x| = 1$ when f itself is perfectly well-behaved for $|x| = 1$ and indeed for all x? (We don't have this query for $1/(1 - x)$ which has a clear problem at $x = 1$.) The answer is clear only when we see this function as $1/(1 + z^2)$ evaluated on the real axis: $R = 1$ because the function has poles at $z = \pm i$.

  Now, it is fairly clear why the series must break down at a pole: the function diverges there. But it must also break down at a branch point around which the function is not single valued. The breakdown of the series is mandated here because when the series is convergent, it is given by a finite single valued sum of single valued terms of the form z^n which cannot reproduce the multiple valuedness around the branch point.

- Here is another mystery from the real axis. Recall the function e^{-1/x^2}. It has a well-defined limit and all its derivatives have a well-defined limit at $x = 0$, namely zero. Thus the origin seems to be a nonsingular point. Yet we cannot develop the function in a Taylor series there: the function and all its derivatives vanish. The existence of such functions means that two different functions, $f(x)$ and $f(x) + e^{-1/x^2}$, can have the same derivatives (of all orders) at the origin. Once again we see the true picture only in the complex plane. There we are assured that every function has a Taylor expansion about any analytic point. What about the function e^{-1/z^2}, you say? How come it cannot be Taylor expanded at the origin, which seems like a very nice point? In turns out the origin is far from being a very nice point, it is home to one of the nastiest singularities, as can be seen in many equivalent ways. First, note that the function is defined by the series

$$e^{-1/z^2} = \sum_{0}^{\infty} \frac{(-1)^n}{z^{2n} n!}. \tag{6.5.9}$$

  This is an expansion in the variable $1/z^2$ and the ratio test tells us it converges for all $1/z^2$ that are finite. In particular it does not converge at the origin: *the origin is an essential singularity*. While we see no sign of this along the real axis, we see upon setting $z = iy$ that the function blows up as e^{1/y^2} as we approach the origin from the imaginary direction. The same goes for all its derivatives along this axis. Thus the function really has no unique limit at the origin and its derivatives are likewise highly sensitive to the direction of approach. The vanishing x-derivatives are not the *bona fide* direction-independent derivatives of an analytic function at a point of analyticity. Any notion that we can have a legitimate Taylor series at the origin is clearly a mistake that comes from looking along just one of infinitely possible directions in the complex plane. The origin is a singular point, and one has to go to the complex plane to appreciate that. In short, there are no exceptions to the rule that an analytic function can be expanded in a Taylor series about any point of analyticity. In particular, a function with all derivatives equal to zero at a point of analyticity is identically zero. For this reason, no two distinct functions can have the same Taylor series: it they did, their difference would have a vanishing Taylor series and would equal zero everywhere, i.e., the functions would be identical, contrary to the assumption.

- Consider the question of defining a function outside the radius of convergence of the Taylor series, taking as an example $f(z) = 1/(1-z)$ whose Taylor series $1+z+z^2+\ldots$ converged only inside the unit disc, shown by C_1 in Fig. 6.6.

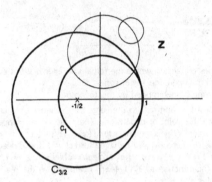

Figure 6.6. Analytic continuation.

Inside the unit circle C_1, the series converges. It also reproduces the given function: since the Taylor series is a geometric series we can sum it (for $|z| < 1$) and reproduce $1/(1-z)$. We know it diverges at some point(s) on the unit circle. Peeking into the closed form result, we know this happens at $z = 1$. If we go outside the unit circle C_1, the series has no meaning, while we know the function itself has a meaning everywhere except at $z = 1$. We generate the function outside the unit circle as follows. We first go to another point within the unit circle, say $z = -1/2$. Since we know f (from the series) in the neighborhood of this point, we can compute in principle its derivatives and launch a new Taylor series centered at this point. The radius for this new series will be determined by the singularity nearest to this point. Peeking into the closed form solution, we know this is still the pole at $z = 1$ which now lies 1.5 units away. Thus the expansion will converge within a circle $C_{3/2}$ of radius 1.5 centered at the new point. (If we did not have the closed form answer at hand, the ratio test would give us the new radius, but not the location of the singularity on the circumference.) The new series will agree with the old one within the unit circle and define the function in the nonoverlap region. We have clearly defined the function in a bigger patch of the complex plane, armed with just the Taylor series at the origin. But will this definition agree with any other, say the closed form expression? Yes, because the series centered at the origin, and the closed form expression $1/(1-z)$, are numerically equal in the entire unit circle and in particular at $z = -1/2$, *as are all their derivatives*. (Imagine extracting the derivatives by surrounding the point $z = -1/2$ by a circle and using Eqn. (6.4.23). Thus the series representing the function and the closed form for it will generate the same Taylor series at the point $z = -1/2$ which therefore agree throughout the circle of radius 3/2 centered there. This process can be continued indefinitely, as shown by a few other circles centered at various points. We will find in the end that f can be defined everywhere except at $z = 1$. In particular if we work our way to $z = 2$, we will get the value $f(2) = -1$. Of course in the present problem we need not go to all this trouble: the closed form $f(z) = 1/(1-z)$ defines it for all $z \neq 1$. But the main point has been to show that the Taylor series at the origin contains all the information that the closed form expression did, albeit in not such a compact and convenient form. This is very important since there are many functions which we know only through their series about some point.

Here are some details on the Taylor expansion centered at $z = -1/2$. First let us use the closed form to generate the Taylor series. There is no need to compute all its derivatives there, we simply rewrite it· to obtain the following Taylor expansion in the variable $z - (-\frac{1}{2}) =$

$z + \frac{1}{2}$:

$$\frac{1}{1-z} = \frac{1}{\frac{3}{2} - (z + \frac{1}{2})} \tag{6.5.10}$$

$$= \frac{\frac{2}{3}}{1 - \frac{2}{3}(z + \frac{1}{2})} \tag{6.5.11}$$

$$= \frac{2}{3}\left[1 + \frac{2}{3}\left[z + \frac{1}{2}\right] + \left[\frac{2}{3}(z + \frac{1}{2})\right]^2 + \cdots \right.$$

$$+ \left. \left[\frac{2}{3}(z + \frac{1}{2})\right]^n + \cdots\right], \tag{6.5.12}$$

which converges for $\frac{2}{3}|z + \frac{1}{2}| < 1$ which is the circle of radius $3/2$ centered at $z = -\frac{1}{2}$ alluded to.

Now we compare this to what we get from the power series at the origin. From the series for $f(z)$ we find its derivative

$$f^{(1)}(z) = \frac{d}{dz}[1 + z + z^2 + z^3 + \cdots] = 1 + 2z + 3z^2 + 4z^3 + \ldots + nz^{n-1} + \ldots \tag{6.5.13}$$

The ratio test says the series for the derivative converges for

$$|z| < R = \lim_{n \to \infty} \frac{n}{n+1} = 1, \tag{6.5.14}$$

i.e., again within the unit circle. Its value at $z = -\frac{1}{2}$ is

$$f^{(1)}(-1/2) = 1 - 1 + \frac{3}{4} - \frac{4}{8} + \frac{5}{16} - \frac{6}{32} + \frac{7}{64} \ldots = \frac{4}{9}. \tag{6.5.15}$$

This coincides with the first derivative in the expansion Eqn. (6.5.12). In other words, the derivative of the series for the function is the series for the derivative of the function within the unit circle. This will keep happening for every derivative. (We are simply using the result that the series for a function may be differentiated term by term to reproduce the series for its derivative, both series having the same radius of convergence.) Thus both schemes (based on the Taylor series at the origin and the closed form expression) will produce the same function inside $C_{3/2}$ and in all other continuations.

- We consider now the notion called the *Permanence of Functional Relations*. Why does a relation like $\sin^2 z + \cos^2 z = 1$ holds off the real axis given that it does on the real axis? To see this consider the function $f(z) = \sin^2 z + \cos^2 z - 1$. It vanishes everywhere along the x-axis. Therefore so do all its derivatives along the x-axis, in particular, at the origin. But the origin is a point of analyticity of this function (which is made of the sine and cosine) and the x-derivatives in question are the direction independent z-derivatives. Therefore f has a Taylor expansion and this expansion gives identically zero in the complex plane. Therefore f will be identically zero, i.e., $\sin^2 z$ will continue to equal $1 - \cos^2 z$ in the complex plane.

More generally, if f and g are two functions which agree on a line segment which lies entirely within the domain of analyticity of both functions, their continuation outside this domain will also agree since the difference $f - g$ has a vanishing Taylor series (the derivatives being taken along the segment) and therefore vanishes. Equivalently, if we construct the Taylor series for f and g starting at some point on the segment, the series will be identical since the functions are numerically identical and hence possess identical derivatives. For this reason, the two series will have the same radius of convergence and run into the same singularities at the boundary. If we now move to another point in this circle and set up a new series to effect an analytical continuation, these new series will again be identical for the same reason as before, and the two functions will forever be equal. Thus given a relation like $\cosh x = (e^x + e^{-x})/2$ on the real axis, it follows that $\cosh z = (e^z + e^{-z})/2$ for all z.

Let us note that the relation in question must be between *analytic functions* before it can be analytically continued. Here is an example that is not analytic. Recall that for real x we had

$$\cos x = \frac{\left[e^{ix}\right] + \left[e^{ix}\right]^*}{2}. \tag{6.5.16}$$

This is not a relation that is preserved upon continuation. In other words

$$\cos z \neq \frac{\left[e^{iz}\right] + \left[e^{iz}\right]^*}{2} \tag{6.5.17}$$

for z off the real axis.

Problem 6.5.2. *Compare the two sides of the above relation on the imaginary axis $z = iy$ and show that they do not match.*

The problem is that the above relation involves complex conjugation which converts the allowed variable z to the forbidden one z^*. Does this mean we cannot relate the exponential to the sines and cosines after continuation into the complex plane? Not so, as we have already seen. Instead of thinking of e^{-ix} as the complex conjugate of $f(x) = e^{ix}$, think of it as $f(-x)$. Then we can view the above relation as

$$\cos z = \frac{e^{iz} + e^{-iz}}{2} \tag{6.5.18}$$

evaluated on the real axis. But now we have a relation between the analytic functions $\cos z$, $e^{\pm iz}$, valid in the complex plane.

Problem 6.5.3. *On the real axis we had $e^{ix} = \cos x + i \sin x$. Is this an analytic relationship which will survive the continuation to complex z?*

Problem 6.5.4. *On the real axis we have $\left[e^{ix}\right]\left[e^{ix}\right]^* = 1$. Is it true that $\left[e^{iz}\right]\left[e^{iz}\right]^* = 1$? Give reasons for your answer and then the supporting calculation.*

6.6. Summary

The highlights of this chapter are as follows.

- The function $f(x, y) = u(x, y) + iv(x, y)$ is analytic in a domain $\mathcal{D}$ if the first partial derivatives of u and v are continuous and obey the Cauchy–Riemann equations:

$$u_x = v_y \qquad u_y = -v_x.$$

These merely ensure that $f(x, y) = f(x + iy)$. The functions u and v are then harmonic:

$$u_{xx} + u_{yy} = v_{xx} + v_{yy} = 0.$$

- An analytic function can have singularities at isolated points. At a pole it behaves as

$$f(z) \to \frac{R(z_0)}{z - z_0},$$

where the residue is given by

$$R(z_0) = \lim_{z \to z_0} (z - z_0) f(z).$$

You must know about the existence of higher-order poles, essential singularities and branch points.

- To differentiate a function $f(z)$ treat z just like x of real variable calculus. Thus $dz^n/dz = nz^{n-1}$. If f is given in terms of x and y, take the partial derivative with respect to x or iy.

- The power series

$$\sum_{n=0}^{\infty} a_n z^n$$

converges within a radius

$$R = \lim_{n \to \infty} \left| \frac{a_n}{a_{n+1}} \right|.$$

Inside this circle of convergence it defines a function $f(z)$. The series may be differentiated any number of times to yield the series for the corresponding derivative of $f(z)$. It may likewise be integrated.

-

$$e^z = \sum_{0}^{\infty} \frac{z^n}{n!} \qquad [|z| < \infty]$$

You must know the first few terms of this series as well as the series for hyperbolic and trigonometric functions which are related to the exponential exactly as in real variables.

- The logarithm is the inverse of the exponential

$$z \equiv e^{\ln z}$$

$$\ln(1 + z) = z - \frac{z^2}{2} + \frac{z^3}{3} + \cdots$$

- The N-th root of $z = re^{i\theta}$ has N values given by

$$z^{1/N} = r^{1/N} \exp[\frac{1}{N}(\theta + 2\pi m)] \qquad m = 0, 1, ..N - 1,$$

- You must know how $\int f(z)dz$ along a path P is defined (as the limit of a sum over tiny segments of P), and how to evaluate such integrals by parametrizing the path.

- Main properties of complex integrals:

$$\int_{z_1}^{z_2} f(z)dz \text{ is path independent}$$

for paths lying entirely in the domain of analyticity $\mathcal{D}$.

$$\int^z f(z)dz = F(z) + c,$$

where $F(z)$ is any integral of f and c is a constant depending on the lower limit. Thus for example

$$\int^z z^n dz = \frac{z^{n+1}}{n+1} + c.$$

You must know Cauchy's Residue Theorem for a meromorphic function:

$$\oint f(z)dz = 2\pi i \sum_i R(z_i),$$

where the contour goes counterclockwise and enclosed poles labeled i at points z_i with residues $R(z_i)$. A special case of the above is the result

$$\oint \frac{f(z)}{z - z_0} dz = 2\pi i f(z_0)$$

if the contour lies in $\mathcal{D}$, the domain of analyticity of $f(z)$. Thus a function is fully determined by its values on a contour surrounding the point z_0. By taking derivatives of both sides with respect to z_0 one can relate the derivatives at z_0 to integrals on the surrounding contour as per

$$\frac{n!}{2\pi i} \oint_{C \in D} \frac{f(z)dz}{(z - z_0)^{n+1}} = \frac{d^n f(z_0)}{dz_0^n} \equiv f^{(n)}(z_0).$$

- To evaluate real integrals, try to first get the contour to run from $-\infty$ to ∞, using symmetries. For example the integral of an even function from 0 to ∞ is half the integral from $-\infty$ to ∞. Then try to add on the semicircle at infinity. If only powers of z are involved in the integrand, do a power counting to see if it is allowed. If the exponential is present, make sure it converges in the half-plane where you are adding on a semicircle. If trigonometric functions are involved, break them into exponentials and do the integrals in pieces or relate the original problem to the real or imaginary part of an exponential.

- Every function can be expanded in a Taylor series about a point of analyticity. The series will converge inside the circle that extends to the nearest singularity. Functional relations satisfied in a line segment (which lies within the region of analyticity) will continue to hold when the functions are analytically continued. Equivalently, if two analytic function coincide in such a segment, they coincide everywhere. Thus for example

$$e^{iz} = \cos z + i \sin z \qquad \sin(-z) = -\sin z \qquad \sin^2 z + \cos^2 z = 1$$

and so on because these relations are true on the real axis.

VECTOR CALCULUS

In this chapter we will combine elementary notions from vector analysis and calculus. We begin with a review of the former topic.

7.1. Review of Vectors Analysis

In Chapter 9 we will discuss vectors in great detail and generality. For the present we need just the primitive notions you must be familiar with. Let us begin with the notion of a vector as an arrow, that is, an object with a magnitude and direction. The example we will often use is that of a displacement vector. Let us assume first that the vectors lie in a plane, the $x - y$ plane.

The next notion is that of adding two vectors **A** and **B**. The rule for defining their sum comes naturally if we recall how two displacements combine and is as follows: place the tail of **B** at the tip of **A** and draw a vector which runs from the tail of **A** to the tip of **B** to obtain their sum **A** + **B**.

The next notion is that of multiplying a vector by a number c. Consider first **A** + **A**. By our rule this is a vector that is parallel to **A** and twice as long. Since it is natural to call the new vector 2**A**, it is natural to define c**A** to be **A** stretched out by a factor c if $c > 0$. We will discuss shortly the case $c < 0$.

We define the *null vector* **0** to be vector of zero length. It is clear that adding **0** to any vector does not change the vector. Consider now the vector −**A** defined to be the reversed version of **A**, with tail and tip interchanged. Our addition rule tells us that **A** + (−**A**) = **0**. In view of this it is natural to identify reversing the vector with multiplication by −1. It follows that multiplying by $c < 0$ reverses the vector and rescales it by $|c|$.

Consider now two *unit vectors*: **i** and **j** pointing along the x and y axes and of unit length. We say **i** and **j** form an *orthonormal basis* meaning they are mutually orthogonal, their *norm* or length is unity. Any vector in the plane can be written in terms of them:

$$\mathbf{A} = A_x\mathbf{i} + A_y\mathbf{j}. \tag{7.1.1}$$

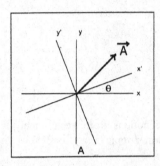 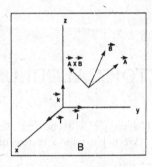

Figure 7.1. A. Two coordinate systems in two dimensions for describing the vector **A**. Not shown are unit vectors i and j along x and y axes and i′ and j′ along x' and y' axes. **B.** A right handed system in three dimensions for defining the cross product.

Clearly **A** is a vector obtained by adding a piece of length A_x along the x-axis with a piece of length A_y along the y-axis. We refer to A_x and A_y as the *components of* **A**. Note that the components of **A** are sensitive to the choice of coordinate axes. For example if we switch to $x' - y'$ axes, as in Fig. 7.1A, rotated with respect to the $x - y$ axes by an angle θ in the counterclockwise direction, with unit vectors i′ and j′ along x' and y' axes, (not shown), the same vector will have different components:

$$A'_x = A_x \cos \theta + A_y \sin \theta \qquad (7.1.2)$$

$$A'_y = -A_x \sin \theta + A_y \cos \theta. \qquad (7.1.3)$$

On the other hand, the length of the vector, denoted by $|A|$, and determined by the Pythagoras theorem, will be unaffected by the rotation:

$$|A|^2 = A_x^2 + A_y^2 = (A'_x)^2 + (A'_y)^2. \qquad (7.1.4)$$

Problem 7.1.1. *Prove Eqns. (7.1.2-7.1.4). First draw a sketch of the coordinate systems and a vector **A** starting from the origin and terminating at the point (A_x, A_y) in the first system to obtain the coordinate transformation in the first two equations. Now check the last one algebraically.*

These ideas generalize to three dimensions. We introduce a third unit vector k perpendicular to the other two and oriented along the positive z-axis of a *right handed coordinate system* shown in Fig. 7.1B.

The adjective *right-handed* means that if you grab the z-axis with your right hand, your thumb will point along k and the other four fingers will curl from i to j.

So far we have discussed addition and scalar multiplication of vectors. We now consider two kinds of products between a pair of vectors: the *dot product* which yields a scalar and the *cross product* which yields a vector.

The dot product of two vectors $\mathbf{A} \cdot \mathbf{B}$ is a number whose value is given by

$$\mathbf{A} \cdot \mathbf{B} = |A||B|\cos\theta_{AB} \qquad (7.1.5)$$

where $|W|$ is the length of vector $\mathbf{W}$ and θ_{AB} is the angle between $\mathbf{A}$ and $\mathbf{B}$, measured in the plane containing both. (It does not matter to the cosine whether you measure the angle from $\mathbf{A}$ to $\mathbf{B}$ or vice versa.) The subscript on the angle will be suppressed if there is no room for confusion.

The dot product is equivalently given as the length of one vector times the projection of the other in its direction

$$\mathbf{A} \cdot \mathbf{B} = |A|_B|B| \qquad (7.1.6)$$

where $|A|_B = |A|\cos\theta_{AB}$ stands for the projection of $\mathbf{A}$ along $\mathbf{B}$.

Note that the dot product is a *scalar*: its value is unaffected by rotating the coordinate system. Under such a rotation the components of $\mathbf{A}$ and $\mathbf{B}$ will change, but $|A|, |B|$, and $\cos\theta_{AB}$ will not.

A special case of the dot product where the vectors are equal gives us:

$$\mathbf{A} \cdot \mathbf{A} = |A|^2. \qquad (7.1.7)$$

The dot product has a very important *distributive property*:

$$\mathbf{A} \cdot (\mathbf{B} + \mathbf{C}) = \mathbf{A} \cdot \mathbf{B} + \mathbf{A} \cdot \mathbf{C}. \qquad (7.1.8)$$

Problem 7.1.2. *Draw a sketch to verify that the projection of* $\mathbf{B} + \mathbf{C}$ *along* $\mathbf{A}$ *equals the sum of the projections of* $\mathbf{B}$ *and* $\mathbf{C}$ *along* $\mathbf{A}$. *Proceed to prove the distributive property of the dot product.*

The distributive property allows us to express the dot product of two vectors in terms of their components. Given

$$\mathbf{i} \cdot \mathbf{i} = \mathbf{j} \cdot \mathbf{j} = 1 \qquad (7.1.9)$$

and

$$\mathbf{i} \cdot \mathbf{j} = 0, \qquad (7.1.10)$$

we find

$$\begin{aligned}\mathbf{A} \cdot \mathbf{B} &= (A_x\mathbf{i} + A_y\mathbf{j}) \cdot (B_x\mathbf{i} + B_y\mathbf{j}) \qquad (7.1.11)\\ &= A_xB_x + A_yB_y. \qquad (7.1.12)\end{aligned}$$

Notice that of the four possible terms that could arise on opening up the brackets, only two survive due to the orthogonality of the basis vectors.

In three dimensions, where every vector has an expansion of the form

$$\mathbf{A} = A_x \mathbf{i} + A_y \mathbf{j} + A_z \mathbf{k}, \qquad (7.1.13)$$

the result generalizes to

$$\mathbf{A} \cdot \mathbf{B} = A_x B_x + A_y B_y + A_z B_z. \qquad (7.1.14)$$

If the vectors are chosen to be the same we get as a special case,

$$\mathbf{A} \cdot \mathbf{A} = |A|^2 = A_x^2 + A_y^2 + A_z^2 \qquad (7.1.15)$$

in agreement with the Pythagoras theorem.

Suppose we want to write some vector $\mathbf{A}$ in the plane in terms of a different set of basis vectors $\mathbf{i}'$ and $\mathbf{j}'$ associated with a rotated system of coordinates as in Fig. 7.1A. From the figure it is clear that

$$\mathbf{i}' = \mathbf{i} \cos\theta + \mathbf{j} \sin\theta \qquad (7.1.16)$$
$$\mathbf{j}' = -\mathbf{i} \sin\theta + \mathbf{j} \cos\theta. \qquad (7.1.17)$$

Using the dot product we can easily find out the new components. Starting with the assumed expansion

$$\mathbf{A} = A_x' \mathbf{i}' + A_y' \mathbf{j}' \qquad (7.1.18)$$

we get, on taking the dot product of both sides with $\mathbf{i}'$,

$$\mathbf{i}' \cdot \mathbf{A} = A_x' \mathbf{i}' \cdot \mathbf{i}' + A_y' \mathbf{i}' \cdot \mathbf{j}' = A_x', \qquad (7.1.19)$$

where we have used the orthonormal nature of the basis vectors. Thus we find

$$A_x' = \mathbf{i}' \cdot \mathbf{A} = (\mathbf{i} \cos\theta + \mathbf{j} \sin\theta) \cdot (A_x \mathbf{i} + A_y \mathbf{j}) = A_x \cos\theta + A_y' \sin\theta \quad (7.1.20)$$

as in Eqn. (7.1.2) and a similar result for A_y, as given in Eqn. (7.1.3). The generalization to three (or more) dimensions is obvious. This simple idea will be central to a lot of manipulations we will perform later on, and you must make sure you understand it well.

When you encounter a new concept such as the dot product you should not only learn all the definitions and rules, but also ask why they are natural, i.e., why or how would anyone think of them. In some cases, such as the invention of analytic functions, the answer is the deep mathematical intuition of the inventors. However, in most of the cases we study in this book, the origin can be traced to some problem in physics where the concept arises naturally. In some cases, such as matrix theory, it is both: the mathematicians invented them from their intuition and physicists rediscovered them independently to describe a physics problem.

As an example of how the dot product arises in physics, let us begin by considering how $T = \frac{1}{2}mv^2$, the kinetic energy of a body of mass m moving in one dimension with velocity v, changes with time when a force F acts:

$$\frac{dT}{dt} = mv\frac{dv}{dt} = vF, \tag{7.1.21}$$

where we have used the one-dimensional version of Newton's Second Law $\mathbf{F} = m(d\mathbf{v}/dt)$. Consider next motion in two dimensions. The kinetic energy is

$$T = \frac{1}{2}mv^2 = \frac{1}{2}m(v_x^2 + v_y^2)$$

and its rate of change is

$$\frac{dT}{dt} = m(v_x\frac{dv_x}{dt} + v_y\frac{dv_y}{dt}) = \mathbf{v} \cdot \mathbf{F}. \tag{7.1.22}$$

Notice how the final formula for the rate of change of kinetic energy very naturally involves the dot product. One implication of the formula is that if a force acts perpendicular to the velocity, the kinetic energy is unaffected. An example is the force a string exerts on a mass as the latter is twirled in a circle.

Later in this chapter we will run into other examples of the dot product that naturally arise in physics.

Consider next the *cross product of two vectors:* $\mathbf{A} \times \mathbf{B}$, defined to be a vector of magnitude

$$|\mathbf{A} \times \mathbf{B}| = |A||B||\sin\theta_{AB}| \tag{7.1.23}$$

and direction perpendicular to the plane containing $\mathbf{A}$ and $\mathbf{B}$. (From now on the subscript on θ_{AB} will be dropped.) Since the plane containing the vectors can have two perpendiculars differing by a sign, we introduce the *screw rule*: turn a screw driver in the sense going from $\mathbf{A}$ to $\mathbf{B}$ and take the direction of advance of the screw as the direction of $\mathbf{A} \times \mathbf{B}$. Now, for a pair of vectors $\mathbf{A}$ and $\mathbf{B}$ there are always two ways to rotate from $\mathbf{A}$ to $\mathbf{B}$. *It will be understood that the shortest route, i.e., the smallest rotation angle is chosen.* This makes the sense of rotation unique and one will never have to rotate by more than $180°$. If the vectors are at $180°$ there is an ambiguity, but the cross product vanishes in this case.

This rule is also called the *right hand rule*: if you curl your four fingers from $\mathbf{A}$ to $\mathbf{B}$, (again along the quickest way to go from $\mathbf{A}$ to $\mathbf{B}$), your thumb points along their cross product. Note that while the dot product is proportional to the projection of one vector along the other, the cross product is proportional to the projection of one vector perpendicular to the other.

The cross product changes sign when we reverse the order of the factors:

$$\mathbf{A} \times \mathbf{B} = -\mathbf{B} \times \mathbf{A} \qquad (7.1.24)$$

which follows from the right hand rule or screw rule in the definition.

The cross product also is distributive:

$$\mathbf{A} \times (\mathbf{B} + \mathbf{C}) = \mathbf{A} \times \mathbf{B} + \mathbf{A} \times \mathbf{C} \qquad (7.1.25)$$

which follows from the fact that the projection of $(\mathbf{B}+\mathbf{C})$ *perpendicular* to $\mathbf{A}$ is the sum of the projections of $\mathbf{B}$ and $\mathbf{C}$ separately in that direction.

Once again classical mechanics illustrates the naturalness of the concept in question. Suppose you are trying to open a door by pulling on the knob. It is clear that the further the knob is from the hinge, the more effective any applied force will be. It is also clear that any applied force will be most effective if it is applied perpendicular to the plane of the door (unless you are trying to rip it off the hinges). Thus the real turning ability of the force is measured by the *torque*, which is the product of the force, the distance from the knob to the hinge and the sine of the angle between the force and the vector connecting the hinge and knob. This leads us to define the torque as the vector

$$\tau = \mathbf{r} \times \mathbf{F}. \qquad (7.1.26)$$

The magnetic force on a moving charge provides another example. Experiments tell us that the magnetic field is experienced only by moving charges. Thus the magnetic force $\mathbf{F}$ depends on the charge q, its velocity $\mathbf{v}$ and the magnetic field $\mathbf{B}$. So are seeking a vector $\mathbf{F}$ that is given by two other vectors $\mathbf{v}$ and $\mathbf{B}$. Experiments further tell us that

$$\mathbf{F} = q\mathbf{v} \times \mathbf{B} \qquad (7.1.27)$$

showing once again that the cross product is not just a sterile definition but a preferred notion singled out by nature to express her laws. Notice also that many implications follow from the force above: for example, since the magnetic force is perpendicular to the velocity (involving as it does the cross product of the velocity and the field) it does not change the kinetic energy of the particle.

Consider finally a parallelogram with adjacent sides given by vectors $\mathbf{A}$ and $\mathbf{B}$. Elementary geometry tells us the area of the figure is $|A||B||\sin\theta|$. If we define a vector $\mathbf{S} = \mathbf{A} \times \mathbf{B}$, its length is the area of the figure bounded by these vectors. How about the fact that $\mathbf{S}$ also has a direction (pointing perpendicular to the plane of the area)? The fact that the area appears as the length of a *vector* suggests that it is more natural to associate a vector and not a scalar with an area, at least a planar one. This is indeed the right thing to do and we shall exploit this possibility later in the chapter. But you must note that only in three dimensions is it possible to define a product of two vectors that equals another, or to associate an area with a vector. The reason is that only in three dimensions do two non-coplanar vectors define a unique direction, namely the direction perpendicular to the plane they define. For example in four dimensions, with coordinates (x, y, z, t); given two vectors in the $x - y$ plane, any vector in the $z - t$ plane is perpendicular to them.

Problem 7.1.3. *Consider now three vectors* a, b, c *not in the same plane. Show that the* box product *(also called the* scalar triple product*)* $c \cdot a \times b$ *gives the volume of the parallelepiped with these vectors as three adjacent edges. At least verify this for the case when the* c *is perpendicular to the* a − b *plane. Show that* $a \cdot b \times c = b \cdot c \times a = c \cdot a \times b$ *either geometrically or algebraically. What happens to the box product when two of the vectors are parallel?*

We now turn to the practical computation of cross products. Recall that in the case of the dot product, we could either compute the product of the lengths and the cosine of the angle between the vectors, or if the components were known in any orthonormal basis, write

$$A \cdot B = A_x B_x + A_y B_y + A_z B_z \qquad (7.1.28)$$

where the result follows from expanding the vectors in the chosen basis and using the distributive property of the dot product. We now wish to do the same for the cross product. Let us now compute the cross product of the basis vectors. Eqn. (7.1.23) gives

$$i \times j = k \; et \; cycl \qquad (7.1.29)$$

where *et cycl* means in cyclic permutation: that is we replace i by j, j by k, and k by i and get two more relations.

Using the distributive property of the cross product property we can write

$$
\begin{aligned}
A \times B &= (A_x i + A_y j + A_z k) \\
&\times \; (B_x i + B_y j + B_z k) \qquad (7.1.30) \\
&= i(A_y B_z - A_z B_y) + et \; cycl. \qquad (7.1.31)
\end{aligned}
$$

7.2. Time Derivatives of Vectors

Consider a particle moving in the $x - y$ plane along some trajectory shown by the thick line in Fig. 7.2.

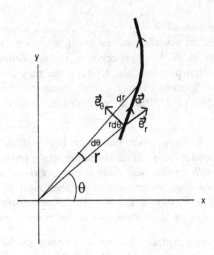

Figure 7.2. Kinematics in two dimensions.

In terms of the unit vectors **i** and **j**, we may write the particle's *position vector* $\mathbf{r}(t)$ as

$$\mathbf{r}(t) = x(t)\mathbf{i} + y(t)\mathbf{j} \tag{7.2.1}$$

at time t and

$$\mathbf{r}(t + dt) = x(t + dt)\mathbf{i} + y(t + dt)\mathbf{j} \equiv \mathbf{r} + \mathbf{\Delta r} \tag{7.2.2}$$

at time $t + dt$.

We define the rate of change or *velocity vector* as

$$\mathbf{v} = \lim_{\Delta t \to 0} \frac{\mathbf{\Delta r}}{\Delta t} = \tag{7.2.3}$$

$$= \frac{dx}{dt}\mathbf{i} + \frac{dy}{dt}\mathbf{j}. \tag{7.2.4}$$

To obtain this result more quickly we simply differentiate both sides of Eqn. (7.2.1) remembering that the unit vectors are constant in time and only the components vary.

Sometimes we like to use a different set of unit vectors in place of **i** and **j**. In polar coordinates one uses unit vectors $\mathbf{e}_r$ and $\mathbf{e}_\theta$ which point in the direction of increasing r and θ at each point, as shown in Fig. 7.2.

Unlike **i** and **j**, these vectors vary from point to point since the directions of increasing r and θ do (unlike the directions of increasing x and y). The position vector of the particle looks very simple in this system:

$$\mathbf{r} = r\mathbf{e}_r. \tag{7.2.5}$$

How about the velocity? There are two ways to proceed. First we argue that if the coordinates change by dr and $d\theta$ in time dt, then the position vector change is a sum of two parts, (see Fig. 7.2), one in the radial and one in the angular direction:

$$\mathbf{dr} = dr\,\mathbf{e}_r + r\,d\theta\,\mathbf{e}_\theta \quad \text{so that} \tag{7.2.6}$$

$$\frac{\mathbf{dr}}{dt} = \frac{dr}{dt}\mathbf{e}_r + r\frac{d\theta}{dt}\mathbf{e}_\theta. \tag{7.2.7}$$

Another way to obtain this result is to go about differentiating Eqn. (7.2.5), taking into account that the unit vector $\mathbf{e}_r$ itself changes with the particle's position which changes with time. Thus

$$\frac{\mathbf{dr}}{dt} = \frac{dr}{dt}\mathbf{e}_r + r\frac{\mathbf{de}_r}{dt}. \tag{7.2.8}$$

To proceed we must determine $\frac{\mathbf{de}_r}{dt}$. Consider two nearby points. The unit vectors in the radial direction are different only because of the difference $d\theta$; had the points differed only in r, the radial direction and hence unit vectors would have been the same. As for the change due to $d\theta$, it is clear that the new unit vector is obtained by rotating the old one by angle $d\theta$. This clearly amounts to adding an infinitesimal vector of length $1 \cdot d\theta$ in the angular direction:

$$\mathbf{e}_r(r + dr, \theta + d\theta) = \mathbf{e}_r(r, \theta) + d\theta\,\mathbf{e}_\theta \tag{7.2.9}$$

so that

$$\frac{\mathbf{de}_r}{dt} = \frac{d\theta}{dt}\mathbf{e}_\theta \quad \text{and} \tag{7.2.10}$$

$$\frac{\mathbf{dr}}{dt} = \frac{dr}{dt}\mathbf{e}_r + r\omega\mathbf{e}_\theta = \mathbf{v} \tag{7.2.11}$$

where

$$\omega = \frac{d\theta}{dt}. \tag{7.2.12}$$

To obtain the acceleration we need to differentiate once more and invoke the additional relation

$$\frac{\mathbf{de}_\theta}{dt} = -\frac{d\theta}{dt}\mathbf{e}_r \tag{7.2.13}$$

which can be derived the same way we derived Eqn. (7.2.10). The result is

$$\mathbf{a} = \frac{\mathbf{dv}}{dt} \tag{7.2.14}$$

$$= (\frac{d^2r}{dt^2} - \omega^2 r)\mathbf{e}_r + (2\omega\frac{dr}{dt} + r\frac{d\omega}{dt})\mathbf{e}_\theta. \tag{7.2.15}$$

The physics behind the various terms is left to the physics courses.

Problem 7.2.1. *Fill in the missing steps in the derivation of the preceding formula.*

I have assumed that the topics discussed so far are not new to you, and hence not many exercises were given. Here are a few, just to let you know where you stand. If you have problems with these, you should check out a book on this topic and deal with this material.

Problem 7.2.2. *Show that* $V = i - j + 2k$, $W = 4j + 2k$ *and* $Z = -10i - 2j + 4k$ *are mutually orthogonal. Find* $|V \times W|$, $|W \times Z|$, *and* $|Z \times V|$ *using Eqn. (7.1.31) to compute the cross products.*

Problem 7.2.3. *Show that the triple product of* $i + 2k$, $4i + 6j + 2k$, *and* $3i + 3j - 6k$ *is* -54.

Problem 7.2.4. *A particle has a position vector* $r = i \cos \omega t + j \sin \omega t$. *Describe its motion in cartesian coordinates with a sketch, and in words. Compute its velocity and acceleration. Find the magnitudes of both. Now switch to polar coordinates. What are* r *and* θ *for this problem? What is* a *as per Eqn. (7.2.15)?*

7.3. Scalar and Vector Fields

We just reviewed the dynamics of a single particle, described by the single vector $r(t)$. This notion is readily extended to many particles, labeled by some index i so that r_i describes particle numbered i. Now we turn to a more complicated object called a *field*. A field has infinite degrees of freedom. Here is a simple example: to each point (x, y, z) is associated a temperature $T(x, y, z)$. The function $T(x, y, z)$ is a *scalar field*. The term "scalar" signifies that at each point is defined just one number (as compared to say a vector) and this number does not change as we rotate our axes. If we rotate our axes, the point that used to be called (x, y, z) will now bear a new name (x', y', z') but the *numerical value* attached to that physical point will still be the same, say $87°$ in the case of the temperature field. Other examples of scalar fields are $P(x, y, z)$, the pressure at each point; or $N(x, y, z)$, the density of pollen spores at each point. *Note that unlike in the case of particle dynamics,* (x, y, z) *do not stand for the dynamical degrees of freedom, but are simply labels for the field variable, like the label* i *on* r_i. Thus for example, the pressure P is a dynamical variable, and it is labeled by its values at each point in space and in particular has the value $P(x, y, z)$ at the point (x, y, z).

We will assume throughout this chapter that the coordinates go over all of three-dimensional space, or if we restrict ourselves to two dimensions, to all of the $x - y$ plane. More fancy three-dimensional regions like three dimensional space minus some points will not be discussed. Likewise when we speak of a surface, we will mean something like the surface of a sphere or hemisphere; fancier surfaces like a Möbius strip or Klein bottle will not be discussed. If you do not know what these are, you are in good shape since you will not get into trouble trying to apply the results of this chapter to these fancy cases.

Consider now a *vector field* $\mathbf{W}(x, y, z)$, which associates with each point a vector, say the wind velocity or the strength of the electric or magnetic field. Not only do we associate three numbers at each point now; when described in a rotated system, the three components associated with one and the same point will change in numerical value. For example the velocity vector which was all in the old x-direction could now have an x and y component in the new system.

We will now learn how to characterize and describe various scalar and vector fields. The fields could also depend on time in addition to the three spatial coordinates, though we will emphasize the latter here.

7.4. Line and Surface Integrals

In this section we define and illustrate the very important ideas of line and surface integrals of vector fields.

Line integrals

Let $\mathbf{F}$ be a vector field. It assigns to each point in space a vector. Take for definiteness the following field in two dimensions:

$$\mathbf{F} = 2xy^2\mathbf{i} + x^2\mathbf{j} \qquad (7.4.1)$$

which assigns, for example, the value $8\mathbf{i} + \mathbf{j}$ to the point $(1, 2)$. In a physics problem, this could be a force field. Let us now pick a path P joining end points $\mathbf{r}_1 \equiv 1$ and $\mathbf{r}_2 \equiv 2$ as in Fig. 7.3.

We imagine this path as made up of tiny arrows $[\mathbf{dr}_i | i = 1 \ldots n]$ laid tip to tail. We then form the dot product $\mathbf{F}(\mathbf{r}_i) \cdot \mathbf{dr}_i$ at each segment. (The vector $\mathbf{F}$ is assumed to be a constant over the tiny arrow.) In the physics example this would be the work done by the force over this tiny segment.

Definition 7.1. *The* Line Integral *of* $\mathbf{F}$ *along* P *joining points* 1 *and* 2 *(where*

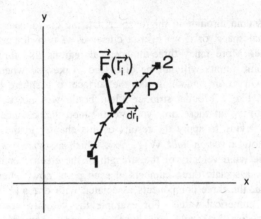

Figure 7.3. Line integral of **F** along path P.

1 *and* 2 *are short for* $\mathbf{r}_1$ *and* $\mathbf{r}_2$ *respectively) is defined as*

$$\int_1^2 \mathbf{F}(\mathbf{r}) \cdot d\mathbf{r} = \lim_{n \to \infty} \sum_{i=1}^{n} \mathbf{F}(\mathbf{r}_i) \cdot d\mathbf{r}_i, \qquad (7.4.2)$$

i.e., in the limit in which each segment is of vanishing size.

The evaluation of line integrals is very similar to the evaluations of integrals in the complex plane: one must parametrize the path and reduce everything to an integral over the parameter t. Here is an example from two dimensions.

Let once again

$$\mathbf{F} = 2xy^2\mathbf{i} + x^2\mathbf{j} \qquad (7.4.3)$$

be the vector field and let us integrate it from the origin to the point $(1,1)$ along the curve $x = y^2$.

The first step is to *parametrize* the path with a variable t, i.e., make up functions $x(t)$ and $y(t)$ such that as t varies, the locus of points traces out the curve in question. If you wish you may pretend that t is really the time and the curve describes the trajectory of some fictitious particle moving in the $x-y$ plane. In our problem the following choice will do:

$$y = t \qquad x = t^2 \qquad [0 \le t \le 1] \qquad (7.4.4)$$

which describes motion along the curve $x = y^2$. This choice now leads to

$$\mathbf{r}(t) = \mathbf{i}t^2 + \mathbf{j}t \qquad (7.4.5)$$

$$\frac{d\mathbf{r}}{dt} = 2t\mathbf{i} + \mathbf{j} \tag{7.4.6}$$

$$\int_{\mathbf{r}_1}^{\mathbf{r}_2} \mathbf{F} \cdot d\mathbf{r} = \int_{t_1}^{t_2} \mathbf{F} \cdot \frac{d\mathbf{r}}{dt} dt \tag{7.4.7}$$

$$= \int_{t=0}^{t=1} (2t^2 t^2 \mathbf{i} + t^4 \mathbf{j}) \cdot (2t\mathbf{i} + \mathbf{j}) dt \tag{7.4.8}$$

$$= \int_0^1 (4t^5 + t^4) dt \tag{7.4.9}$$

$$= \frac{13}{15}. \tag{7.4.10}$$

Note that even though the curve lives in high dimensions, it is still one dimensional and the line integral can therefore be mapped into an ordinary integral in the variable t. It is the integrand that is calculated from higher dimensional considerations, i.e., by evaluating the dot product of the vector field with the displacement. In the physics case the parametrization has a direct meaning. We are trying to find the work done by the force as it drags some particle from 1 to 2 along P. If we imagine the motion as taking place in real time, $\mathbf{F} \cdot \frac{d\mathbf{r}}{dt}$ stands for the *power* expended by the force and its integral over time is the work done. Thus the line integral becomes the ordinary integral over t of the power $\mathbf{F} \cdot \frac{d\mathbf{r}}{dt}$ between $t_1 = 0$ and $t_2 = 1$.

The extension of this notion to vector fields in three dimensions is straightforward.

As with integrals in the complex plane, integrals over a closed curve will be denoted by $\oint_P$.

The line integral is generally path dependent. For example if we join the same end points by a line at $45°$, i.e., the line $x = y$, the result will be $5/6$. The details are left to the following exercise.

Problem 7.4.1. *Verify that the line integral of* $\mathbf{F} = 2xy^2\mathbf{i} + x^2\mathbf{j}$ *between the origin and* $(1, 1)$ *along the line* $x = y$ *is 5/6. Note that you can use* x *itself as a parameter. Show the integral along a path that first goes horizontally (i.e.,* $d\mathbf{r} = \mathbf{i}dx$, $y = 0$*) to* $(1, 0)$ *and then straight up to* $(1, 1)$*,* $(d\mathbf{r} = \mathbf{j}dy$, $x = 1)$ *is 1. Show the answer is the same for a path that first goes up to* $(0, 1)$ *and then horizontally to* $(1, 1)$*.*

Problem 7.4.2. *Show that the line integral of* $\mathbf{F} = \mathbf{i}y^2 + \mathbf{j}xy + \mathbf{k}xz$ *from the origin to the point* $(1, 1, 1)$ *along a straight line connecting the end points equals unity and along the curve* $x = t$, $y = t^2$, $z = t^3$ *equals* $\frac{36}{35}$*.*

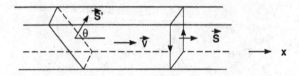

Figure 7.4. The flux.

Definition 7.2. *Given any vector field* **F***, if the line integral is dependent on just the end points and not the path we say the field is* conservative. *An equivalent definition of a conservative field is that*

$$\oint \mathbf{F} \cdot d\mathbf{r} = 0 \tag{7.4.11}$$

for any closed path.

The equivalence of the two definitions is exactly as it was in the case of integration in the complex plane.

Problem 7.4.3. *(i) Calculate the line integral of* $\mathbf{F} = \mathbf{i}2xy^2 + \mathbf{j}x^2$ *between* $(0,0)$ *and* $(1,1)$ *but along* $y = x^2$. *(ii)What is the line integral around a unit circle centered at the origin? First make a sensible choice for the parameter t that will move the coordinates along the circle. Note that even for a nonconservative field the line integral along* certain *closed paths can vanish.*

Surface integrals

Imagine a steady stream of particles moving via a pipe of rectangular cross section, along the x-axis with a velocity $\mathbf{v} = \mathbf{i}v_x$, as shown in Fig. 7.4.

Let the density be ρ particles per unit volume. The *current density* is defined as

$$\mathbf{j} = \rho\mathbf{v}, \tag{7.4.12}$$

and is given by

$$\mathbf{j} = \mathbf{i}j_x \tag{7.4.13}$$

in our problem. Consider an area (denoted by **S** in the figure) of size S, that lies transverse to this flow, i.e., in the $y - z$ plane. The number of particles going through the area *in one second* is called Φ, the *flux* through the area. To find it, let us follow the particles that cross it at $t = 0$. At the end of this period, they would have gone a distance $L = v_x$. Thus all the particles that crossed the area in the last one second lie inside a parallelepiped of base S and length $L = v_x$. The flux is then

$$\Phi = \rho v_x S = j_x S. \tag{7.4.14}$$

Now the flux is a scalar, a number. We are obtaining it from a vector, $\mathbf{j}$, and an area. Since one usually gets a scalar from taking the dot product of two vectors, we ask if we can represent the area as a vector. (I have already hinted at this possibility earlier.) Given a planar area as in this case, we can associate a vector with it of magnitude equal to the numerical value of the area and direction perpendicular to the plane of the area, i.e., the normal. But we can draw two normals to any plane. So we assign an *orientation to the area as follows*. We draw arrows around the perimeter of the area in such a way that as a person walks around the area along the arrows, she sees the area always to her left or right. For each choice, we choose the normal vector to point as per the screw rule: the normal points along the direction of advance of a screw as the screw driver is turned in the same sense as the arrows along the perimeter.[1] Thus the area vector in the figure is given by an arrow parallel to the current for the given choice of arrows along its edges. (If you stand upstream of the area and look at it, the arrows along the edges will run counterclockwise.) Using this convention for representing areas as vectors, we can write the area vector as

$$\mathbf{S} = \mathbf{i}S \qquad (7.4.15)$$

and the flux as

$$\Phi = Sj_x = \mathbf{S} \cdot \mathbf{j} \qquad (7.4.16)$$

using the fact that Sj_x is really the dot product of two vectors which happen to lie entirely in the x-direction. But the above equation implies more than that. Suppose we tilt the area so that it no longer lies transverse to the flow of particles, i.e., so that its area vector does not lie parallel to the velocity, but enlarge it so that it still intercepts all the particles that flow in the pipe. The new area called S' in the figure is bigger by a factor $1/\cos\theta$, θ being the tilt. By construction the flux through it is the same as through S. Equation (7.4.16) would say the flux through S' is $S' \cdot \mathbf{j}$, which is correct, since the increased area of S' is exactly offset by the cosine of the angle between the area vector and the current vector.

    Thus the dot product of a current density vector and the area vector gives the flux of particles penetrating the area. We use the word flux for the dot product of any vector with an area even if it does not stand for the flow of anything. For example if $\mathbf{B}$ is the magnetic field vector, its dot product with an area is called the *magnetic flux*.

---

[1] An area without this orientation specified is like a vector from which we have erased the head and tail. Note also that only in three space dimensions can we represent an area as a vector, using that fact the area (assumed planar) has a unique direction perpendicular to it (up to a sign) which we can assign to the vector. In four dimensions for example an area in the $x - y$ plane can have an infinity of distinct vectors perpendicular to it. An analogous statement is that in two dimensions we can define a unique normal to a curve at each point (up to a sign) while in three dimensions the normal can rotate in the plane perpendicular to the line at that point.

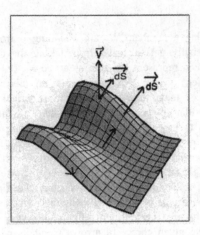

Figure 7.5. An undulating surface built out of smaller planar patches. The orientation of the patches is specified by the normals instead of arrows running along the perimeter. If you looked at the surface from above, these arrows would run counterclockwise.

Note that the flux is given by the above formula only if the area is planar and the current a constant over its extent. But we can generalize the notion of flux as follows. First, we note that any infinitesimal area can be approximated by a plane and specified by the above convention. It will be referred to as $d\mathbf{S}$. For example an area of size 10^{-8} m^2, sitting at the north pole of a sphere centered at the origin, with its normal pointing outward, will be denoted by $d\mathbf{S} = 10^{-8}\mathbf{k}$. Next we learn how to build up a macroscopic area by patching together little ones. Recall how the path P was built out of oriented line segments placed end-to-end. Each arrow was a directed segment with a beginning (tail) and an end (tip), i.e., with two boundaries. Whenever two arrows were so joined tip-to-tail, one boundary of the first arrow (its tip) was neutralized by one of the next (its tail) leading to a segment whose boundaries were the tail of the first and the tip of the second. We similarly glue areas, each with its boundaries marked by arrows running around in a definite sense. We bring together two tiny areas and place them next to each other so that on the adjacent edges the orienting arrows run oppositely as in Fig. 6.2B. (That is, the area vectors are roughly parallel and not antiparallel.) Thus the common perimeter gets neutralized and we get a bigger area with a new boundary that is the "sum" of the old boundaries (where in the sum the common edges have canceled). In this fashion we build up an undulating surface in three dimensions whose boundary is made up of the exposed edges of the pieces that tile it. A sample is shown in Fig. 7.5.

We now define the surface integral:

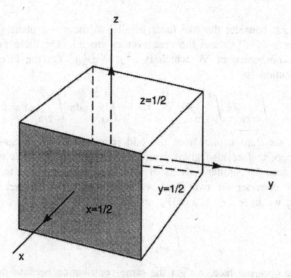

Figure 7.6. The unit cube centered at the origin. The faces we can see are at $x = 1/2$, $y = 1/2$, or $z = 1/2$. The other three are at minus these values.

Definition 7.3. *The surface integral of a vector field* **V** *over the surface S is given by*

$$\int_S \mathbf{V} \cdot \mathbf{dS} = \lim_{n \to \infty} \sum_{i=1}^{n} \mathbf{V}(\mathbf{r}_i) \cdot \mathbf{dS}_i \qquad (7.4.17)$$

where i labels the patches, all of which become vanishingly small as $i \to \infty$.

Note that even though the integration is over a two-dimensional region we use only one integration symbol. This is a simplification that should not cause any confusion. If the surface in question is closed (like a sphere) we will use the symbol $\oint_S$.

Imagine a point source of alpha particles at the origin. If we surround it with a surface and calculate the particle flux, i.e., the surface integral of the current density, (with area vectors pointing outward) we will get a nonzero number, determined by the emission rate. On the other hand, if we calculate the net flux out of a surface not enclosing the origin, we will get zero, since what comes in goes out. Likewise if the origin has a sink which absorbs particles, we will get a negative surface integral.

Let us now work out an example: the surface integral of

$$\mathbf{W} = \mathbf{i}x^3 y + \mathbf{j}y^2 x + \mathbf{k}z \qquad (7.4.18)$$

over a unit cube centered at the origin, as in Fig. 7.6.

First consider the two faces parallel to the $y - z$ plane. These have coordinates $x = \pm 1/2$ and the area vectors are $\pm \mathbf{i}$. On these faces we need just the x-component of $\mathbf{W}$ which is $x^3 y = \pm \frac{1}{8} y$. On the face with area $\mathbf{i}$, the contribution is

$$\int_{-1/2}^{1/2} dy \int_{-1/2}^{1/2} dz \; \mathbf{W} \cdot \mathbf{i} = \frac{1}{8} \int_{-1/2}^{1/2} y \, dy \int_{-1/2}^{1/2} dz = 0, \qquad (7.4.19)$$

where the zero comes from the odd integrand in the y-integration. Likewise the opposite face also gives zero. If we consider the faces with areas $\pm \mathbf{j}$, we get a zero for similar reasons of symmetry, this time due to the x-integration. Finally consider the two faces with areas $\pm \mathbf{k}$. On the face with area up the z-axis, we have $z = 1/2 = W_z$ and the surface integral is

$$\int_{-1/2}^{1/2} dx \int_{-1/2}^{1/2} dy \cdot \frac{1}{2} = \frac{1}{2}. \qquad (7.4.20)$$

On the opposite face, we get the same contribution because there are two extra minus signs that compensate, one due to the area vector and the other due to the change in sign of W_z. The total contribution is

$$\oint \mathbf{W} \cdot \mathbf{dS} = 1. \qquad (7.4.21)$$

Sometimes the field is more naturally given in non-cartesian coordinates. Consider thus the field of a point charge q at the origin:

$$\mathbf{E} = \frac{q}{4\pi \varepsilon r^2} \mathbf{e}_r \qquad (7.4.22)$$

where $\mathbf{e}_r$ is a unit vector in the radial direction and $\varepsilon_0 = 8.85419 \cdot 10^{-12}$ Coulombs^2/(Newton $\cdot$ m^2) is a constant called the *permittivity of free space*. Let us find the surface integral over a sphere of radius R also centered at the origin. Consider a tiny patch of size dS on the sphere. The vector corresponding to it is $\mathbf{e}_r dS$ and the flux through it is $\mathbf{E} \cdot \mathbf{dS}$. The total flux out of the sphere is

$$\oint_S \mathbf{E} \cdot \mathbf{dS} = \oint_S \frac{q}{4\pi \varepsilon_0 R^2} \mathbf{e}_r \cdot \mathbf{e}_r dS \qquad (7.4.23)$$

$$= \frac{q}{4\pi \varepsilon_0} \int_S \frac{dS}{R^2} \qquad (7.4.24)$$

$$= \frac{q}{\varepsilon_0} \qquad (7.4.25)$$

where we have used the fact that the area of the sphere is $4\pi R^2$.

Problem 7.4.4. *Show that the surface integral of* $\mathbf{V} = \mathbf{i}yz + \mathbf{j}xz + \mathbf{k}xy$ *over a unit cube in the first octant, with one corner at the origin, is zero.*

7.5. Scalar Field and the Gradient

Let ϕ be a scalar field in three dimensions, say the temperature. The first question one asks is how it varies from point to point, i.e., its rate of change. The change in ϕ from one point (x, y, z) to a nearby point $(x + dx, y + dy, z + dz)$ is given to first order by

$$d\phi = \frac{\partial \phi}{\partial x}dx + \frac{\partial \phi}{\partial y}dy + \frac{\partial \phi}{\partial z}dz. \tag{7.5.1}$$

Compare this to Eqn. (7.1.12) and notice that it looks like a dot product of two vectors. We can see that in the present case

$$d\phi = \frac{\partial \phi}{\partial x}dx + \frac{\partial \phi}{\partial y}dy + \frac{\partial \phi}{\partial z}dz \equiv \nabla\phi \cdot \mathbf{dr} \text{ where} \tag{7.5.2}$$

$$\nabla\phi = \mathbf{i}\frac{\partial \phi}{\partial x} + \mathbf{j}\frac{\partial \phi}{\partial y} + \mathbf{k}\frac{\partial \phi}{\partial z}, \text{ and} \tag{7.5.3}$$

$$\mathbf{dr} = \mathbf{i}dx + \mathbf{j}dy + \mathbf{k}dz \tag{7.5.4}$$

is the displacement vector. One refers to $\nabla\phi$ as the *gradient of* ϕ or more often as "grad ϕ", where "grad" rhymes with "sad". *At each point in space* $\nabla\phi$ *defines a vector field built from the various derivatives of the scalar field.* For example if

$$\phi = x^2y + y + xyz \tag{7.5.5}$$

then

$$\nabla\phi = \mathbf{i}(2xy + yz) + \mathbf{j}(x^2 + 1 + xz) + \mathbf{k}(xy). \tag{7.5.6}$$

The gradient contains a lot of information. We just saw that it tells us how much ϕ changes (to first order) when we move in *any* direction: we simply take the dot product of the gradient with the displacement vector. Next, if we write the dot product in the equivalent form as

$$d\phi = |\nabla\phi||dr|\cos\theta \tag{7.5.7}$$

we see that for a given magnitude of displacement, $|dr|$, we get the maximum change if the displacement is parallel to the vector $\nabla\phi$. *Thus* $\nabla\phi$ *points in the direction of greatest increase of* ϕ *and* $|\nabla\phi|$ *gives the greatest rate of change.*

Let us pause to understand the significance of this result. Suppose you were in a temperature field $T(x, y)$ in two dimensions and wanted to find the greatest rate of change. Since the rate of change is generally dependent on direction, you may think that you would have to move in every possible direction, compute and tabulate the rates of change, and choose from your list the winner. Not so! You simply measure the rates of change in two perpendicular directions and make a vector out of those directional derivatives. This vector, the gradient, gives the magnitude and direction of the greatest rate of change. To obtain the rate of change in any other direction, you take the projection of the gradient in that direction: as per Eqn. (7.5.7),

$$\frac{d\phi}{dr} = |\boldsymbol{\nabla}\phi|\cos\theta, \qquad (7.5.8)$$

where θ is the angle between the gradient and the direction of interest.

In particular, if we move perpendicular to $\boldsymbol{\nabla}\phi$, ϕ does not change to first order. If we draw a surface on which ϕ is constant, called a *level surface*, $\boldsymbol{\nabla}\phi$ will be perpendicular to the surface at each point. Let us consider a simple function which measures the square of the distance from the origin:

$$\phi = x^2 + y^2 + z^2. \qquad (7.5.9)$$

Thus the surfaces of constant ϕ are just spheres centered at the origin. For example the surface $\phi = 25$ is a sphere of radius 5. Clearly

$$\boldsymbol{\nabla}\phi = 2x\mathbf{i} + 2y\mathbf{j} + 2z\mathbf{k} \qquad (7.5.10)$$

points in the radial direction (since its components are proportional to that of the position vector) which is indeed the direction of greatest increase. The rate of change in this direction is $|\boldsymbol{\nabla}\phi| = 2\sqrt{x^2 + y^2 + z^2} = 2r$, which is correct given that $\phi = r^2$. Finally the gradient vector is clearly perpendicular to the (spherical) surface of constant ϕ or r.

Thus suppose you find yourself in a temperature field $\phi = T$. If you feel too hot where you are, you must quickly compute the gradient at that point and move opposite to it; if you are too cold you must move along the gradient, and if T is just right and you feel like moving around, you should move perpendicular to the gradient. Since the gradient changes in general from point to point, you must repeat all this after the first step. Here is an example. Let the temperature be given by

$$T = y^2 - x^2 \qquad \text{with} \quad \boldsymbol{\nabla}T = -2x\mathbf{i} + 2y\mathbf{j}. \qquad (7.5.11)$$

The temperature and its gradient are depicted in Fig. 7.7.

For another example of the gradient, let us revisit the Lagrange multiplier method. When $f(x, y, z)$ is to be extremized with respect to the constraint

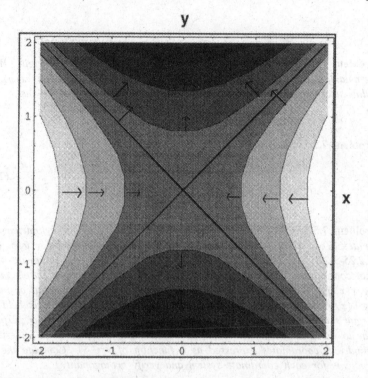

Figure 7.7. The plot of the temperature $T = y^2 - x^2$: darker regions are at higher temperature. A few gradient vectors (not exactly to scale) are shown.

$g(x, y, z) = 0$, we write the condition as

$$\frac{\partial f}{\partial x} = \lambda \frac{\partial g}{\partial x} \qquad (7.5.12)$$

and similarly for the other two coordinates. These can clearly be written as one vector equation:

$$\nabla f = \lambda \nabla g. \qquad (7.5.13)$$

This means that at the extremum, the normals to the surfaces $g = 0$ and $f = c$ are parallel, or that the surfaces are in tangential contact. We can understand this as follows: if ∇f had a component along the constant g surface ($g = 0$) we could move along the surface and vary f to first order. All this will be clearer if you go back and look at Fig. 3.1 in Chapter 3. As we move along the constraint circle of radius 2, we cross contours of fixed f. This means that by moving a little further we change the value of f. When we get to the extremum, the contour is tangent to the circle and there is no first order change in f.

Problem 7.5.1. *Consider a sphere of radius R centered at the origin. Write a formula for $h(x,y)$, the height of the hemisphere above the point (x,y). Calculate its gradient and compare the results against your expectations.*

Problem 7.5.2. *Show that*

$$\nabla(\phi\chi) = \phi\nabla\chi + \chi\nabla\phi. \tag{7.5.14}$$

Problem 7.5.3. (Very instructive). *Recall the equations for cylindrical coordinates ρ, ϕ, z (Eqns. (3.2.20-3.2.22)), and spherical coordinates, r, θ, ϕ, (Eqns. (3.2.25-3.2.27)) in terms of cartesian coordinates. Verify that these are orthogonal coordinates, which means for example that the direction of purely increasing $r(x,y,z)$, at fixed θ and ϕ, is perpendicular to the direction of increasing $\theta(x,y,z)$ and $\phi(x,y,z)$ with the other two coordinates similarly held fixed. (Argue that the direction in which just one coordinate increases is perpendicular to the gradients of the other two coordinates (with all three spherical or cylindrical coordinates expressed as a function of x, y, z). Compute the three gradients for each coordinate system and verify orthogonality.)*

Consider now the line integral of $\mathbf{W} = \nabla\phi$, between points 1 and 2 (which stand for $\mathbf{r}_1$ and $\mathbf{r}_2$ respectively). It follows from the definition of the gradient (see Eqn. (7.5.2)) that

$$\int_1^2 \nabla\phi \cdot d\mathbf{r} = \phi(2) - \phi(1). \tag{7.5.15}$$

In the present case the integral does not depend on the path: as we move along any chosen path, the integral keeps adding the tiny increments $d\phi$ and these must necessarily add up to $\phi(2) - \phi(1)$ when we are finished. For example let $\phi = h(x,y)$ be the height above sea level at a point labeled (x,y) on a map. Let us take a walk from point 1 to point 2 along any path. At each step let us add the change in altitude, which is what $\nabla h \cdot d\mathbf{r}$ represents. It is clear that no matter which path we take between fixed end points, the total will add up to the height difference $h(2) - h(1)$.

Thus the gradient is a conservative field. Shortly we will find a diagnostic for conservative fields in general and the gradient will pass the test.

7.5.1. Gradient in other coordinates

We now derive an expression for the gradient in polar coordinates in two dimensions. We begin with

$$d\phi = \frac{\partial \phi}{\partial r}dr + \frac{\partial \phi}{\partial \theta}d\theta \qquad (7.5.16)$$

$$= \frac{\partial \phi}{\partial r}dr + \frac{1}{r}\frac{\partial \phi}{\partial \theta}rd\theta, \qquad (7.5.17)$$

where we rewrote the second piece with some canceling factors of r because

$$\mathbf{dr} = \mathbf{e}_r dr + \mathbf{e}_\theta rd\theta \qquad (7.5.18)$$

has an angular component $rd\theta$ and not $d\theta$ and we want to write $d\phi$ as $\nabla\phi\cdot\mathbf{dr}$. It is now clear upon inspection that

$$\nabla\phi = \frac{\partial \phi}{\partial r}\mathbf{e}_r + \frac{1}{r}\frac{\partial \phi}{\partial \theta}\mathbf{e}_\theta. \qquad (7.5.19)$$

Notice that the significance of the gradient is still the same: it is the sum over orthogonal unit vectors times the spatial rate of change in the corresponding directions.

In the case of other coordinates or three dimensions the gradient is given by

$$\nabla\phi = \sum_i \mathbf{e}_i \frac{1}{h_i}\frac{\partial \phi}{\partial u_i}, \qquad (7.5.20)$$

where h_i is the scale factor which converts the coordinate shift du_i to the corresponding displacement in space and $\mathbf{e}_i$ is the unit vector in the direction of increasing u_i.

Problem 7.5.4. *Write down $\nabla\psi$ in cylindrical and spherical coordinates in three dimensions. Show that x, y, z form an orthogonal system of coordinates by evaluating their gradients (after expressing them in terms of cylindrical and spherical coordinates). (This is just the converse of Problem(7.5.3.)).*

Problem 7.5.5. *(i) Given that $\phi = x^2 y + xy$ find the gradient in cartesian and polar coordinates. (ii) Evaluate $|\nabla\phi|$ at $(x = 1, y = 1)$ in both coordinates and make sure they agree. (iii) Find the rate of change at this point in the direction parallel to $2\mathbf{i} + 3\mathbf{j}$. (iv) Integrate $\nabla\phi$ from the origin to $(x = 1, y = 1)$ along a $45°$ line. What should this equal? Did you get that result?*

Problem 7.5.6. *Show that the work done in displacing a pendulum of length L, bob mass m, by an angle θ from the vertical is $W = mgL(1 - \cos\theta)$, g being the acceleration due to gravity.*

Problem 7.5.7. *You are on a hot volcanic mountain where the temperature is given by* $T(x, y) = x^2 + xy^3$. *(i) If you are located at* $(x = 1, y = 1)$, *in which direction will you run to beat the heat? (ii) If your steps are 1/10 units long, by how much will the temperature drop after the first step? (Work to first order.)*

Problem 7.5.8. *Find the directional derivative of the following fields in the radial direction at the point* $(1, 1, 1)$: *(i)* $f = x^2 + y^2 + z^2$, *(ii)* $f = e^{x+y+z}$, *(iii)* xyz. *(iv) Repeat for the direction parallel to* $\mathbf{i} + \mathbf{j} - 2\mathbf{k}$. *Can you interpret the results?*

7.6. Curl of a Vector Field

For a general vector field, the line integral around a closed path will not be zero and is called the *circulation* around that loop. The circulation measures the tendency of the field to go around some point. Consider for example the flow of water as you drain the tub: the velocity vector $\mathbf{V}$ has a large component in the angular direction. Let us say it is $\mathbf{V} = V_r \mathbf{e}_r + V_\theta \mathbf{e}_\theta$. Consider a contour that goes around the origin at fixed radius R. Then $d\mathbf{r} = R d\theta \mathbf{e}_\theta$ and the circulation, $\int_0^{2\pi} V_\theta R d\theta$ will be nonzero. Likewise the magnetic field $\mathbf{B}$ due to an infinitely long conductor carrying current I coils around the wire: if the current is along the z-axis and we use cylindrical coordinates,

$$\mathbf{B} = \mathbf{e}_\phi \frac{\mu_0 I}{2\pi \rho}, \tag{7.6.1}$$

where ρ and ϕ are coordinates in the plane perpendicular to the z-axis and $\mu_0 = 4\pi 10^{-7}$ Newtons/Amp^2 is a constant called the *permeability of free space*. The line integral of $\mathbf{B}$ along a contour encircling the wire and lying in the plane perpendicular to the wire will be nonzero. We shall have more to say about this later.

In contrast consider a field $\mathbf{W} = W_r \mathbf{e}_r$ in the $x - y$ plane. This purely radial field has no tendency to go around any point and the circulation around a circular path centered at the origin vanishes (since $\mathbf{W}$ and $d\mathbf{r}$ are perpendicular) in agreement.

Consider a loop C in the $x - y$ plane and the line integral of $\mathbf{W}$ taken in the counterclockwise direction. Just as in the case of the complex integral (Fig. 6.1) we can argue that this integral is the sum over integrals around smaller nonoverlapping tiles or *plaquettes* that fully cover the area S enclosed

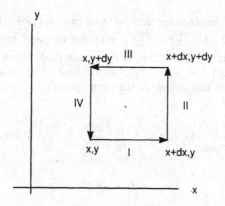

Figure 7.8. The tiny contour for the curl formula.

by the given contour: the contributions from edges shared by any two adjacent tiles cancel, leaving behind just the exposed edges, whose union is just C. Consider therefore the tiny rectangle from Fig. 7.8

The contributions from edges I and III are computed as in the complex plane:

$$I + III = \int_x^{x+dx} W_x(x,y)dx - \int_x^{x+dx} W_x(x,y+dy)dx \quad (7.6.2)$$

$$= \int_x^{x+dx} \left(-\frac{\partial W_x(x,y)}{\partial y} \right) dy\, dx \quad (7.6.3)$$

$$= -\frac{\partial W_x(x,y)}{\partial y}dx\ dy + O(\text{cubic}). \quad (7.6.4)$$

Adding now the similar contribution from sides II and IV, we find

$$\oint = I + III + II + IV = \quad (7.6.5)$$

$$= \left(\frac{\partial W_y}{\partial x} - \frac{\partial W_x}{\partial y} \right) dx dy. \quad (7.6.6)$$

Note that the final contribution is proportional to the area of the plaquette and the neglected terms are of higher order. Let us recall what we have done so far. We wanted the line integral around a macroscopic contour C. We argued that it is the sum of integrals around tiny plaquettes that tile the area bounded by C. We now find that each plaquette makes a contribution equal to its area times $\left(\frac{\partial W_y}{\partial x} - \frac{\partial W_x}{\partial y} \right)$. Consequently as we increase the number of tiles enclosed and decrease the area of each one, we will obtain in the limit the double integral of $\left(\frac{\partial W_y}{\partial x} - \frac{\partial W_x}{\partial y}\right)$ over the enclosed area. The neglected terms (called "cubic" in the derivation above and proportional to higher powers of the infinitesimals) will not survive in the limit. The reason must be quite familiar from earlier discussions of the same question.

It is clear that

$$\oint_{C=\partial S} \mathbf{W} \cdot \mathbf{dr} = \int\int_S \left(\frac{\partial W_y}{\partial x} - \frac{\partial W_x}{\partial y}\right) dx dy \qquad (7.6.7)$$

where

$$C = \partial S \qquad (7.6.8)$$

means that C is the boundary of the region S. This result is called *Green's Theorem.* It allows us to express the line integral of a vector field around a closed loop as the area integral of the object $\frac{\partial W_y}{\partial x} - \frac{\partial W_x}{\partial y}$ called its *curl.* In particular if the vector field is conservative, this integral must vanish on any region S which in turn means that

$$\mathbf{W} \text{ is conservative } \rightarrow \left(\frac{\partial W_y}{\partial x} - \frac{\partial W_x}{\partial y}\right) = 0. \qquad (7.6.9)$$

Consider for example our friend $\mathbf{F} = 2xy^2\mathbf{i} + x^2\mathbf{j}$. This better not be conservative since its line integral from the origin to $(1,1)$ was found to be different on different paths. We find indeed

$$\left(\frac{\partial F_y}{\partial x} - \frac{\partial F_x}{\partial y}\right) = 2x - 4xy, \qquad (7.6.10)$$

On the other hand we have seen that any gradient is a conservative vector field since its integral depends only on the difference of the scalar field at the two ends of the path. It must then be curl-free. If we apply Eqn. (7.6.9) to $\mathbf{W} = \mathbf{i}\frac{\partial \phi}{\partial x} + \mathbf{j}\frac{\partial \phi}{\partial y}$ we find

$$\frac{\partial^2 \phi}{\partial x \partial y} - \frac{\partial^2 \phi}{\partial y \partial x} \equiv 0. \qquad (7.6.11)$$

In other words the curl of a gradient vanishes identically. We shall have more to say on this later.

Problem 7.6.1. *Consider the complex integral $\oint f(z)dz$, where $f = u + iv$ and $dz = dx + idy$ over a contour that lies in the domain of analyticity of f. Write the real and imaginary parts as circulations of two real vector fields and verify that the CRE ensure that both fields are conservative.*

Problem 7.6.2. *Test each of the following fields to see if it is conservative:* (i) $\mathbf{W} = \mathbf{i} y \sin x + \mathbf{j} \cos y$, (ii) $\mathbf{W} = \mathbf{i} 2xy^3 + \mathbf{j} 3y^2 x^2$, (iii) $\mathbf{W} = \mathbf{i} \sin x + \mathbf{j} \cos y$, (iv) $\mathbf{W} = \mathbf{i} \cosh x \cosh y + \mathbf{j} \sinh x \sinh y$.

We saw that the vector field $\nabla \phi$ is conservative. *The converse is also true: if a field is conservative, it may be written as a gradient of some scalar field.* The corresponding scalar field is not unique: given one member, we can generate a whole family by adding a constant without changing the gradient. To find any one member of this family corresponding to a given conservative field $\mathbf{W}$, pick some point 1 and set $\phi(1) = 0$. Assign to ϕ at any other point $\mathbf{r}$ the value

$$\phi(\mathbf{r}) = \int_1^{\mathbf{r}} \mathbf{W} \cdot d\mathbf{r} \qquad (7.6.12)$$

where the path connecting 1 and $\mathbf{r}$ is arbitrary. If you change the upper limit infinitesimally, you will see from the integral that $d\phi = \mathbf{W}(\mathbf{r}) \cdot d\mathbf{r}$. But this must equal $\nabla \phi(\mathbf{r}) \cdot d\mathbf{r}$. It follows that $\mathbf{W} = \nabla \phi$ as promised.

As mentioned earlier, in classical mechanics if $\mathbf{F}$ is a force field, the work done by $\mathbf{F}$ is given by its line integral between the starting and ending points of some path. If the work done is path-independent, the force is conservative and the corresponding $-\phi$ is called the *potential* corresponding to the force. (The minus sign in the definition is merely a matter of convention.)

Suppose you are asked to find the line integral of

$$\mathbf{F} = \mathbf{i}(\cos x \cos y) - \mathbf{j}(\sin x \sin y)$$

between the origin and the point $(\pi/4, \pi/4)$ along the curve $y = x^2$. You could do what we did earlier, i.e., parametrize the curve in terms of some t, and do the integral. But a better way would be to observe that $\mathbf{F}$ is curl free. It is therefore the gradient of some ϕ. Rather than use Eqn. (7.6.12), we can easily guess that $\phi = \sin x \cos y$. In that case, by Eqn. (7.5.15),

$$\int_{0,0}^{\pi/4,\pi/4} \mathbf{F} \cdot d\mathbf{r} = \sin x \cos y \big|_{0,0}^{\pi/4,\pi/4} = \frac{1}{2}. \qquad (7.6.13)$$

Here is an illustration of what you must do if you cannot guess the scalar field. Consider $\mathbf{W} = \mathbf{i} 2xy^2 + \mathbf{j} 2x^2 y$ which is given to be a conservative field. To find the potential (up to a constant) we need to evaluate

$$\phi(x, y) = \int_{0,0}^{x,y} (\mathbf{i} 2\bar{x}\bar{y}^2 + \mathbf{j} 2\bar{x}^2\bar{y}) \cdot (\mathbf{i} d\bar{x} + \mathbf{j} d\bar{y}) \qquad (7.6.14)$$

along any path connecting the origin and the point (x, y). (We are calling the dummy variables $(\bar{x}, \bar{y})$ since (x, y) stand for the end points.) Let us choose the straight line path:

$$\bar{x}(t) = tx \qquad \bar{y}(t) = ty \qquad (0 \le t \le 1). \qquad (7.6.15)$$

Thus

$$\phi(x,y) = \int_0^1 (\mathbf{i}2xy^2t^3 + \mathbf{j}2x^2yt^3) \cdot (\mathbf{i}x\,dt + \mathbf{j}y\,dt) = \int_0^1 (4x^2y^2t^3)dt = x^2y^2,$$

(7.6.16)

which you may verify is correct. Note that although we know ϕ only up to an additive constant, the *difference* in ϕ between two points is free of any ambiguity since the constant drops out. Thus the line integral of $(\mathbf{i}2xy^2 + \mathbf{j}2x^2y)$ between any two points is given by the difference of x^2y^2 between those points.

Problem 7.6.3. *(i) Calculate the work done by* $\mathbf{F} = 2xy\mathbf{i} + x^2\mathbf{j}$ *along the straight line path from the origin to the point* $(1,1)$. *(ii) Repeat using a path that goes from the origin to* $(1,0)$ *and then straight up. (iii) Can this force possibly be conservative? (iv) If yes, find the corresponding potential by inspection.*

Problem 7.6.4. *Let* $\mathbf{F} = 3x^2y\mathbf{i} + x^3\mathbf{j}$. *(i) Find its line integral on the curve* $y = x^2$ *between the origin and the point* $(1,1)$. *(ii) Show that* $\mathbf{F}$ *can be the gradient of some scalar and find the latter by inspection. Use this information to regain the result for the line integral.*

Problem 7.6.5. *Let* $\mathbf{W} = \mathbf{i}x + \mathbf{j}xy$. *Evaluate its line integral around a unit square with its lower left vertex at the origin and edges parallel to the axes. Can this be the gradient of some scalar?*

Problem 7.6.6. *If* $u + iv$ *is an analytic function, show that the curves of constant* u *intersect the curves of constant* v *orthogonally at each point. (Bring in the gradient to solve this.) For the case* $f(z) = z^2$, *sketch the curves of constant* u *and* v *and check this claim.*

We would now like to extend Green's Theorem to three dimensions. So we ask what the circulation is around a tiny contour $C = \partial S$ which bounds a tiny area suspended in three dimensions. Let us begin by returning to the tiny rectangle in the $x - y$ plane bounded by a contour C running counterclockwise. The infinitesimal area vector corresponding to this is

$$d\mathbf{S} = \mathbf{k}\,dx\,dy.$$

(7.6.17)

Since the circulation around this loop was $\left(\frac{\partial W_y}{\partial x} - \frac{\partial W_x}{\partial y}\right)dx\,dy$, it is natural to think of this as the dot product of $d\mathbf{S} = \mathbf{k}\,dx\,dy$ with a vector whose z-component is $\left(\frac{\partial W_y}{\partial x} - \frac{\partial W_x}{\partial y}\right)$. The complete vector, called the *curl of* $\mathbf{W}$, is defined as follows:

$$\boldsymbol{\nabla} \times \mathbf{W} = \mathbf{i} \left(\frac{\partial W_z}{\partial y} - \frac{\partial W_y}{\partial z} \right) + \mathbf{j} \left(\frac{\partial W_x}{\partial z} - \frac{\partial W_z}{\partial x} \right) + \mathbf{k} \left(\frac{\partial W_y}{\partial x} - \frac{\partial W_x}{\partial y} \right).$$
(7.6.18)

Notice the cyclic fashion in which the indices run. The i-th component of the curl is the j-th derivative of the k-th component of the field minus the other cross derivative. To get the j-th component, move all the indices up by one notch cyclically: $i \rightarrow j \rightarrow k \rightarrow i$. To get the k-th component do this one more time. Thus we say that for a tiny loop we considered,

$$\oint \mathbf{W} \cdot \mathbf{dr} = \boldsymbol{\nabla} \times \mathbf{W} \cdot \mathbf{dS}.$$
(7.6.19)

The beauty of writing the result in vector notation is that this formula works for an area not necessarily in the $x-y$ plane. This is because a dot product of two vectors, depending as it does only on rotationally invariant quantities such as the lengths or angles between vectors, is invariant under a rotation of axes. If we therefore rotate our axes so that the tiny patch in the old $x - y$ plane has some general orientation now, the circulation will still be given by the dot product of the area and the curl, though both would have different components now. Thus we come to the general result that

The circulation of $\mathbf{W}$ *around a tiny loop bounding an area* $\mathbf{dS}$ *is* $\boldsymbol{\nabla} \times$ $\mathbf{W} \cdot \mathbf{dS}$.

To see if this formula is indeed correct, you can look at infinitesimal loops in the other two planes, as indicated in the following exercise. If you want to know how exactly it works for a loop in some arbitrary direction, you must work out Problem (7.6.8.). It will help you understand in what sense an area not along the principal planes, (or an area vector not along one of the axes), is a sum of areas (or area vectors) along the principal planes (axes).

Problem 7.6.7. *By considering the circulation around loops in the other two planes confirm the correctness of the other two components of the curl.*

Problem 7.6.8. *This exercise will give you a feeling for how many of the above ideas regarding areas as vectors really work in detail. Consider a planar area in the first octant* ($x > 0, y > 0, z > 0$) *that cuts the three axes at the points* $(a, 0, 0)$, $(0, a, 0)$, *and* $(0, 0, a)$. *Sketch it and show its area is*

$$A = \frac{\sqrt{3}a^2}{2}.$$
(7.6.20)

Choose the area vector to point away from the origin. Given that the area vector points in a direction that is an equal admixture of the three unit vectors and has the above magnitude, show

$$\mathbf{A} = \frac{a^2}{2}(\mathbf{i} + \mathbf{j} + \mathbf{k}). \tag{7.6.21}$$

Imagine now integrating a vector $\mathbf{W}$ *counterclockwise around this area. Show through a sketch that this contour is the sum of three contours: one in the $x - y$ plane, one in the $y - z$ plane and one in the z-x plane. (Hint: The area in the x-y plane is bounded by the two axes and the line $x + y = a$.) Show that the vectors corresponding to the areas bound by these contours are precisely $\frac{a^2}{2}\mathbf{i}$, $\frac{a^2}{2}\mathbf{j}$ and $\frac{a^2}{2}\mathbf{k}$ so that Eqn. (7.6.21) is an expression of this equivalence. Now show that the circulation around each of these three areas is given by the three terms in the dot product in Eqn. (7.6.19).*

There is an analogy between the gradient and curl. The former gives the magnitude and direction of the greatest range of change (per unit displacement) of the scalar field and the latter gives the magnitude and direction of the greatest circulation per unit area of the vector field. At each point if you compute $\nabla\phi$, you know that that is the direction in which the rate of change is a maximum, given by $|\nabla\phi|$. If you move in any other direction, the rate of change will be diminished by the cosine of the angle between $\nabla\phi$ and the displacement vector. Likewise, if you know $\nabla \times \mathbf{W}$ at some point, you know that the circulation per area will be greatest around a loop that lies in the plane perpendicular to $\nabla \times \mathbf{W}$ (i.e., the area vector is parallel to $\nabla \times \mathbf{W}$)and that the circulation will equal the product of the area and $|\nabla \times \mathbf{W}|$. The circulation around any other differently oriented area will be reduced by the cosine of the angle between the area vector and $\nabla \times \mathbf{W}$.

Let us take one more look at Eqn. (7.6.19). It says that the circulation around a tiny loop is the product of the area of the loop times the component of curl parallel to the area vector. *Conversely, the component of curl in any direction is the circulation per unit area in a plane perpendicular to that direction.*

Consider now a surface S in three dimensions. For definiteness imagine this is a hemisphere. The boundary $C = \partial S$ of this surface is the equator. If we now tile the hemisphere with little areas it follows that the circulation around the equator is the sum of circulations around the tiles, as shown in Fig. 7.9 (Only a few representative tiles are shown.)

Since the latter are given by Eqn. (7.6.19) we obtain

$$\oint_{C=\partial S} \mathbf{W} \cdot d\mathbf{r} \simeq \sum_{\text{tiles}} \nabla \times \mathbf{W} \cdot d\mathbf{S}. \tag{7.6.22}$$

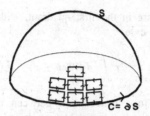

Figure 7.9. An example of Stokes' Theorem. The line integral of a vector field along the equator of the hemisphere is the sum of line integrals around the plaquettes that tile it. The latter sum becomes the surface integral of the curl in the limit of vanishingly small plaquettes.

We do not use the equality sign since as long as the tiles are of finite size, the contour can only be approximated by the tiles and the question of where in each tile to evaluate the various derivatives will plague us. However if we consider the limit of infinitely many tiles of vanishing size, the result becomes exact. In this limit the right-hand side becomes the *surface integral* of $\nabla \times \mathbf{W}$ over the surface S and

$$\oint_{C=\partial S} \mathbf{W} \cdot d\mathbf{r} = \int_S \nabla \times \mathbf{W} \cdot d\mathbf{S} \qquad (7.6.23)$$

which is called *Stokes' Theorem*.

Thus Green's Theorem is Stokes' Theorem restricted to the plane. Note that Stokes' Theorem relates the line integral of a vector around a *closed loop* to the surface integral of its curl over *any* surface bounded by the closed loop. There is no Stokes' Theorem for the line integral over an open curve P_{12} with boundaries 1 and 2.

Note also that for a general vector field, the surface integral can depend on the details of the surface given a fixed boundary. On the other hand, if the vector field is the curl of another vector field, the answer depends on just the boundary $C = \partial S$. This should remind you of line integrals of vector fields that are generally sensitive only to the path, but depend only on the boundary (end points) of the path when the vector field is a gradient. We shall have more to say on this later.

Let us consider two examples of Green's Theorem. Suppose we want to integrate

$$\mathbf{F} = \mathbf{i}\, xy + \mathbf{j}\, x^2 y \qquad (7.6.24)$$

counterclockwise around some contour in the x-y plane. The theorem tells us

$$\oint \mathbf{F} \cdot d\mathbf{r} = \int\int dx\,dy \left(\frac{\partial(x^2 y)}{\partial x} - \frac{\partial(xy)}{\partial y} \right) = \int\int dx\,dy(2xy - x). \quad (7.6.25)$$

Say the contour is a unit square *centered* at the origin. Then the double integral is zero since the functions being integrated are odd in x or y or both.

Consider then a unit square in the first quadrant, with its lower left-hand corner at the origin. In that case

$$\int\int dx\, dy(2xy - x) = \int_0^1 dx \int_0^1 dy(2xy - x) = \frac{1}{2}(x^2\big|_0^1 \, y^2\big|_0^1 - x^2\big|_0^1) = 0$$

$$(7.6.26)$$

where we get zero due to a cancellation between the two terms. Had the square been of side two, the result would have been

$$\frac{1}{2}(x^2\big|_0^2 \, y^2\big|_0^2 - x^2\big|_0^2) = 6. \qquad\qquad (7.6.27)$$

As the second example consider the two dimensional field $\mathbf{F} = 2xy^2\mathbf{i} + x^2\mathbf{j}$ whose curl was $2x - 4xy$. (See Eqn. (7.6.10) in the worked example.) Since the curl is nonzero, not only do we know the answer is path-dependent, we can also relate the difference (in the line integral) between any two paths to the surface integral of the curl over the area trapped between the paths. Let us recall that the line integral was 1 if we went along the x-axis to $(1, 0)$ and then up to $(1, 1)$, while it was equal to 5/6 if one went along $x = y$. It follows that if we went to $(1, 1)$ along the first route and *came back* along the second, the contribution from this closed counter-clockwise path would be $1 - 5/6 = 1/6$. Let us verify that this is indeed what we get by integrating the curl within the triangular region between the paths:

$$\int_0^1 dx \int_0^x dy[2x - 4xy] = \frac{2}{3} - \frac{1}{2} = \frac{1}{6}. \qquad (7.6.28)$$

Problem 7.6.9. *Recall that the line integral of $\mathbf{F} = 2xy^2\mathbf{i} + x^2\mathbf{j}$ between the origin and $(1, 1)$ was equal to 1 along paths that first went up and then to the right and vice versa. Integrate the curl over the region in question and reconcile this fact with Green's Theorem. Suppose we go to $(0, 1)$, then straight up to $(1, 1)$ and back along $x = y^2$. From the worked example (in the introduction to line integrals) deduce what we get for this path. Reconcile with Green's Theorem by integrating the curl over the appropriate area. Finally reconcile the fact that the line integral has the same value on paths $y = x^2$ and $x = y^2$ by integrating the curl in the region between the two curves.*

Problem 7.6.10. *Show that the line integral of $\mathbf{W} = -\frac{1}{2}(y - \frac{1}{2}\sin 2y)\mathbf{i} + x\cos^2 y\mathbf{j}$ counterclockwise around a circle of radius π centered at the point $(x = e, y = e^\pi)$ equals π^3.*

Problem 7.6.11. *(i) Find the work done by the force $\mathbf{F} = \mathbf{i}x + \mathbf{j}y$ as one moves along the semicircle of radius 2 centered at the origin starting from $(1, 0)$ and* *ending at* $(-1, 0)$. *(ii) How much work is done in returning to the starting point along the diameter?* *(iii) What does all this suggest about the force? Verify your conjecture.*

Problem 7.6.12. *Show that the area enclosed by a counterclockwise curve* C *in the plane is given by* $\frac{1}{2} \oint_C (x\, dy - y\, dx)$.

Problem 7.6.13. *Evaluate* $\oint (y\, dx - x\, dy)$ *clockwise around the circle* $(x-2)^2 + y^2 = 1$, *any way you want.*

Problem 7.6.14. *Show that the line integral of* $\mathbf{F} = \mathbf{i}2y + \mathbf{j}x + \mathbf{k}z^2$ *from the origin to* $(1, 1, 1)$ *along the path* $x^2 + y^2 = 2z$, $x = y$ *is* $\frac{11}{6}$. *Visualize the path and parametrize it using* x. *Repeat the integral along the straight line joining the points. Show that this vector field is not curl free. How do we reconcile this with the above two results?*

Problem 7.6.15. *Evaluate* $\oint ((x - 2y)\, dx - x\, dy)$ *around the unit circle first as a parametrized line integral and then by the use of Green's theorem and show that it equals* π.

7.6.1. Curl in other coordinate systems

Let us say u_1, u_2, u_3 are three orthogonal coordinates, such as the spherical coordinates. Let h_1, h_2, h_3 be the scale factors that convert the differentials du_i to distances. The curl can then be written as follows:

$$\nabla \times \mathbf{W} = \mathbf{e}_1 \frac{1}{h_2 h_3} \left[\frac{\partial (W_3 h_3)}{\partial u_2} - \frac{\partial (W_2 h_2)}{\partial u_3} \right] + et\ cycl. \qquad (7.6.29)$$

Let us try to understand this. Consider the component of curl along $\mathbf{e}_1$. This is given by the circulation per unit area on a loop that lies in the plane spanned by coordinates u_2, u_3. Consider the roughly rectangular loop bounded by the lines $u_2, u_2 + du_2, u_3, u_3 + du_3$. By pairing together line integrals along opposite edges, we will get as in cartesian coordinates,

$$\text{circulation} = \left[\frac{\partial W_3 h_3}{\partial u_2} - 3 \leftrightarrow 2 \right] du_2 du_3, \qquad (7.6.30)$$

where the scale factors multiply W_i because the corresponding infinitesimal line elements are $h_i du_i$. We must then divide this by the area of the loop, $h_2 du_2 h_3 du_3$, to get the curl.

Problem 7.6.16. *Write the expression for the curl in two dimensional polar coordinates, spherical and cylindrical coordinates.*

Problem 7.6.17. *Consider once again $\oint((x - 2y)dx - xdy)$ around the unit circle from Problem(7.6.15.). Given that this is the line integral of a vector field, extract the vector field in cartesian coordinates and then express it in polar coordinates. (Start by writing down i and j in terms of e_r and e_θ.) Now do the line integral in polar coordinates. Check using Green's theorem in polar coordinates. Repeat for the following: $\oint(xy\ dx + x\ dy)$ counter-clockwise around the unit circle centered at the origin.*

7.7. The Divergence of a Vector Field

Imagine a closed surface S which is the boundary of some volume V: $S = \partial V$. Our goal is now to compute the surface integral of **W** over S. Imagine filling the volume V with tiny parallelepipeds, called cubes to save space and avoid spelling mistakes. (If they are infinitesimal, they can fill any shape.) Each cube has its area vector pointing outward on each face. Thus adjacent faces of neighboring cubes have their area vectors oppositely directed. Suppose we know how to compute the surface integral over any such tiny cube. Then we can express the surface integral over S as the sum of surface integrals over the tiny cubes that fill it. The reason is just like in Stokes' Theorem. Consider any cube in the interior. *Every one of its six faces is shared by some other cube. Consequently the surface integral over this cube is annulled by the contribution from these very same faces when considered as parts of the surrounding cubes.* The reason is clear: we are integrating the same vector field in the two cases, but with oppositely pointing normals: the outward pointing normal on our cube is inward pointing according to its neighbors. If the neighboring cubes themselves have other neighbors, they will experience a similar cancellation and all we will have in the end is the contribution from the exposed faces, i.e., the surface S.

To summarize, the surface integral of the vector field over the closed surface S is the sum of surface integrals over tiny cubes that fill the enclosed volume V in the usual limit of infinite cubes of vanishing size. So our first job is to compute the surface integral over an infinitesimal cube shown in Fig. 7.10.

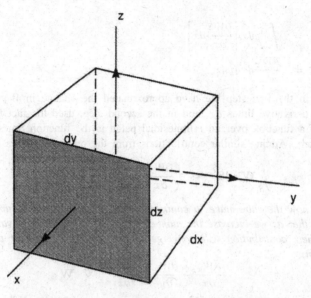

Figure 7.10. The infinitesimal cube at the origin on whose faces we compute the surface integral.

It is centered at the origin (which is a convenience but not a necessity) and has linear dimensions dx, dy, and dz. Its surface is the union of six planar areas. Each is represented by a vector normal to the plane and pointing outward. Let us imagine that this cube is immersed in the vector field $\mathbf{W}$ and calculate the flux coming out of this cube by adding the flux on all six faces, i.e., do the surface integral of $\mathbf{W}$ over this cube. First imagine that $\mathbf{W}$ is constant. Then the total flux will vanish. This is because whatever we get from one face will be exactly canceled by the face opposite to it since the two normals are opposite in direction while the vector is constant. If the flux were indeed due to flow of particles, this would mean that whatever number flows into the cube from one face flows out of the opposite face. Thus, for us to get a nonzero flux, the field will have to be nonuniform.

Consider two of the cube's faces that lie in the $y - z$ plane. Their areas are given by $\pm i\,dy\,dz$, where the plus/minus signs apply to the faces with $x > 0$ and $x < 0$ respectively. (The area vectors point along the outward normals to the surface.) The flux for these two faces comes from just the x-component of $\mathbf{W}$ and is given by

$$\Phi_{yz} = \int_{x=dx/2} dy\,dz\,[W_x(dx/2, y, z)] - \int_{x=-dx/2} dy\,dz\,[W(-dx/2, y, z)]$$

$$= \int dy dz \left[\frac{\partial W_x}{\partial x} dx \right] \qquad (7.7.1)$$

$$= \frac{\partial W_x}{\partial x} dx dy dz, \qquad (7.7.2)$$

where in the first step we have approximated the change in W_x by the first partial derivative times dx, and in the second step, used the fact that the integral of a function over an infinitesimal patch is the function times the area of the patch. Adding similar contributions from the other four faces we get

$$\oint_S \mathbf{W} \cdot d\mathbf{S} = \left(\frac{\partial W_x}{\partial x} + \frac{\partial W_y}{\partial y} + \frac{\partial W_z}{\partial z} \right) dx dy dz. \qquad (7.7.3)$$

Notice how the cube makes a contribution proportional to its volume. This will ensure that as we increase the number of cubes, in the limit of vanishing cube size, their contribution will converge to the volume integral of the following function:

$$\frac{\partial W_x}{\partial x} + \frac{\partial W_y}{\partial y} + \frac{\partial W_z}{\partial z} \equiv \boldsymbol{\nabla} \cdot \mathbf{W} \qquad (7.7.4)$$

called the *divergence* of $\mathbf{W}$ and often pronounced "div W" where "div" rhymes with "give."

By this definition the flux out of the infinitesimal cube or the surface integral over the cube is the volume integral of the divergence. Conversely, the divergence is the outgoing flux per unit volume.

As stated earlier, if we patch together many infinitesimal cubes, we can build a macroscopic volume V whose boundary surface $S = \partial V$ is the union of the unshared or exposed faces. Thus

$$\oint_{S=\partial V} \mathbf{W} \cdot d\mathbf{S} = \int_V \boldsymbol{\nabla} \cdot \mathbf{W} dx dy dz \qquad (7.7.5)$$

where it is assumed that we have taken the limit of zero cube size. The above is called *Gauss's Theorem*. Notice that the theorem relates the surface integral of a vector over a *closed* surface to the integral of its divergence over the volume enclosed. An open surface, like a hemisphere, does not enclose a volume and there is no Gauss' Theorem for integrals over it.

Note also another curious fact. In general the integral of a scalar over a solid volume V cannot be reduced to an integral of anything over just its boundary surface $S = \partial V$. But if the integrand is the divergence of a vector field, this is possible. Notice once again the analogy to line integrals of gradients and surface integrals of curls both of which get contributions from the boundary. We shall return to this point.

Let us now work out an example: the surface integral of

$$\mathbf{W} = \mathbf{i} x^3 y + \mathbf{j} y^2 x + \mathbf{k} z \qquad (7.7.6)$$

over a unit cube centered at the origin as in Fig. 7.10. We have already done this in the introduction to surface integrals and found the answer equals unity.

If we use Gauss's theorem, the divergence is

$$\nabla \cdot \mathbf{W} = 3x^2y + 2xy + 1 \qquad (7.7.7)$$

from the three partial derivatives. So the surface integral equals

$$\int_{-1/2}^{1/2} dx \int_{-1/2}^{1/2} dy \int_{-1/2}^{1/2} dz (3x^2y + 2xy + 1) = 1 \qquad (7.7.8)$$

where only the third term contributes and the others vanish by symmetry.

Problem 7.7.1. *Find the divergence of* $\mathbf{W} = \mathbf{i}\cos x + \mathbf{j}\sin z + \mathbf{k}z^2$.

Problem 7.7.2. *Consider the surface in Fig. 7.9 with a boundary at the equator. Slice a piece off the top at the Tropic of Cancer. Now the surface has two boundaries. Express the integral of the curl over this S to its boundaries by drawing little tiles again. Do the similar thing with Gauss' Theorem by considering a solid region with a piece scooped out of its insides so that it too has two boundaries. Pay attention to the signs of the two boundary contributions. Compare to the simple result*

$$\int_1^2 \nabla\phi \cdot d\mathbf{r} = \phi(2) - \phi(1), \qquad (7.7.9)$$

where 1 and 2 are as usual short for $\mathbf{r}_1$ *and* $\mathbf{r}_2$, *respectively.*

Problem 7.7.3. *Find the surface integral of* $\mathbf{W} = \mathbf{i}x^3y + \mathbf{j}y^2x + \mathbf{k}z$ *over a unit cube centered at the origin, with edges parallel to the axes, by direct computation and by Gauss's Theorem. Use symmetries to save yourself some work.*

Problem 7.7.4. *Find the surface integral of* $\mathbf{W} = \mathbf{i}x\sin^2 y + \mathbf{j}x^2 + \mathbf{k}z\cos^2 y$ *over a sphere of radius* π *centered at the point* (e, π^e, e^π).

Problem 7.7.5. *Show that the surface integral of* $\mathbf{F} = \mathbf{i}y + \mathbf{j}x + \mathbf{k}z^2$ *over a unit cube in the first octant with one of its corners at the origin and three of its edges aligned with the axes is unity, without using Gauss's theorem. Check the result using Gauss's theorem. Show likewise that if the closed surface is instead the upper hemisphere of a unit sphere and the* $x - y$ *plane, the answer is* $\frac{\pi}{2}$, *and check using Gauss's theorem. For the hemisphere, note that the unit radial vector can be written as* $(\mathbf{i}x + \mathbf{j}y + \mathbf{k}z)/r$.

7.7.1. Divergence in other coordinate systems

The formula for divergence in any orthogonal system of coordinates is

$$\boldsymbol{\nabla} \cdot \mathbf{W} = \frac{1}{h_1 h_2 h_3} \left[\frac{\partial}{\partial u_1}(W_1 h_2 h_3) + \frac{\partial}{\partial u_2}(W_2 h_3 h_1) + \frac{\partial}{\partial u_3}(W_3 h_1 h_2) \right] . \quad (7.7.10)$$

To see why this is correct, recall that the divergence is the flux per unit volume. Pairing the opposite faces of a cube-like object bounded by the level surfaces $u_1, u_1 + du_1, u_2, u_2 + du_2, u_3, u_3 + du_3$, we see that the net contribution from the two opposite faces in the 2-3 planes is given by

$$\frac{\partial W_1 h_2 h_3}{\partial u_1} du_1 du_2 du_3,$$

where the scale factors now multiply W_1 because the area element that enters the flux calculation is $h_2 du_2 h_3 du_3$. Finally when we add the contribution from the other four faces and divide by the volume element $h_1 du_1 h_2 du_2 h_3 du_3$ we obtain Eqn. (7.7.10).

Problem 7.7.6. *(i) Find the surface integral of* $\mathbf{F} = \mathbf{i}x + \mathbf{j}y + \mathbf{k}z$ *over a sphere of radius R centered at the origin without using Gauss's theorem. Check it using Gauss's theorem. Use the right set of coordinates. (ii) Repeat the calculation for the closed surface bounded by the upper hemisphere and the $x - y$ plane. (iii) Repeat for the case where the closed surface is a unit cube centered at the origin.*

Problem 7.7.7. *Write down the expressions for curl, gradient and divergence in cylindrical and spherical coordinates.*

7.8. Differential Operators

Let us look at the formulae for curl and divergence in cartesian coordinates, Eqn. (7.6.18, 7.7.4). For example the divergence is

$$\boldsymbol{\nabla} \cdot \mathbf{W} = \frac{\partial W_x}{\partial x} + \frac{\partial W_y}{\partial y} + \frac{\partial W_z}{\partial z}. \quad (7.8.1)$$

This looks like the dot product of the following object, called "del"

$$\boldsymbol{\nabla} = \mathbf{i}\frac{\partial}{\partial x} + \mathbf{j}\frac{\partial}{\partial y} + \mathbf{k}\frac{\partial}{\partial z} \quad (7.8.2)$$

with $\mathbf{W}$. Likewise the curl looks like the cross product of the two objects ∇ and $\mathbf{W}$. (It is to be understood that the vector $\mathbf{W}$ always stands to the right so that the dot or cross product is a sum of various derivatives of the components of $\mathbf{W}$.)

The object in Eqn. (7.8.2) is an example of a *vector differential operator*. We want to know what it means. Let us begin with an easier object, a differential operator

$$D \equiv \frac{d}{dx}. \tag{7.8.3}$$

Clearly D is not a number valued object. It looks like a derivative with its head cut off: what happened to the f in $\frac{df}{dx}$? Here is an analogy. We all know what $\sin x$ means. It is a function, which gives us a definite number for any given x: the number in question is

$$\sin x = x - \frac{x^3}{3!} + \frac{x^5}{5!} + \cdots \tag{7.8.4}$$

But what meaning do we attach to just the expression "sin"? This is not a number valued object. On the other hand given an x, it will spit out a number as per the above formula. But even without the x, sin stands for a sequence of instructions that will be performed on the given x, namely the infinite sum shown above. The sequence of instructions coded under the name "sin" is called an *operator*. It needs an *operand*, namely x to yield a number. We say the operator sin acts on the operand x to give the number $\sin x$.

(As an aside let us recall that even in physics we sometimes split a single entity into two parts and at first we feel uneasy working with half the team. Take for example the electrostatic force between two charges separated by $\mathbf{e}_r r$ given by Coulomb's law:

$$\mathbf{F} = \frac{q_1 q_2 \mathbf{e}_r}{4\pi \varepsilon_0 r^2} \tag{7.8.5}$$

where $\mathbf{e}_r$ is a unit vector going from point 1 to point 2 and ε_0 is a constant determined from electrostatic measurements. We then split this expression into two parts: an electric field

$$\mathbf{E} = \frac{q_1 \mathbf{e}_r}{4\pi \varepsilon_0 r^2} \tag{7.8.6}$$

due to charge q_1 at the location of charge q_2 which will cause charge q_2 to experience a force $q_2 \mathbf{E}$ when placed there. Thus the field is a result of just one charge and it runs counter to the simpler notion that it takes two charges for anything to happen. The field is a force waiting to happen. But we give it a meaning even before the other charge is put there. To get a force out of the field, you have to introduce the other charge q_2.)

The object D is similar. *Whereas sin takes in a number and spits out a number, D takes in a function and spits out a function. The function it spits out is the derivative of the input function.* Thus

$$D \text{ acting on } f = \frac{df}{dx}. \tag{7.8.7}$$

Thus D acting on the function x^2 gives the function $2x$; acting on the function $\sin x$ gives $\cos x$ and so on. One calls D the *derivative operator*. It is usually written as $\frac{d}{dx}$ to make its action very transparent. You can use any symbol you like, but what it does is what it does: take the derivative of what you feed it.

The derivative operator is a linear operator:

$$D(af(x) + bg(x)) = aDf(x) + bDg(x). \tag{7.8.8}$$

That is to say, the action on a linear combination is the sum of the action on the pieces. By contrast, sin is not linear because $\sin(x+y) \neq \sin x + \sin y$. However if you were always going to let x and y be small, or if you were somehow restricted to small x and y, then you could come to the conclusion that sin is a linear operator since then $\sin x \simeq x$ and $\sin(x + y) \simeq x + y$. Many relations in nature that are really nonlinear appear linear at first sight since experiments initially probe the situation in a limited range.

The vector differential operator, "del," in Eqn. (7.8.2) is within our grasp now: acting on a scalar ϕ it gives a vector field $\nabla\phi$, acting on a vector field via a dot product, it gives a scalar, the divergence, while acting via the cross product, it gives another vector field, the curl.

Problem 7.8.1. *Can you define the operator D^2? Is it linear?*

The vector differential operator can be written in other orthogonal coordinate systems as

$$\nabla = \mathbf{e}_1 \frac{1}{h_1} \frac{\partial}{\partial u_1} + \mathbf{e}_2 \frac{1}{h_2} \frac{\partial}{\partial u_2} + \mathbf{e}_3 \frac{1}{h_3} \frac{\partial}{\partial u_3}. \tag{7.8.9}$$

7.9. Summary of Integral Theorems

Let us recall the three integral theorems established in the preceding pages:

$$\int_{P_{12}} \nabla\phi \cdot \mathbf{dr} = \phi|_{\partial P} = \phi(2) - \phi(1), \tag{7.9.1}$$

$$\int_S \nabla \times \mathbf{A} \cdot \mathbf{dS} = \oint_{P=\partial S} \mathbf{A} \cdot \mathbf{dr} \text{ (Stokes' Theorem)}, \tag{7.9.2}$$

$$\int_V \nabla \cdot \mathbf{A} \, dV = \oint_{S=\partial V} \mathbf{A} \cdot \mathbf{dS} \text{ (Gauss's Theorem)}. \tag{7.9.3}$$

Note that in all three cases

- On the left-hand side some sort of derivative of a field is integrated over a one-, two-, or three-dimensional region.

- On the right-hand side is the contribution of the field from just the boundary of the region in question in the left hand side.

In general, the line, surface, and volume integrals will not receive their contribution from just the boundaries of the integration regions. This is so only for the gradient, curl, or divergence in the three respective cases shown above. It follows for these special fields that if the integration region on the left-hand side is over a region with no boundary, we will get zero. This will lead to some identities we discuss in the next section.

7.10. Higher Derivatives

So far we have computed some first-order derivatives of vector and scalar fields: $\nabla\phi$, $\nabla \cdot \mathbf{W}$ and $\nabla \times \mathbf{W}$. Let us move on to higher-order derivatives of fields. For example, we can explore the divergence of the curl or the gradient of a divergence and so on. We will examine a few second-order derivatives. We start with two that are identically zero:

$$\nabla \times \nabla \phi \equiv 0 \qquad (7.10.1)$$

$$\nabla \cdot \nabla \times \mathbf{W} \equiv 0, \qquad (7.10.2)$$

which you may easily verify by straightforward differentiation. All you will need is the equality of mixed partial derivatives: $f_{ij} = f_{ji}$, where $i, j = x, y$ or z.

Problem 7.10.1. *Verify these identities.*

We can understand these identities in a deeper way. Consider the line integral of a gradient. It is the same on any two paths with the same end points since all the contribution comes from the end points. The fact that we get the same answer on two paths joining the same end points means that *the difference*, the integral on a *closed path* (obtained by going up one path and coming back counter to the other) is zero. But by Green's or Stokes' Theorems, the surface integral of the curl (of this gradient) had to vanish on the area bounded by this closed path. Since the paths and hence enclosed area were arbitrary, and could be infinitesimal, the integrand of the surface integral, which is the curl of the grad, had to vanish everywhere.

Consider now the surface integral of a curl. It is the same on any two surfaces with the same boundary since the answer comes only from the boundary. In other words the *difference* of the two surface integrals is zero. This difference is then a surface integral over a *closed surface*. Let us make sure we understand the last point, by considering an example. Consider a unit circle on the $x - y$ plane, with its circumference running counterclockwise as seen from above the $x - y$ plane. It is the boundary of the unit disc in the plane and also of a unit hemisphere sitting above the $x - y$ plane. In computing surface integrals, we are instructed by the orientation of the loop to choose the outward normal on the hemisphere and the upward normal on the disc. In the difference of integrals, we will use the downward normal for the disc. The difference is then the surface integral over a closed surface formed by the hemisphere and disc, with all normals pointing outward. With this fact clear, we now use Gauss's Law to relate the integral over the closed surface of the curl to the integral of the divergence (of the curl) over the enclosed volume. Since this is to vanish for any closed volume, even an infinitesimal one, the divergence of the curl must vanish.

7.10.1. Digression on vector identities

There is yet another way to understand these identities, another way that tells you the identities had to be true.[2] As a prelude to this we require the following theorem stated without proof:

The boundary of a boundary is zero.

Here are some examples to illustrate this. Consider a path P_{12} in the $x - y$ plane going from point 1 to point 2. Its boundaries are its end points. But the path itself is not the boundary of anything. But now, let the two extremities be brought together making it a closed curve. Two things have simultaneously happened:

- The path has become the boundary of a two dimensional region. (In two dimensions this region is unique. In three dimensions there are many surfaces bounded by this path. Think of the closed path as the rim of a butterfly net. As the net wiggles in three dimensions, that rim continues to be its boundary.)

- The path no longer has a boundary.

We summarize this by saying

$$\text{If } P \; = \; \partial S \text{ then} \tag{7.10.3}$$
$$\partial P = \partial \partial S \; \equiv \; 0. \tag{7.10.4}$$

Consider next the case of S, the butterfly net. It has a boundary $\partial S = P$, where P is the rim. Shrink the boundary to a point. Two things happen now:

- The surface S is now closed and is the boundary of a volume V: $S = \partial V$.

- The surface no longer has a boundary!

Once again we see

$$\text{If } S \; = \; \partial V \text{ then} \tag{7.10.5}$$
$$\partial S = \partial \partial V \; \equiv \; 0. \tag{7.10.6}$$

All these cases are summarized by one powerful identity:

$$\partial^2 \equiv 0. \tag{7.10.7}$$

If you want to go beyond our intuitive treatment of this result, you must look it up under the heading *Homology*.

It turns out that the identities Eqn. (7.10.1,7.10.2) are related to this identity. Let us begin with

$$\int_1^2 \boldsymbol{\nabla}\phi \cdot \mathbf{dr} = \phi(2) - \phi(1) \equiv \phi|_{\partial P} \tag{7.10.8}$$

(where 1 and 2 are short for $\mathbf{r}_1$ and $\mathbf{r}_2$ respectively) which states that the contribution to the integral comes entirely from the boundary of the region of integration. Let us now join the end points so as to obtain a closed path. Let S be any surface with this closed path as its boundary. Once the path is closed, two things happen:

- The end point contribution is zero since there is no end point.

- We can now use Stokes' Theorem relating the line integral to the surface integral over the enclosed surface.

---

[2]This section is purely for your amusement.

Thus

$$\int_S (\nabla \times \nabla \phi) \cdot d\mathbf{S} \;=\; \oint_{P=\partial S} \nabla \phi \cdot d\mathbf{r} \qquad (7.10.9)$$

$$= \phi|_{\partial P=\partial\partial S\equiv 0} \qquad (7.10.10)$$

$$\equiv 0. \qquad (7.10.11)$$

Now S is any surface with a boundary. We can take then a very tiny infinitesimal surface $d\mathbf{S}$ on which the vector field $(\nabla \times \nabla \phi)$ is essentially constant. In that case the surface integral is just $\nabla \times \nabla \phi \cdot d\mathbf{S}$. If this is to vanish *for any orientation of the surface,* $(\nabla \times \nabla \phi)$ must itself vanish giving us the identity Eqn. (7.10.1).

Consider next Stokes' Theorem

$$\int_S \nabla \times \mathbf{W} \cdot d\mathbf{S} = \oint_{P=\partial S} \mathbf{W} \cdot d\mathbf{r}, \qquad (7.10.12)$$

where once again the entire contribution comes from the boundary of $\mathbf{S}$. Let us now shrink the boundary $P = \partial S$ to a point. This closes the surface and makes it the boundary of a volume V: $S = \partial V$. Evaluating the surface integral above using Gauss's Theorem we find

$$\int_V \nabla \cdot \nabla \times \mathbf{W}\,dx\,dy\,dz \;=\; \oint_{S=\partial V} \nabla \times \mathbf{W} \cdot d\mathbf{S} \qquad (7.10.13)$$

$$= \oint_{P=\partial S=\partial\partial V\equiv 0} \mathbf{W} \cdot d\mathbf{r} \qquad (7.10.14)$$

$$\equiv 0. \qquad (7.10.15)$$

In the last step all we are saying is that the circulation on a closed surface (such as the surface of a sphere) must vanish since every tiny tile is fully surrounded by other tiles so that the contribution from any edge of any tile to the line integral is canceled by an adjoining tile with which it shares that edge.

Since the volume integral must vanish for any volume, the integrand itself must vanish, giving us the identity Eqn. (7.10.2).

To summarize, in Eqn. (7.10.9) we are taking the surface integral of a curl (of a gradient). So the answer comes entirely from the boundary, as the line integral of the gradient. The line integral of the gradient once gives all its contribution at the boundary of the boundary, which is zero. In Eqn. (7.10.13) we are doing this again. We are integrating the divergence (of a curl) over a volume. So the contribution comes from the boundary surface. But on the boundary we are integrating a curl, which gives all its contribution from the boundary of the bounding surface, which is zero.

Notice that in Eqns. (7.10.9-7.10.11) and Eqns. (7.10.13-7.10.15) we start with the integral of a (scalar or vector) field that has been differentiated twice. At the first step we trade one derivative on the field for one ∂ acting on the integration region, giving its boundary. Then we do it again, trading one more derivative on the integrand for one more ∂ acting on the integration region. But this is one too many boundary operations, for two strikes and you are out in this game.

The advantage of the last way of thinking about vector identities is that it is very general and applies to all kinds of manifolds in all dimensions.

We now return to our study of higher-order operators.

A second derivative that does not vanish and plays a significant role is called the *Laplacian*. It is defined as follows:

$$\nabla \cdot \nabla \phi \;=\; \frac{\partial^2 \phi}{\partial x^2} + \frac{\partial^2 \phi}{\partial y^2} + \frac{\partial^2 \phi}{\partial z^2} \qquad (7.10.16)$$

$$\equiv \; \nabla^2 \phi. \qquad (7.10.17)$$

A function that obey

$$\nabla^2 \phi = 0 \qquad (7.10.18)$$

is said to obey Laplace's equation, or to be a harmonic function. (Recall that the real and imaginary parts of analytic function are harmonic in two dimensions.) The differential operator

$$\nabla^2 = \frac{\partial^2}{\partial x^2} + \frac{\partial^2}{\partial y^2} + \frac{\partial^2}{\partial z^2} \qquad (7.10.19)$$

is called the *Laplacian*.

By combining the expressions for gradient and divergence in other orthogonal systems, Eqns. (7.5.19, 7.7.10), we can obtain the Laplacian in other coordinates:

$$\nabla^2 \phi = \frac{1}{h_1 h_2 h_3} \left[\frac{\partial}{\partial u_1} \left(\frac{h_2 h_3}{h_1} \frac{\partial \phi}{\partial u_1} \right) + \frac{\partial}{\partial u_2} \left(\frac{h_3 h_1}{h_2} \frac{\partial \phi}{\partial u_2} \right) + \frac{\partial}{\partial u_3} \left(\frac{h_1 h_2}{h_3} \frac{\partial \phi}{\partial u_3} \right) \right].$$
$$(7.10.20)$$

There are only two other second derivatives:

$$\nabla(\nabla \cdot \mathbf{W}), \qquad (7.10.21)$$

$$\nabla \times \nabla \times \mathbf{W} \;=\; \nabla(\nabla \cdot \mathbf{W}) - \nabla^2 \mathbf{W}. \qquad (7.10.22)$$

The first has no name and rarely comes up. The second equation is true in cartesian coordinates and will be invoked later in this chapter.

Problem 7.10.2. *Verify that there are no other second derivatives we can form starting with scalar and vector fields and the "del" operator.*

7.11. Applications from Electrodynamics

In this section we will discuss some ideas from electrodynamics that will illustrate how the ideas developed above arise naturally in physics.

We begin with the flow of electric charge. Let ρ be the density of charge and $\mathbf{v}$ its velocity at some point. First consider a wire along the x-axis and ignore the vectorial character of the velocity.[s] The current through the wire, I, is given by the amount of charge crossing the cross section of the wire per unit time. By arguments given earlier in this chapter,

$$I = \rho v S \qquad (7.11.1)$$

where S is the cross-sectional area of the wire. The current density j is given as the current per unit area and is given by

$$j = \rho v. \qquad (7.11.2)$$

In general

$$\mathbf{j} = \rho \mathbf{v} \qquad (7.11.3)$$

is the current density at a point where the charge density is ρ and velocity is $\mathbf{v}$.

Consider some volume V bounded by a surface S. The total charge leaving the surface is given by the surface integral of j:

$$\text{charge leaving } W = \oint_S \mathbf{j} \cdot \mathbf{dS}. \qquad (7.11.4)$$

Now it turns out that the total charge in the universe is conserved, i.e., does not change with time. Furthermore, the conservation is local: we do not see charge disappear in one region and instantaneously reappear elsewhere. Instead we see a more detailed accounting: any charge decrease in a volume V is accounted for by the flow out of that volume. Indeed without this local accounting of charge, its conservation is quite empty and unverifiable: if some charge is suddenly lost in our part of the universe, it is pointless to argue over whether this charge is really gone or hiding in some remote and inaccessible part of our universe. Using the fact that the total charge in V is the volume integral of ρ, and the above equation for the flow out of V, we may express charge conservation as follows:

$$\frac{\partial}{\partial t} \int_V \rho(x, y, z, t) d^3 r = -\oint_S \mathbf{j} \cdot \mathbf{dS} \qquad (7.11.5)$$

$$= -\int_V \nabla \cdot \mathbf{j} d^3 r \quad \text{(Gauss's Theorem)} \qquad (7.11.6)$$

where $d^3 r$ is shorthand for $dx\,dy\,dz$ in this case and for whatever is the element of volume in other coordinate systems, such as $r^2 \sin\theta d\theta d\phi$ in the spherical system. Since the volume in the above equation is arbitrary, the integrands on both sides must be equal, giving us

$$\frac{\partial}{\partial t} \rho(x, y, z, t) + \nabla \cdot \mathbf{j} = 0, \qquad (7.11.7)$$

which is called the *continuity equation*. This equation also appears in other contexts such as fluid mechanics, where it expresses the conservation of matter.

We now turn to Maxwell's electrodynamics. The cause of every field is the electric charge, at rest or in motion. The effect is also felt by the charges, at rest or in motion. But we see it as a two stage process: charges produce fields and fields lead to forces felt by charges. As mentioned earlier, the electrostatic force

$$\mathbf{F} = \frac{q_1 q_2}{4\pi\varepsilon_0 r^2} \mathbf{e}_r \qquad (7.11.8)$$

between two static charges is seen as a two stage process: the first charge produces a field

$$\mathbf{E} = \frac{q_1}{4\pi\varepsilon_0 r^2} \mathbf{e}_r \qquad (7.11.9)$$

at the location of the second charge and this field causes a force on the second given by

$$\mathbf{F} = q_2 \mathbf{E}. \qquad (7.11.10)$$

Electric charges in motion also experience another force, called the magnetic force, which was discussed earlier. Thus the total force on a charge is given by the *Lorentz force law*:

$$\mathbf{F} = q(\mathbf{E} + \mathbf{v} \times \mathbf{B}). \tag{7.11.11}$$

To find the electric and magnetic fields at a point, we first put a static charge there and measure the force (by measuring the acceleration times mass), giving us $\mathbf{E}$. Then we put moving charges with different velocities there and measure the force they feel to find out $\mathbf{B}$.

Problem 7.11.1. *Give a set of measurements that will determine* $\mathbf{B}$.

This then is one half of the story: how the charges respond to the fields. The other half specifies the relation between the fields and the charges that produce them.

Let us begin with the electric field due to a static point charge q at the origin in MKS units:

$$\mathbf{E} = \frac{q}{4\pi\varepsilon_0 r^2} \mathbf{e}_r \tag{7.11.12}$$

where $\mathbf{e}_r$ is a unit radial vector. If we now calculate the flux of $\mathbf{E}$ over a sphere of radius R we find

$$\oint_S \mathbf{E} \cdot d\mathbf{S} = \oint_S \frac{q}{4\pi\varepsilon_0 R^2} \mathbf{e}_r \cdot \mathbf{e}_r dS \tag{7.11.13}$$

$$= \frac{q}{4\pi\varepsilon_0} \int_S \frac{R^2 d\Omega}{R^2} \tag{7.11.14}$$

$$= \frac{q}{\varepsilon_0}, \tag{7.11.15}$$

where we have used the fact that the infinitesimal area vector on a sphere of radius r is everywhere radial and has a magnitude $r^2 \sin\theta d\theta d\phi \equiv r^2 d\cos\theta d\phi \equiv r^2 d\Omega$, i.e.,

$$d\mathbf{S} = \mathbf{e}_r r^2 d\cos\theta d\phi \tag{7.11.16}$$

and that

$$\int d\Omega = 4\pi. \tag{7.11.17}$$

Observe that the surface integral of $\mathbf{E}$ due to the charge, (q/ε_0), is independent of the radius of the sphere. If we increase the radius, the area grows quadratically but the field decays as the inverse square, exactly compensating. Indeed we will now argue that the answer is insensitive to any change of shape of the surface as long as it encloses the origin. Let us go back to Eqn. (7.11.13) but no longer assume S is a sphere. We find

$$\oint_S \mathbf{E} \cdot d\mathbf{S} = \oint_S \frac{q}{4\pi\varepsilon_0 r^2} \mathbf{e}_r \cdot d\mathbf{S} \tag{7.11.18}$$

$$= \frac{q}{4\pi\varepsilon_0} \int_S d\Omega \tag{7.11.19}$$

$$= \frac{q}{\varepsilon_0}, \tag{7.11.20}$$

where we have used the fact that $\mathbf{e}_r \cdot d\mathbf{S}$ is the projection of the area vector in the radial direction (or the projection of the plane of the area in the tangential directions) so that

$$\mathbf{e}_r \cdot d\mathbf{S}/r^2 = d\Omega.$$

Thus the total flux is controlled only by the solid angle enclosed by the surrounding surface and independent of its detailed shape.

All this sounds very much like Cauchy's Theorem: there an integral enclosing the pole told us about the residue of the pole and the answer was independent of the shape of the surrounding contour; here the surface integral tells us about the charge enclosed and is independent of the detailed shape of the surface. Even the proof is similar: there we began with a circle centered at the pole and slowly modified the contour, here we began with a sphere centered on the charge and slowly deformed it.

What if the surface encloses two charges? The *principle of superposition*, which is a physical (and not mathematical) principle based on experiment, tells us that the electric field due to two or more charges is the sum of the electric fields due to each one. It follows that the same goes for the surface integral of **E** due to a collection of charges. Thus we have the result:

$$\oint_{S=\partial V} \mathbf{E} \cdot \mathbf{dS} = \sum_{i \in V} \frac{q_i}{\varepsilon_0}. \tag{7.11.21}$$

We will now convert this macroscopic relation to a microscopic differential relation as follows. We express the surface integral on the left hand side in terms of the volume integral of the divergence using Gauss's Theorem and write the total charge in the right-hand side as the volume integral of the charge density to obtain:

$$\int_V \mathbf{\nabla} \cdot \mathbf{E} d^3 r = \int_V \frac{\rho}{\varepsilon_0} d^3 r. \tag{7.11.22}$$

Equating the integrands on both sides since the result is true for any volume whatsoever, we obtain the first Maxwell equation:

$$\mathbf{\nabla} \cdot \mathbf{E} = \frac{\rho}{\varepsilon_0} \quad \text{(Maxwell I)} \tag{7.11.23}$$

This equation tells us that if the electric field has a divergence anywhere it is due to the presence of charge: every charge is a source of radially outgoing **E** lines, i.e., a field with a divergence at the location of the charge. Note that in the radial field of a point charge, the divergence is nonzero only at the origin: any other surface not enclosing the origin will have as much flux entering it as leaving it. (What is the corresponding statement in the complex plane regarding a pole?)

The integral form of Gauss's theorem

$$\oint_{S=\partial V} \mathbf{E} \cdot \mathbf{dS} = \int_V \frac{\rho}{\varepsilon_0} d^3 r \tag{7.11.24}$$

is called Gauss's law in electrodynamics. Together with some symmetry arguments, it can be used to solve problems that would otherwise be very difficult. Consider first the electric field due to a ball of charge of radius R and constant charge density ρ at a distance $r > R$ from the center. Since we know from Coulomb's law the field due to a point charge, we can divide the sphere into tiny cubes each containing charge $\rho dx dy dz$, find the force due to each cell, and add all the contributions. This is very hard since each cube is a different distance from the point where we want the field and each cube contributes a vector that must be added to all the other vectors. Gauss's law allows to solve the problem as follows. We first argue that the field must be radial by symmetry: if there is a reason for it to swerve to one side, there is just as good a reason to swerve to the other, the spherically symmetric charge distribution allows no room for any kind of preference. Here is another way to see it. The radially outgoing field distribution has the feature that if I rotate the whole configuration around the origin, nothing changes. Suppose you came up with a configuration that looked different after a rotation. We would have a problem, since the charge distribution will look the same after the rotation and we will have a situation where the same distribution produces two different field configurations! Continuing, we also know the strength of the field can depend only on r. So we know $\mathbf{E} = \mathbf{e}_r f(r)$. Choosing as our surface a sphere of radius $r > R$, concentric with the charge distribution, we get

$$\oint_{S=\partial V} \mathbf{E} \cdot \mathbf{dS} = \oint_{S=\partial V} \mathbf{e}_r f(r) \cdot \mathbf{e}_r dS \qquad (7.11.25)$$

$$= 4\pi r^2 f(r) \qquad (7.11.26)$$

$$= \int_V \frac{\rho}{\varepsilon_0} d^3 r \qquad (7.11.27)$$

$$= \frac{4\pi R^3}{3\varepsilon_0} \rho \qquad (7.11.28)$$

from which follows

$$f(r) = \frac{\frac{4\pi R^3}{3}\rho}{4\pi r^2 \varepsilon_0}, \qquad (7.11.29)$$

which is the field you would get if all the charge in the ball concentrated at the origin. You should convince yourself that

- The result can be generalized even if $\rho = \rho(r)$.

- If ρ has angular dependence, so will $\mathbf{E}$ and given its integral over a sphere, we cannot work our way to the values at all points on it. Gauss's law gives us just one equation per surface and if only one thing is unknown about the field (such as its magnitude at some distance) it may be determined.

Problem 7.11.2. *Consider an infinite wire along the z-axis with charge λ per unit length. Give arguments for why the field will be radially outward and a function only of ρ, the axial distance to the wire in cylindrical coordinates. By considering a cylindrical surface of length L, coaxial with the wire, show that $\mathbf{E} = \mathbf{e}_\rho \frac{\lambda}{2\pi r \varepsilon_0}$:*

Problem 7.11.3. *Consider an infinite plane of charge with density σ per unit area. By considering a cylinder that pierces the plane with axis normal to the plane, show that the field is perpendicular to it, has no dependence on distance from the plane and is of magnitude $E = \frac{\sigma}{2\varepsilon_0}$.*

Problem 7.11.4. *Consider the radial field $\mathbf{E} = \frac{\mathbf{e}_r}{4\pi\varepsilon_0 r^2}$. Show that its divergence is zero at any point other than the origin by using the divergence formula in spherical coordinates. Next calculate the surface integral on a sphere of radius r. Argue that if the divergence is zero at the origin as well Gauss's Theorem will be violated. Thus the divergence must be given by a function that has support only at the origin, but have so much punch there that its integral is finite. A function with this property is called Dirac delta function. It is denoted by the symbol $\delta^3(\mathbf{r} - \mathbf{r}_0)$ and it has the following properties:*

$$\delta^3(\mathbf{r} - \mathbf{r}_0) = 0 \ \textit{for} \ \mathbf{r} \neq \mathbf{r}_0 \qquad (7.11.30)$$

$$\int_{\mathbf{r}_0 \in V} \delta^3(\mathbf{r} - \mathbf{r}_0) d^3 r = 1. \qquad (7.11.31)$$

Thus $\delta^3(\mathbf{r} - \mathbf{r}_0)$ is nonzero only at $\mathbf{r}_0$ but so singular there that its volume integral is unity. Express the divergence of $\mathbf{E}$ in this problem in terms of the Dirac delta functions. We shall return to a fuller study of this function in Chapter 9.

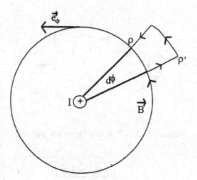

Figure 7.11. Ampere's Law for one current.

Consider next the magnetic field **B**. It is an empirical fact that there are no magnetic charges producing radial magnetic fields. (How then do we have any magnetic fields? Because moving electric charges, i.e., currents, can produce a magnetic field. More on this shortly.) The absence of magnetic charges is expressed by the second Maxwell equation:

$$\nabla \cdot \mathbf{B} = 0 \quad \text{(Maxwell II)} \tag{7.11.32}$$

Now we ask about the magnetic field produced by currents. Consider an infinite wire carrying a current I up the z-axis. In cylindrical coordinates, we can summarize the experiments by the equation

$$\mathbf{B} = \mathbf{e}_\phi \frac{\mu_0 I}{2\pi\rho} \tag{7.11.33}$$

where μ_0 is a constant determined by measuring magnetostatic (time-independent) phenomena. The line integral of **B** around a circle of radius ρ (i.e., $\mathbf{dr} = \mathbf{e}_\phi \rho d\phi$) centered on the wire and lying in the $x - y$ plane is given by

$$\oint \mathbf{B} \cdot \mathbf{dr} = \int_0^{2\pi} \frac{\mu_0 I \rho d\phi}{2\pi\rho} = \mu_0 I. \tag{7.11.34}$$

Notice that the result is independent of the circle radius. Indeed it is independent of the detailed shape of the contour, as long as it encloses the z-axis.

To see this, first deform the circle as follows (see Fig. 7.11): within an interval $d\phi$, cut out the arc, pull it radially outward, rescaling it so it subtends the same angle at the center but forms a part of a circle of radius $\rho' > \rho$. Thus we first move along the old circle until we come to the modified sector, go radially outward from ρ to ρ', move along the new radius by $d\phi$, come in radially to the old radius, and continue as before. This change of contour does not change the line integral because the newly added radial segments are normal to the field (which is in the angular direction) while the increased radius on the curved section is exactly offset by the $1/\rho$ in the field strength. Once we can make such a blimp and get away with it, we can do it any number of times and deform the contour as long as it does not cross the wire.

To see this another way, consider a general closed path lying in the $x - y$ plane and surrounding the wire. Decompose each segment $\mathbf{dr}$ into a radial part $\mathbf{e}_\rho d\rho$ and an angular part $\mathbf{e}_\phi \rho d\phi$. Only the angular part contributes to the line integral with **B**:

$$\mathbf{B} \cdot \mathbf{dr} = \frac{\mu_0 I \rho d\phi}{2\pi\rho} \tag{7.11.35}$$

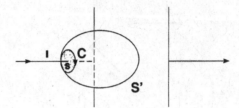

Figure 7.12. The displacement current.

$$= \frac{\mu_0 I d\phi}{2\pi} \tag{7.11.36}$$

and this is *proportional to the angle subtended at the origin*. The total line integral is then sensitive only to the fact that the contour surrounds the origin and insensitive to its detailed shape.

If several currents pierce the surface we can use the Principle of Superposition to add their fields, paying attention to the direction of flow to obtain

$$\oint \mathbf{B} \cdot \mathbf{dr} = \mu_0 \sum I, \tag{7.11.37}$$

which is called *Ampere's Law*.

Problem 7.11.5. *Can you find the field at the center of a single loop in the $x - y$ plane carrying a clockwise current I using the integral theorem above?*

What if the loop is not planar? We can still break up **dr** into its components and only the component along $\mathbf{e}_\phi$ will contribute; this contribution will again be proportional to $d\phi$.

Let us now use Stokes' Theorem on the left-hand side of Eqn. (7.11.37) and write the currents on the right hand side as the integral of the current density to obtain

$$\int_S (\nabla \times \mathbf{B}) \cdot \mathbf{dS} = \mu_0 \int_S \mathbf{j} \cdot \mathbf{dS}. \tag{7.11.38}$$

Notice that S is not a unique surface, it is any surface with the loop as its boundary.

Since the surface is arbitrary, we may equate the integrands to obtain

$$(\nabla \times \mathbf{B}) = \mu_0 \mathbf{j}. \tag{7.11.39}$$

This is however not a Maxwell equation since it is incorrect as it stands for situations which are not static. Consider an ac circuit with a capacitor, as shown in Fig. 7.12.

The current I oscillates with time. So does Q, the charge on the capacitor, oscillate as per $I = dQ/dt$. Let us consider the line integral of **B** around a contour C that lies in a plane that cuts the wire just before it meets one of the capacitor plates. This must equal the surface integral of $\nabla \times \mathbf{B} = \mu_0 \mathbf{j}$ on *any* surface with the loop as its boundary. First consider a surface S on the plane just alluded to. Since the wire passes through it, **j** is nonzero on this surface and makes a contribution that equals the line integral. Imagine now pulling the surface out of the plane (keeping the boundary fixed at the original loop) until it passes between the plates of the capacitor and becomes the surface S'. Note that still $\partial S' = C$. Now no **j** passes this surface! Yet the line integral is the line integral and does not care about what surface you plan to use in Stokes' Theorem. It follows that Eqn. (7.11.39) is incomplete, $(\nabla \times \mathbf{B})$ must equal $\mu_0 \mathbf{j}$ plus an additional term $\mu_0 \mathbf{W}$ which kicks in when you cross the region between the capacitor plates. Equation (7.11.39) has another problem: it violates the continuity equation. If we take the divergence of both sides, the left vanishes identically, while the right does not. We will first address the latter problem by enforcing the continuity equation. Let us take the divergence of both sides of

$$\nabla \times \mathbf{B} = \mu_0 \mathbf{j} + \mu_0 \mathbf{W}. \tag{7.11.40}$$

The left-hand side vanishes identically so that the right-hand side must satisfy

$$\nabla \cdot \mathbf{j} + \nabla \cdot \mathbf{W} = 0 \tag{7.11.41}$$

But we know

$$\nabla \cdot \mathbf{j} = -\frac{\partial \rho}{\partial t} \quad \text{continuity equation} \tag{7.11.42}$$

$$= -\frac{\partial \nabla \cdot \varepsilon_0 \mathbf{E}}{\partial t} \quad \text{Maxwell I} \tag{7.11.43}$$

$$= -\nabla \cdot \frac{\partial \varepsilon_0 \mathbf{E}}{\partial t} \tag{7.11.44}$$

Comparing to Eqn. (7.11.41) we learn that the choice

$$\mathbf{W} = \varepsilon_0 \frac{\partial \mathbf{E}}{\partial t} \tag{7.11.45}$$

respects the continuity equation. The added term $\mathbf{W} = \varepsilon_0 \frac{\partial \mathbf{E}}{\partial t}$ is called the *displacement current* was put in by hand by Maxwell to arrive the correct equation

$$\nabla \times \mathbf{B} = \mu_0 \mathbf{j} + \mu_0 \varepsilon_0 \frac{\partial \mathbf{E}}{\partial t} \quad \text{(Maxwell III)}. \tag{7.11.46}$$

While it is necessary that the Maxwell equation respect the continuity equation, we must also ensure that the displacement current, when integrated over S' gives the same answer as $\mathbf{j}$ when integrated over S. To verify this you must calculate $\mathbf{E}$ between the capacitor plates as a function of the accumulated charge, and relate the latter to the current by the continuity equation. The details are relegated to the following exercise.

Problem 7.11.6. *Provide the missing steps in this verification. To make life easier, choose S' so that it completely encloses the left plate and is planar and parallel to the plates between the plates. Assume $\mathbf{E}$ is perpendicular to the plates in this region and zero outside.*

The final Maxwell equation comes from Lenz's law which states that the line integral of the electric field (i.e., emf) around any closed loop equals minus the rate of change of the magnetic flux through it:

$$\oint_{C=\partial S} \mathbf{E} \cdot d\mathbf{r} = -\frac{\partial}{\partial t} \int_S \mathbf{B} \cdot d\mathbf{S}. \tag{7.11.47}$$

Using Stokes' Theorem on the left hand side, and equating integrands we get

$$\nabla \times \mathbf{E} = -\frac{\partial \mathbf{B}}{\partial t} \quad \text{(Maxwell IV)}. \tag{7.11.48}$$

Notice that taking the divergence of both sides, we run into no problems.

Let us now look at the four Maxwell equations:

$$\nabla \cdot \mathbf{E} = \frac{\rho}{\varepsilon_0} \qquad \text{Maxwell I,} \tag{7.11.49}$$

$$\nabla \cdot \mathbf{B} = 0 \qquad \text{Maxwell II,} \tag{7.11.50}$$

$$\nabla \times \mathbf{B} = \mu_0 \mathbf{j} + \mu_0 \varepsilon_0 \frac{\partial \mathbf{E}}{\partial t} \qquad \text{Maxwell III,} \tag{7.11.51}$$

$$\nabla \times \mathbf{E} = -\frac{\partial \mathbf{B}}{\partial t} \qquad \text{Maxwell IV.} \tag{7.11.52}$$

All of classical electrodynamics is contained in these equations and the Lorentz force law Eqn. (7.11.11). Our main focus has been on the fact that vector calculus is the natural language to express these equations, just as calculus is the natural language to use to describe Newton's laws.

Vector calculus also helps us solve these equations. We will look at it at two levels of detail. First you will be given some instant gratification for plodding through all this vector calculus by showing how we can unearth a remarkable consequence of Maxwell's equations. Consider them in free space, where all charges and currents vanish. Take the curl of the third equation to get

$$\nabla \times \nabla \times \mathbf{B} = \mu_0 \varepsilon_0 \nabla \times \frac{\partial \mathbf{E}}{\partial t} \tag{7.11.53}$$

$$\nabla(\nabla \cdot \mathbf{B}) - \nabla^2 \mathbf{B} = -\mu_0 \varepsilon_0 \frac{\partial^2 \mathbf{B}}{\partial t^2} \tag{7.11.54}$$

$$\mu_0 \varepsilon_0 \frac{\partial^2 \mathbf{B}}{\partial t^2} - \nabla^2 \mathbf{B} = 0, \tag{7.11.55}$$

where we have recalled Eqn. (7.10.22) and the other Maxwell equations. An identical equation exists for $\mathbf{E}$.

Problem 7.11.7. *Derive the corresponding equation for* $\mathbf{E}$.

Equation (7.11.55) has the form well-known in classical physics, called the wave equation. For example if a string of length L, the points in which are labeled by $0 \leq x \leq L$, vibrates from its equilibrium position by an amount $\psi(x, t)$ at time t, then ψ obeys the wave equation

$$\frac{1}{v^2} \frac{\partial^2 \psi}{\partial t^2} - \frac{\partial^2 \psi}{\partial x^2} = 0 \tag{7.11.56}$$

v being the velocity of propagation. Thus Maxwell was able to deduce that electromagnetic waves exist in free space and that can travel with a velocity

$$v = \frac{1}{\sqrt{\mu_0 \varepsilon_0}}. \tag{7.11.57}$$

Great excitement followed when the numbers were put in for μ_0 and ε_0 and one obtained

$$v = \frac{1}{\sqrt{8.85419 \cdot 10^{-12} \cdot 4\pi \cdot 10^{-7}}} = 2.997925 \cdot 10^8 \text{ m/s} \tag{7.11.58}$$

the velocity of light! Thus light and electromagnetic waves were probably the same! (This was confirmed soon.) This is one of the finest illustrations of how, when we condense our experiments into equations and go on to solve the latter, some *physical* predictions emerge from the *mathematics*. In other words, once we write down the Maxwell equations, by fiddling with charges and wires etc., it is a purely mathematical consequence (of vector calculus, which you have just learned) that there exist in this theory waves traveling at the speed of light.

We now return for a second look at the solution of Maxwell's equations to find further applications of vector calculus. Look at number II. We know the divergence of the curl is identically zero. The converse is also true in three dimensional space (with no points excluded etc.): if the divergence of a field $\mathbf{B}$ is zero everywhere, we can find a vector $\mathbf{A}$ such that

$$\mathbf{B} = \nabla \times \mathbf{A}. \tag{7.11.59}$$

In physics literature $\mathbf{A}$ is called the *vector potential*. The above equation has the same content as Maxwell II. If we feed this into Maxwell IV, we get

$$\nabla \times (\mathbf{E} + \frac{\partial \mathbf{A}}{\partial t}) = 0. \tag{7.11.60}$$

Now we use the converse of the other identity, that curl of a gradient is identically zero, to write

$$\mathbf{E} + \frac{\partial \mathbf{A}}{\partial t} = -\nabla\phi \tag{7.11.61}$$

where ϕ is called the *scalar potential* and the minus sign in the right-hand side is conventional. The above equation is more commonly written as

$$\mathbf{E} = -\nabla\phi - \frac{\partial \mathbf{A}}{\partial t}. \tag{7.11.62}$$

It is obvious that $\mathbf{B}$ and $\mathbf{E}$ given by functions of the form Eqn. (7.11.59, 7.11.61) satisfy Maxwell II and IV identically, but the point is that this is the most general solution to these equations. So the idea is to say the electric and magnetic fields are derivatives of scalar and vector potentials as prescribed above (which takes care of Maxwell II and IV) and then go on to determine the potentials using the other two Maxwell equations.

In looking for the potentials, we must be aware that they are not unique. That is to say, you can walk into a room and say this or that is the electric or magnetic field at a point (by watching the response of test charges and invoking the Lorentz force equation) but you cannot say this or that is the potential. The reason is that if $\mathbf{A}$ leads to a certain $\mathbf{B}$, so does $\mathbf{A} + \nabla\Lambda$ for any Λ, since the curl operation will kill the Λ term identically. Likewise the scalar potential is also not unique. In general the electric and magnetic fields in Eqns. (7.11.59,7.11.61) are insensitive to the change

$$\mathbf{A} \quad \rightarrow \quad \mathbf{A} + \nabla\Lambda \tag{7.11.63}$$

$$\phi \quad \rightarrow \quad \phi - \frac{\partial\Lambda}{\partial t} \tag{7.11.64}$$

called a *gauge transformation*. It is beyond the scope of this book to show that using this freedom, the potentials can always be chosen (whatever be the fields) to obey the *Lorentz gauge condition*:

$$\nabla \cdot \mathbf{A} + \mu_0\varepsilon_0 \frac{\partial\phi}{\partial t} = 0. \tag{7.11.65}$$

Problem 7.11.8. *Show that even with this restriction the potentials are not unique, we can still perform the gauge transformation with Λ's obeying*

$$\nabla^2\Lambda + \mu_0\varepsilon_0 \frac{\partial^2\Lambda}{\partial t^2} = 0 \tag{7.11.66}$$

Let us now solve Maxwell I and III in free space, away from charges. Writing the fields in terms of the potentials, we get from these two

$$\nabla \cdot (-\nabla\phi - \frac{\partial\mathbf{A}}{\partial t}) \quad = \quad 0 \tag{7.11.67}$$

$$\nabla \times \nabla \times \mathbf{A} \quad = \quad \mu_0\varepsilon_0 \frac{\partial}{\partial t}(-\nabla\phi - \frac{\partial\mathbf{A}}{\partial t}) \tag{7.11.68}$$

If we now use the Lorentz gauge conditions and recall the identity Eqn. (7.10.22)

$$\nabla \times \nabla \times \mathbf{A} = \nabla(\nabla \cdot \mathbf{A}) - \nabla^2\mathbf{A} \tag{7.11.69}$$

we get two beautiful equations for the potentials:

$$\mu_0\varepsilon_0 \frac{\partial^2\phi}{\partial t^2} - \nabla^2\phi \quad = \quad 0. \tag{7.11.70}$$

$$\mu_0\varepsilon_0 \frac{\partial^2\mathbf{A}}{\partial t^2} - \nabla^2\mathbf{A} \quad = \quad 0. \tag{7.11.71}$$

Problem 7.11.9. *Derive the equations for the potentials given above.*

As stated earlier, the aim of this chapter has not been so much to teach you electrodynamics, as to show you that the language of vector calculus you are learning here is the one chosen by nature to express some of her most basic laws.

7.12. Summary

The highlights of this chapter are as follows.

- Know the dot product, its definition in terms of components in an orthonormal basis and its properties such as linearity. Ditto for the cross product.

- If $\mathbf{r}(t)$ is a function of time then know that

$$\mathbf{r} = \mathbf{i}x(t) + \mathbf{j}y(t) = \mathbf{e}_r r$$

$$\frac{d\mathbf{r}}{dt} = \mathbf{i}\dot{x}(t) + \mathbf{j}\dot{y}(t) = \mathbf{e}_r \dot{r} + r\omega \mathbf{e}_\theta,$$

which comes from using

$$d\mathbf{e}_r/dt = -\dot{\theta}\mathbf{e}_\theta = -\omega\mathbf{e}_\theta,$$

Using a similar formula for $d\mathbf{e}_\theta/dt$ one can get a formula for the acceleration.

- The line integral of a vector field $\mathbf{F}(\mathbf{r})$ is defined as follows:

$$\int_1^2 \mathbf{F}(\mathbf{r}) \cdot d\mathbf{r} = \lim_{n \to \infty} \sum_{i=1}^n \mathbf{F}(\mathbf{r}_i) \cdot d\mathbf{r}_i$$

(where 1 and 2 are short for $\mathbf{r}_1$ and $\mathbf{r}_2$ respectively) and is found by parametrizing the path. The answer in general depends on the path and not just the end points. If the answer is path independent, the field is conservative. Equivalently

$$\oint \mathbf{F} \cdot d\mathbf{r} = 0$$

for a conservative field.

- The surface integral of a vector field $\mathbf{V}(\mathbf{r})$ over a surface S is defined as

$$\int_S \mathbf{V} \cdot d\mathbf{S} = \lim_{i \to \infty} \sum_i \mathbf{V}(\mathbf{r}_i) \cdot d\mathbf{S}_i,$$

where the infinitesimal areas $d\mathbf{S}_i$ have a magnitude equal to size of areas and direction perpendicular to plane of areas. (The sign is determined by the sense of arrows running around the edges.)

- The gradient $\nabla\phi$ enters as follows:

$$d\phi = \frac{\partial\phi}{\partial x}dx + \frac{\partial\phi}{\partial y}dy + \frac{\partial\phi}{\partial z}dz \equiv \nabla\phi \cdot \mathbf{dr} \text{ where}$$

$$\nabla\phi = \mathbf{i}\frac{\partial\phi}{\partial x} + \mathbf{j}\frac{\partial\phi}{\partial y} + \mathbf{k}\frac{\partial\phi}{\partial z}, \text{ and}$$

$$\mathbf{dr} = \mathbf{i}dx + \mathbf{j}dy + \mathbf{k}dz.$$

Since $d\phi = \nabla\phi \cdot \mathbf{dr} = |\nabla\phi||\mathbf{dr}|\cos\theta$, the gradient gives the direction of greatest rate of change and equals in magnitude that rate of change.

- The integral of the gradient is path independent:

$$\int_1^2 \nabla\phi \cdot \mathbf{dr} = \phi(2) - \phi(1) = \phi|_{\partial P},$$

(where 1 and 2 are short for r_1 and r_2, respectively).

- In non-cartesian coordinates

$$\nabla\phi = \sum_i \mathbf{e}_i \frac{1}{h_i}\frac{\partial\phi}{\partial u_i}$$

- Green's Theorem says if $\mathbf{W}$ and the loop C lie in a plane (say the $x-y$ plane),

$$\oint_{C=\partial S} \mathbf{W} \cdot \mathbf{dr} = \int\int_S \left(\frac{\partial W_y}{\partial x} - \frac{\partial W_x}{\partial y}\right)dxdy$$

-

$$\mathbf{W} \text{ is conservative } \rightarrow \left(\frac{\partial W_y}{\partial x} - \frac{\partial W_x}{\partial y}\right) = 0.$$

- The curl is given by

$$\nabla \times \mathbf{W} = \mathbf{i}\left(\frac{\partial W_z}{\partial y} - \frac{\partial W_y}{\partial z}\right) + \mathbf{j}\left(\frac{\partial W_x}{\partial z} - \frac{\partial W_z}{\partial x}\right) + \mathbf{k}\left(\frac{\partial W_y}{\partial x} - \frac{\partial W_x}{\partial y}\right).$$

If a field is conservative, its curl vanishes everywhere and vice versa. In general coordinates the curl has a more complicated expression. Look it up when needed.

- Stokes' Theorem:

$$\oint_{C=\partial S} \mathbf{W} \cdot \mathbf{dr} = \int_S \nabla \times \mathbf{W} \cdot \mathbf{dS}$$

- Gauss's Law:

$$\oint_{S=\partial V} \mathbf{W} \cdot d\mathbf{S} = \int_V \boldsymbol{\nabla} \cdot \mathbf{W} dx dy dz,$$

where $\boldsymbol{\nabla} \cdot \mathbf{W} = \frac{\partial W_x}{\partial x} + \frac{\partial W_y}{\partial y} + \frac{\partial W_z}{\partial z}$ is the divergence of $\mathbf{W}$.

- Remember the identities

$$\boldsymbol{\nabla} \times \boldsymbol{\nabla}\phi \equiv 0$$

$$\boldsymbol{\nabla} \cdot \boldsymbol{\nabla} \times \mathbf{W} \equiv 0,$$

which come from the fact that the boundary of a boundary is zero or that mixed partial derivatives are independent of the order of differentiation.

- The Laplacian ∇^2 is defined as follows:

$$\boldsymbol{\nabla} \cdot \boldsymbol{\nabla}\phi = \frac{\partial^2 \phi}{\partial x^2} + \frac{\partial^2 \phi}{\partial y^2} + \frac{\partial^2 \phi}{\partial z^2}$$

$$\equiv \nabla^2 \phi.$$

In general coordinates it has a more complicated expression. Look it up when needed.

MATRICES AND DETERMINANTS

8.1. Introduction

Consider a family with parents named P_1 and P_2 who have very imaginatively named their two children C_1 and C_2. Let us say each month the children get an allowance, also denoted by the symbols C_1 and C_2 which are related to the parents' income (likewise denoted by P_1 and P_2) as follows:

$$C_1 = \frac{1}{10}P_1 + \frac{1}{6}P_2 \tag{8.1.1}$$

$$C_2 = \frac{1}{9}P_1 + \frac{1}{8}P_2. \tag{8.1.2}$$

In other words the father gives $\frac{1}{10}$ of his income to the son and the mother gives $\frac{1}{6}$ of hers to the son. The daughter similarly receives $\frac{1}{9}$ and $\frac{1}{8}$ respectively from her father and mother.

Let us assume that no matter what the incomes of the parents, they will always contribute these *fractions*. In that case we would like to store the unchanging fractions $\frac{1}{10}, \frac{1}{6}, \frac{1}{9}, \frac{1}{8}$ in some form. It is logical to store them in an array or *matrix*

$$M = \begin{bmatrix} \frac{1}{10} & \frac{1}{6} \\ \frac{1}{9} & \frac{1}{8} \end{bmatrix}. \tag{8.1.3}$$

The logic is as follows. The contributions naturally fall into two sets: what the son gets and what the daughter gets. This makes the two rows. In each row, there are again two sets: what the father gives and what the mother gives. These make the two columns. In the general case we can write the parental contributions as

$$M = \begin{bmatrix} M_{11} & M_{12} \\ M_{21} & M_{22} \end{bmatrix} \tag{8.1.4}$$

The *matrix element* M_{ij} resides in row i and column j and stands for the contribution to child i from parent j. We can then write in the general case

$$C_1 = M_{11}P_1 + M_{12}P_2 \tag{8.1.5}$$

$$C_2 = M_{21}P_1 + M_{22}P_2. \tag{8.1.6}$$

Let us generalize this notion of a matrix to an array with m rows and n columns. A matrix with m rows and n columns will be referred to as an m *by* n or $m \times n$ *matrix*. Let us introduce two *2 by 1* matrices:

$$C = \begin{bmatrix} C_1 \\ C_2 \end{bmatrix} \qquad (8.1.7)$$

and

$$P = \begin{bmatrix} P_1 \\ P_2 \end{bmatrix} \qquad (8.1.8)$$

A matrix with $m = n$ is called a *square matrix*, a matrix with $n = 1$ is a called a *column vector*, and a matrix with $m = 1$ is called a *row vector*. *We will be dealing only with these three types of matrices. In addition, with a few exceptions, the name matrix will be used to mean a square matrix, while column and row vectors will be referred to as such.*

We next introduce the notion of adding and multiplying matrices. In the world of pure mathematics one can make up rules as one wants, as long as they are consistent. However some definitions prove to be more useful than others, as is true of the ones that follow.

Definition 8.1. *If M and N are two matrices of the same dimensions (same number of rows and columns) their sum $T = M + N$ has entries $T_{ij} = M_{ij} + N_{ij}$.*

Definition 8.2. *If a is a number, aM is defined as a matrix with entries $(aM)_{ij} = aM_{ij}$.*

Definition 8.3. *If M is an m by n matrix and N is an n by r matrix, their product MN is an m by r matrix with entries*

$$(MN)_{ij} = \sum_{r=1}^{n} M_{ir} N_{rj} \qquad (8.1.9)$$

Whereas the sum is easy to remember, the product takes some practice.

To get the ij matrix element of the product you must take the entries of the i-th row of the first matrix, interpret them as components of a vector, and take the dot product with the j-th column of the second matrix, similarly interpreted.

The product is not defined if the number of columns of the first matrix do not equal the number of rows of the second. Likewise the sum is defined only for matrices of the same dimension. Here are a few examples.

$$\begin{bmatrix} 1 & 2 \\ 3 & 4 \end{bmatrix} \begin{bmatrix} 4 & 2 \\ 1 & 3 \end{bmatrix} = \begin{bmatrix} 6 & 8 \\ 16 & 18 \end{bmatrix} \qquad (8.1.10)$$

$$\begin{bmatrix} 11 & 13 & 15 \\ 17 & 19 & 21 \\ 23 & 25 & 27 \end{bmatrix} = \begin{bmatrix} 1 & 2 & 3 \\ 4 & 5 & 6 \\ 7 & 8 & 9 \end{bmatrix} + \begin{bmatrix} 10 & 11 & 12 \\ 13 & 14 & 15 \\ 16 & 17 & 18 \end{bmatrix} \qquad (8.1.11)$$

$$\begin{bmatrix} 1 & 2 \\ 3 & 4 \end{bmatrix} \begin{bmatrix} 4 \\ 1 \end{bmatrix} = \begin{bmatrix} 6 \\ 16 \end{bmatrix}. \qquad (8.1.12)$$

where, for example, the number 6 which is in the $(1,1)$ place of the 2 by 2 matrix in Eqn. (8.1.10) is obtained as follows: $6 = 1 \cdot 4 + 2 \cdot 1$. Likewise the 16 in the $(2, 1)$ place is obtained as $16 = 3 \cdot 4 + 4 \cdot 1$. You should check the other entries similarly.

We are now ready to write Eqn. (8.1.6) as a matrix equation:

$$\begin{bmatrix} C_1 \\ C_2 \end{bmatrix} = \begin{bmatrix} M_{11} & M_{12} \\ M_{21} & M_{22} \end{bmatrix} \begin{bmatrix} P_1 \\ P_2 \end{bmatrix} \qquad (8.1.13)$$

or more compactly as

$$C = MP \qquad (8.1.14)$$

where C, M, and P are matrices whose entries were defined above.

When you first run into the rule for matrix multiplication you might ask how natural it is, or how anyone would arrive at it. Would it not be more natural to define the ij element of the product as the product of the ij elements of the two factors? Such a definition, while allowed, does not have the same range of applicability as the one quoted above. For example our ability to write the income relations compactly as $C = MP$ (Eqn. (8.1.14)) is predicated on using the present multiplication rule. For another illustration of its virtue, consider the situation where the parents in our example themselves get *all their* income from the grandparents according to

$$P = NG \qquad (8.1.15)$$

where G is the (column vector representing the) grandparents' income and N is their matrix of contributions. From their income so obtained the parents give to their children as per the rule $C = MP$. How do we relate C to G directly? Very simple:

$$\begin{aligned} C &= MP & (8.1.16) \\ &= MNG. & (8.1.17) \end{aligned}$$

where the product MN is evaluated by the rule stated. If we go back to the explicit form such as Eqn. (8.1.6) and express P_1, P_2 in terms of G_1, G_2 and compared the relation to C_1 and C_2 we will find the result agrees with the above. For example

$$C_1 = M_{11}P_1 + M_{12}P_2 \qquad (8.1.18)$$

$$= M_{11}[N_{11}G_1 + N_{12}G_2] + M_{12}[N_{21}G_1 + N_{22}G_2] \qquad (8.1.19)$$
$$= [M_{11}N_{11} + M_{12}N_{21}]G_1 + [M_{11}N_{12} + M_{12}N_{22}]G_2 \qquad (8.1.20)$$
$$= (MN)_{11}G_1 + (MN)_{12}G_2, \qquad (8.1.21)$$

If we think of the parents' income as the input and the children's income as the output, the matrix M relates the two. As the input changes, that is, as the parents' income changes the output changes, but the matrix relating the two stays the same. We also see that the input to this problem, P, can itself be the output of another, i.e., $P = NG$. In this case matrix multiplication provides the matrix for the direct conversion for the two-stage problem, in our example from the grandparents to the children.

Now for yet another example. Consider a point in a plane with coordinates (x_1, x_2) with respect to the usual cartesian axes. Suppose we use a different set of axes, related to the first by a counter-clockwise rotation θ. It is easy to show from elementary trigonometry that the same point will have coordinates (x'_1, x'_2) given by

$$x'_1 = x_1 \cos\theta + x_2 \sin\theta \qquad (8.1.22)$$
$$x'_2 = -x_1 \sin\theta + x_2 \cos\theta \qquad (8.1.23)$$

which we may write compactly as

$$X' = R_\theta X \qquad (8.1.24)$$

in obvious notation, where the *rotation matrix* R_θ is

$$R_\theta = \left[\begin{array}{cc} \cos\theta & \sin\theta \\ -\sin\theta & \cos\theta \end{array} \right]. \qquad (8.1.25)$$

Once again we may think of X and X' as input and output, related by R_θ. The point of the matrix notation is this: under one and the same physical process, namely the rotation of axes, each point in the plane gets new coordinates. The matrix R_θ does not care which point you are interested in, it characterizes the rotation once and for all. Likewise the income the children get varies with the parents' income, but the matrix M characterizes, once and for all, the invariant aspect of their contributions.

Instead of saying that the axes were rotated, we could say the vector with components (x_1, x_2) was rotated *clockwise* by θ into the vector with components (x'_1, x'_2). This is called the *active transformation* as compared to the above, which is called the *passive transformation*. In the active case, the axes are fixed and the vectors are rotated, while in the passive case the vectors are fixed and the axes are rotated in the opposite sense.

In either case, consider a sequence of transformation by angles θ followed by θ' generated by matrices R_θ and $R_{\theta'}$. If X'' is the final result, it is related to X

by the matrix $R = R_{\theta'}R_\theta$. Now you know that the result should be a rotation by $\theta + \theta'$. You should verify that $R_{\theta'}R_\theta$ indeed equals $R_{\theta+\theta'}$.

Problem 8.1.1. *Verify that $R_{\theta+\theta'} = R_{\theta'}R_\theta$.*

Let us embed these two examples (income and rotations) in a more general framework. They are both *linear transformations*: i.e., they involve linear relations between one set of variables and another. For example the new coordinates are given as linear combinations of the old ones; there are no higher or lower powers in sight. Now if a set of variables Z is expressed linearly in terms of a set Y and the latter itself is now written linearly in terms of a set X, it follows that if we take the first linear relation and replace every Y there by a linear combination of X's, the result will be a linear relation between Z and X. *Thus a sequence of two linear transformations is itself a linear transformation.* Now we can assemble the parameters of each linear transformations in the form of a matrix. *The matrix multiplication rule we use has the property that the matrix corresponding to a sequence of transformations is given by the product of the matrices corresponding to the individual transformations.*

Problem 8.1.2. *Recall from Problem(1.6.4.) in Chapter 1 that the relativistic transformation of coordinates when we go from one frame of reference to another is*

$$x' = x\cosh\theta - t\sinh\theta \qquad (8.1.26)$$
$$t' = -x\sinh\theta + t\cosh\theta, \qquad (8.1.27)$$

where θ is the rapidity difference between the two frames. Write this in matrix form. Say we go to a third frame with coordinate x'', t'', moving with rapidity θ' with respect to the one with primed coordinates. Show that the matrix relating the doubly primed coordinates to the unprimed ones corresponds to rapidity $\theta + \theta'$.

We now learn some definitions in matrix theory. The *zero matrix* is like the number 0: adding it to any matrix makes no difference. We denote it by the symbol 0 and it clearly has all its entries equal to 0:

$$0 = \begin{bmatrix} 0 & 0 & \cdots & 0 \\ 0 & 0 & \cdots & 0 \\ \vdots & \vdots & \vdots & \vdots \\ 0 & 0 & \cdots & 0 \\ 0 & 0 & \cdots & 0 \end{bmatrix} \qquad (8.1.28)$$

The *unit matrix* I is like the number 1: multiplying by I makes no difference:

$$IM = M \text{ for all } M. \qquad (8.1.29)$$

Its *diagonal* elements (I_{ii}) are unity and the rest zero:

$$I = \begin{bmatrix} 1 & 0 & \cdots & 0 \\ 0 & 1 & \cdots & 0 \\ \vdots & \vdots & \vdots & \vdots \\ 0 & 0 & \cdots & 0 \\ 0 & 0 & \cdots & 1 \end{bmatrix}. \tag{8.1.30}$$

Problem 8.1.3. *Verify that I has the advertised property by using Eqn. (8.1.9).*

How about multiplication by the matrix

$$aI \equiv \begin{bmatrix} a & 0 & \cdots & 0 \\ 0 & a & \cdots & 0 \\ \vdots & \vdots & \vdots & \vdots \\ 0 & 0 & \cdots & 0 \\ 0 & 0 & \cdots & a \end{bmatrix}? \tag{8.1.31}$$

You may show that this matrix rescales every entry of the matrix it multiplies by a. We will therefore write aIM as simply aM, since this is how we defined multiplication of a matrix by a number.

Matrix multiplication has a feature that is not shared by ordinary multiplication of numbers: *it is generally noncommutative*. This means that in general

$$MN \neq NM. \tag{8.1.32}$$

Take for example the matrix

$$P_1 = \begin{bmatrix} 1 & 0 \\ 0 & 0 \end{bmatrix} \tag{8.1.33}$$

which leaves alone the x_1- component of the column vector it multiplies and kills the other component x_2 (check this) and the matrix R_θ which rotates the vector by an angle θ. (Since P_1 projects out the part of the vector in the 1-direction, it is called the *projection operator* along direction 1.) First convince yourself (without actual computation) that the product of these matrices depends on the order of the factors by considering the fate of a vector with $x_1 = 0$ and $x_2 = 1$ when these two matrices multiply it in the two possible orders. Now multiply the matrices in both orders and note the difference.

Problem 8.1.4. *Perform the calculation listed in the preceding discussion involving R_θ and P_1.*

Noncommutativity makes it sensible to define the *commutator*

$$[M, N] \equiv MN - NM. \tag{8.1.34}$$

Of course it is possible that in some cases the matrices *commute*, i.e., their commutator is zero, as in the case of two rotations R_θ and $R_{\theta'}$. Do you see why this is so?

Problem 8.1.5. *Show that the unit matrix commutes with all matrices.*

Products of matrices differ from products of ordinary numbers in other remarkable ways. For example

$$MN = 0 \quad \text{does not imply} \quad M = 0 \text{ or } N = 0. \tag{8.1.35}$$

Problem 8.1.6. *Consider as an example the matrices P_1 and P_2 which project out vectors in the 1 and 2 directions, respectively. Convince yourself (without calculation) that the product of these two will kill any vector. Next write out the two matrices and show that their product indeed is zero.*

Even more bizarre is the case where

$$M^2 = 0 \quad M \neq 0, \tag{8.1.36}$$

an example of which is

$$\begin{bmatrix} 0 & 1 \\ 0 & 0 \end{bmatrix}. \tag{8.1.37}$$

Problem 8.1.7. *For the pairs M and N given below, find $M + N$, M^2, MN and $[M, N]$.*

$$M = \begin{bmatrix} 1 & 2 \\ 3 & 4 \end{bmatrix} \quad N = \begin{bmatrix} 5 & 6 \\ 7 & 8 \end{bmatrix}$$

8.2. Matrix Inverses

Consider a linear transformation such as

$$C = MP \tag{8.2.1}$$

relating the children's income to the parents'. Suppose we want to express the parents' income in terms of the children's. If C, M, and P were just ordinary numbers, we could divide both sides by M and write a formula for P in terms of C. How are we to do it now? Here is a related problem where the same question comes up. Say we want to solve a pair of simultaneous equations

$$4x_1 - 3x_2 = 14 \qquad (8.2.2)$$
$$x_1 - 6x_2 = -7 \qquad (8.2.3)$$

We can proceed as in our childhood, eliminate, say, x_2 in favor of x_1 by using the second equation, plug that into the first and find that $x_1 = 5$ and $x_2 = 2$. But here is an alternative using matrices. First we cast the two equations in matrix form as

$$MX = C \qquad (8.2.4)$$

where

$$M = \begin{bmatrix} 4 & -3 \\ 1 & -6 \end{bmatrix} \qquad (8.2.5)$$

$$X = \begin{bmatrix} x_1 \\ x_2 \end{bmatrix} \qquad C = \begin{bmatrix} 14 \\ -7 \end{bmatrix} \qquad (8.2.6)$$

The goal here is to find X in Eqn. (8.2.4). Once again if M were a number we would simply divide both sides by M. We have not however defined what it means to divide by a matrix. We do so now. Recall that in the case of ordinary numbers dividing by 4 is the same as multiplying by $1/4$, which is the reciprocal or inverse of 4. The inverse $1/4$ has the property that its product with 4 gives the number 1 which is the unique number with the property that multiplying by it makes no difference. We likewise define M^{-1}, the *inverse of a matrix* M, as matrix with the property

$$M^{-1}M = I, \qquad (8.2.7)$$

where I is the unit matrix which leaves all matrices invariant upon matrix multiplication.

Postponing for the moment the question of how M^{-1} is to be found, let us assume we know it. Then we can solve the simultaneous equations as follows:

$$MX = C \qquad (8.2.8)$$
$$M^{-1}MX = IX = M^{-1}C \qquad (8.2.9)$$
$$X = M^{-1}C. \qquad (8.2.10)$$

Now for the question of how M^{-1} is to be determined.

Let us start with the 2×2 case and then move to bigger things later. I state that for any 2×2 matrix M

$$M^{-1} = \frac{1}{|M|} \begin{bmatrix} M_{22} & -M_{12} \\ -M_{21} & M_{11} \end{bmatrix} \qquad (8.2.11)$$

where the *determinant of the matrix* $|M|$ is given by

$$|M| = M_{11}M_{22} - M_{12}M_{21}. \tag{8.2.12}$$

Thus

$$\begin{bmatrix} 4 & -3 \\ 1 & -6 \end{bmatrix}^{-1} = \frac{-1}{21} \begin{bmatrix} -6 & 3 \\ -1 & 4 \end{bmatrix} \tag{8.2.13}$$

and the solution to Eqns. (8.2.2-8.2.3) is

$$X = \begin{bmatrix} x_1 \\ x_2 \end{bmatrix} = \frac{-1}{21} \begin{bmatrix} -6 & 3 \\ -1 & 4 \end{bmatrix} \begin{bmatrix} 14 \\ -7 \end{bmatrix} = \begin{bmatrix} 5 \\ 2 \end{bmatrix}. \tag{8.2.14}$$

Problem 8.2.1. *Verify that the inverse given in Eqn. (8.2.11) satisfies $M^{-1}M = I$.*

Just as the number 0 has no inverse, some matrices will have no inverses. *It is clear from the formula that the inverse matrix will not exist if the determinant vanishes.* You may verify that this is the case for the projection operators.

Problem 8.2.2. *Find the inverse of the rotation matrix R_θ. Give arguments for why this had to be the answer. Show that the determinant for the rotation matrix will never vanish by explicit computation. Argue that this had to be so since every rotation must have an inverse. What will that inverse do in general?*

Theorem 8.1. *A matrix and its inverse will always commute.*

Proof: We first show that just as in the case of numbers, the inverse of the inverse is the matrix itself:

$$(M^{-1})^{-1} = M. \tag{8.2.15}$$

Let us say $(M^{-1})^{-1} = G$. Then by definition, Eqn. (8.2.7) tells us

$$GM^{-1} = I. \tag{8.2.16}$$

Postmultiplying both sides by M and using $M^{-1}M = I$, we find that indeed $G = M$. Thus

$$M^{-1}M = I = MM^{-1}. \ \blacksquare \tag{8.2.17}$$

Since $M^{-1}M = I$ we see that the inverse of a matrix undoes whatever the matrix does. This is why $R_\theta^{-1} = R_{-\theta}$. The reason P_1 has no inverse is that it kills every vector that lies along the 2-direction and no finite operator can undo this carnage.
    More generally if a matrix kills any nonzero vector, it cannot have an inverse for exactly this reason. We can see this another way. Suppose a matrix M takes a vector X to X'. Then its inverse, if it exists, will take X' back to X in order to satisfy $M^{-1}X' = M^{-1}MX = I \cdot X = X$. Suppose now that M kills a nonzero vector Y. Then $M(X + Y) = MX + MY = MX = X'$. Thus both X and $X + Y$ end up as X'. If there were an M^{-1} it would not know what to do acting

on X': is it to give X or $X + Y$? The answer is of course that M^{-1} does not exist in this case. Projection operators have no inverses since they kill vectors in the direction perpendicular to the projection.

Problem 8.2.3. *Show by computing a matrix inverse that the solution to*

$$x_1 + 2x_2 = 9 \quad 3x_1 + 4x_2 = 23 \tag{8.2.18}$$

is $x_1 = 5$, $x_2 = 2$.

Problem 8.2.4. *Find the inverse of the Lorentz Transformation matrix from Problem (8.1.2.) and the rotation matrix R_θ. Does the answer make sense? (You must be on top of the identities for hyperbolic and trigonometric function to do this. Remember: when in trouble go back to the definitions in terms of exponentials.)*

Problem 8.2.5. *Stare at the equations*

$$2x + 3y = 5$$

$$4x + 6y = 10$$

and argue that a unique solution cannot be found given this information. Double check by showing the appropriate determinant vanishes.

Problem 8.2.6. *Go back to the income problem and consider the case where each child receives from each parent the same fraction of the parental income. Thus if the mother gives 5% to the son and 6% to the daughter so does the father. Argue that in this case the money received by either child depends only on the* total *parental income. Argue on this basis that it must not be possible to find the individual parental incomes from the children's incomes. Verify by computing the appropriate determinant.*

Theorem 8.2. *The inverse of the product is the product of the inverses in reverse.*

$$(NM)^{-1} = M^{-1}N^{-1}. \tag{8.2.19}$$

Proof:

$$M^{-1}N^{-1}NM \; = \; M^{-1}IM. \tag{8.2.20}$$

$$= \; I \; \blacksquare \tag{8.2.21}$$

Problem 8.2.7. *Verify this for a product of three operators.*

We now address the question of inverses for bigger matrices. There is a procedure for that. This however requires some knowledge of determinants of bigger matrices. To this end we digress to study this subject.

8.3.   Determinants

Prior to introducing determinants of larger matrices let us summarize some properties of the 2×2 determinant:

$$|MN| = |M||N| \qquad (8.3.1)$$
$$|M^T| = |M| \qquad (8.3.2)$$
$$|M_{ex}| = -|M| \qquad (8.3.3)$$
$$|M_a| = a|M| \qquad (8.3.4)$$

where M_{ex} is related to M by exchanging any two rows or any two columns, M_a has one of the rows or columns rescaled by a factor a and M^T is the *matrix transpose* defined as follows:

Definition 8.4. *The transpose of a matrix M, denoted by M^T has elements*

$$M_{ij}^T = M_{ji}, \qquad (8.3.5)$$

i.e., the transpose matrix M^T is obtained by exchanging the rows and column of M.

The last two equations (8.3.3-8.3.4) imply that *if two rows or columns of a matrix are proportional the determinant vanishes.* The first implies

$$|M^{-1}| = \frac{1}{|M|}. \qquad (8.3.6)$$

Problem 8.3.1. *Verify all the above mentioned properties of determinants by explicit computation on 2×2 matrices M and N.*

Problem 8.3.2. *We saw that $|M| = 0$ if one row is a times another. What does this mean in the study of simultaneous equations?*

We now seek a generalization of the determinant to bigger matrices which has all the above features.

We will start with the 3×3 case which will have the greatest use for you. At the end the general case will be discussed and it will have no qualitative differences from this one.

Let

$$M = \begin{bmatrix} M_{11} & M_{12} & M_{13} \\ M_{21} & M_{22} & M_{23} \\ M_{31} & M_{32} & M_{33}. \end{bmatrix} \qquad (8.3.7)$$

Definition 8.5. *The* **cofactor matrix** M_C *has elements*

$$(M_C)_{ij} = (-1)^{i+j} determinant\ of\ matrix\ with\ row\ i\ and\ column\ j\ deleted$$
$$(8.3.8)$$

Notice that the cofactor for the 3×3 case involves only 2×2 determinants which we know how to evaluate.

Definition 8.6. *The determinant of a* 3×3 *matrix* M *is*

$$|M| = M_{11}(M_C)_{11} + M_{12}(M_C)_{12} + M_{13}(M_C)_{13}. \qquad (8.3.9)$$

We will state without proof that this definition preserves the properties of the 2×2 determinant that were highlighted. There is one other feature we cannot see in the 2×2 case: *The* 3×3 *determinant is unchanged by a cyclic permutation of its rows or its columns.* Thus if row 1 goes into row 2 and 2 goes to 3 and 3 goes to 1, nothing happens. Thus is because such a cyclic permutation can be effected by an even number of row exchanges and this brings about an even number of sign changes.

Consider as an example of the determinant formula above the case of

$$A = \begin{bmatrix} a & b & c \\ d & e & f \\ g & h & i \end{bmatrix} \qquad (8.3.10)$$

In this case

$$\begin{aligned} |A| &= (-1)^{1+1}a(ei - fh) + (-1)^{1+2}b(di - fg) + (-1)^{1+3}c(dh - eg). \\ &= a(ei - fh) - b(di - fg) + c(dh - eg). \end{aligned} \qquad (8.3.11)$$

So we basically start with a plus sign for the $(1, 1)$ element and alternate the sign as we go along.

Now it happens that you can also focus on the second row and the write the determinant as the sum over the product of each entry in that row times the corresponding cofactor. You can also do it with the entries in a column. But we will be content with the one method described above. The theory of determinants is very beautiful and elaborate. We are only interested in them here in order to obtain inverses. Here is the answer:

$$M^{-1} = \frac{M_C^T}{|M|}. \qquad (8.3.12)$$

Problem 8.3.3. *Apply the notion of the cofactor to a* 2×2 *matrix and get back the definition given in Eqn. (8.2.12). Interpret the determinant of a* 1×1 *matrix as just the number.)*

Problem 8.3.4. *Solve the following simultaneous equations by matrix inversion:*

$$3x - y - z = 2$$
$$x - 2y - 3z = 0$$
$$4x + y + 2z = 4$$

and

$$3x + y + 2z = 3$$
$$2x - 3y - z = -2$$
$$x + y + z = 1.$$

Problem 8.3.5. *For the matrix*

$$\begin{bmatrix} 1 & 2 & 3 \\ 4 & 5 & 6 \\ 7 & 8 & 10 \end{bmatrix} \tag{8.3.13}$$

find the cofactor matrix and the inverse. Verify that your inverse does the job.

Problem 8.3.6. *Show that*

$$\begin{bmatrix} 2 & 1 & 3 \\ 0 & 1 & 2 \\ -1 & 1 & 1 \end{bmatrix}^{-1} = \begin{bmatrix} 1 & -2 & 1 \\ 2 & -5 & 4 \\ -1 & 3 & -2 \end{bmatrix}$$

$$\begin{bmatrix} 2 & 1 & 3 \\ 4 & 1 & 2 \\ 0 & -1 & 2 \end{bmatrix}^{-1} = \frac{1}{12} \begin{bmatrix} -4 & 5 & 1 \\ 8 & -4 & -8 \\ 4 & -2 & 2 \end{bmatrix}$$

Problem 8.3.7. *Show that the cross product* $\mathbf{A} \times \mathbf{B}$ *can be formally written as a determinant as follows:*

$$\mathbf{A} \times \mathbf{B} = \begin{vmatrix} \mathbf{i} & \mathbf{j} & \mathbf{k} \\ A_x & A_y & A_z \\ B_x & B_y & B_z \end{vmatrix}. \tag{8.3.14}$$

I say "formally" since this determinant has vector entries and is itself equal to a vector and is not a number.
    Show likewise that

$$\nabla \times \mathbf{V} = \begin{vmatrix} \mathbf{i} & \mathbf{j} & \mathbf{k} \\ \frac{\partial}{\partial x} & \frac{\partial}{\partial y} & \frac{\partial}{\partial z} \\ V_x & V_y & V_z \end{vmatrix}. \tag{8.3.15}$$

Consider the case where **A** and **B** have lengths $|A|$ and $|B|$ respectively and lie in the x-y plane making angles $0 < \theta_A < \theta_B < 90°$, with respect to the x-axis. Compute the cross product using the right-hand rule and express the answer in terms of lengths and angles of the vectors. Next write out the cartesian components of the vectors, evaluate the cross product using Eqn. (8.3.14), and regain the old answer.

Problem 8.3.8. *Show, by referring to Eqn. (8.3.14) that the "box" or scalar triple product of three vectors can be written as a determinant:*

$$\mathbf{A} \cdot (\mathbf{B} \times \mathbf{C}) = \begin{vmatrix} A_x & A_y & A_z \\ B_x & B_y & B_z \\ C_x & C_y & C_z \end{vmatrix}. \tag{8.3.16}$$

By invoking the invariance of the determinant under cyclic change of rows show that the box or scalar triple product is invariant under the cyclic exchange of the three vectors. The antisymmetry of the determinant (i.e., its change of sign) under exchange of rows corresponds to the antisymmetry of the cross product of two vectors under the exchange. The vanishing of the determinant when two rows are proportional corresponds to the vanishing of the "box" when two of the adjacent edges become parallel.

Problem 8.3.9. *Consider the passage from cartesian coordinates to some general nonorthogonal coordinates u, v, w. What is the Jacobian? Since the coordinates are not orthogonal, the solid bounded by the surfaces u, v, w and $u+du, v+dv, w+dw$ does not have perpendicular edges and we cannot write the volume element as $h_u h_v h_w du dv dw$. But we can use the fact that the box product of three vectors, not necessarily orthogonal, gives the volume of the rectangular parallelepiped defined by them. Argue that in this case the infinitesimal solid bounded by the above-mentioned surfaces has edges given by the infinitesimal vectors:*

$$d\mathbf{r}_u = \left[\mathbf{i} \frac{\partial x}{\partial u} + \mathbf{j} \frac{\partial y}{\partial u} + \mathbf{k} \frac{\partial z}{\partial u} \right] du, \tag{8.3.17}$$

$$d\mathbf{r}_v = \left[\mathbf{i} \frac{\partial x}{\partial v} + \mathbf{j} \frac{\partial y}{\partial v} + \mathbf{k} \frac{\partial z}{\partial v} \right] dv, \tag{8.3.18}$$

$$d\mathbf{r}_w = \left[\mathbf{i} \frac{\partial x}{\partial w} + \mathbf{j} \frac{\partial y}{\partial w} + \mathbf{k} \frac{\partial z}{\partial w} \right] dw. \tag{8.3.19}$$

By forming the box product show that the Jacobian is

$$J \left(\frac{x, y, z}{u, v, w} \right) = \begin{vmatrix} \partial x/\partial u & \partial y/\partial u & \partial z/\partial u \\ \partial x/\partial v & \partial y/\partial v & \partial z/\partial v \\ \partial x/\partial w & \partial y/\partial w & \partial z/\partial w \end{vmatrix} \tag{8.3.20}$$

This means that

$$\int dx\,dy\,dz \rightarrow \int J\left(\frac{xyz}{uvw}\right) du\,dv\,dw. \qquad (8.3.21)$$

There is a geometric interpretation of the formula for the matrix inverse which is interesting. Consider a 3×3 matrix whose elements we purposely name as follows:

$$M = \begin{bmatrix} A_x & A_y & A_z \\ B_x & B_y & B_z \\ C_x & C_y & C_z \end{bmatrix} \qquad (8.3.22)$$

so that each row stands for the components of a vector. These vectors need not be orthogonal or of unit length. Let us define their *reciprocal vectors* $\mathbf{A}^*, \mathbf{B}^*, \mathbf{C}^*$ such that

$$\mathbf{A} \cdot \mathbf{A}^* = 1 \qquad (8.3.23)$$
$$\mathbf{A} \cdot \mathbf{B}^* = \mathbf{A} \cdot \mathbf{C}^* = 0 \qquad (8.3.24)$$

and similarly for the other two. That is, each reciprocal vector has unit dot product with its counterpart in the original set and is perpendicular to the other two. Suppose we have a means of generating these vectors. Then the problem of inverting matrix M is done: M^{-1} is given by a matrix whose *columns* contain the reciprocal vectors. Indeed

$$\begin{bmatrix} A_x & A_y & A_z \\ B_x & B_y & B_z \\ C_x & C_y & C_z \end{bmatrix}\begin{bmatrix} A_x^* & B_x^* & C_x^* \\ A_y^* & B_y^* & C_y^* \\ A_z^* & B_z^* & C_z^* \end{bmatrix} = \begin{bmatrix} \mathbf{A} \cdot \mathbf{A}^* & \mathbf{A} \cdot \mathbf{B}^* & \mathbf{A} \cdot \mathbf{C}^* \\ \mathbf{B} \cdot \mathbf{A}^* & \mathbf{B} \cdot \mathbf{B}^* & \mathbf{B} \cdot \mathbf{C}^* \\ \mathbf{C} \cdot \mathbf{A}^* & \mathbf{C} \cdot \mathbf{B}^* & \mathbf{C} \cdot \mathbf{C}^* \end{bmatrix} \qquad (8.3.25)$$

$$= I. \qquad (8.3.26)$$

Let us now determine the reciprocal vectors and complete the problem. Consider $\mathbf{A}^*$. It has to be normal to $\mathbf{B}$ and $\mathbf{C}$. So we will choose it proportional to $\mathbf{B} \times \mathbf{C}$. As for the actual scale of the vector, we choose it so that its dot product with $\mathbf{A}$ is unity. The answer is clearly

$$\mathbf{A}^* = \frac{\mathbf{B} \times \mathbf{C}}{\mathbf{A} \cdot (\mathbf{B} \times \mathbf{C})} \qquad (8.3.27)$$

$$= \frac{\mathbf{B} \times \mathbf{C}}{|M|}. \qquad (8.3.28)$$

If we put the components of this vector in the first column of a matrix, we see that it agrees exactly with the first column of the inverse given by Eqn. (8.3.12). The other two columns follow analogously. (We must remember the invariance of the box product under cyclic permutations.)

How about higher order determinants and inverses of larger square matrices? The rules for both are exactly as in Eqn. (8.3.8) and Eqn. (8.3.12). Notice that the $N \times N$ determinant is defined in terms of cofactors that involve $(N - 1) \times (N - 1)$ determinants and these in turn are defined in terms of $(N - 2) \times (N - 2)$ determinants, and so on. Thus if you know how to evaluate a 2×2 determinant, you can evaluate anything bigger.

8.4. Transformations on Matrices and Special Matrices

We will now learn about some additional transformations one can perform on a matrix and special matrices that respond to these in a simple way.

Let us begin with the notion of a transpose that was discussed earlier. For any matrix M, the transpose is defined by

$$(M^T)_{ij} = M_{ji} \tag{8.4.1}$$

which just means that M^T is obtained from M by exchanging the rows and columns, i.e., by writing down the first row of M in the first column of M^T and so on. In general then this process will affect the matrix. If however a matrix satisfies

$$M_{ij} = \pm M_{ji}, \tag{8.4.2}$$

i.e., it equals $\pm$ its own transpose it is said to be *symmetric/antisymmetric*. You can tell a symmetric matrix by inspection: its elements will be symmetric with respect to the diagonal. Here is an example

$$M = \begin{bmatrix} 1 & 2 & 3 \\ 2 & 4 & 5 \\ 3 & 5 & 6 \end{bmatrix} = M^T. \tag{8.4.3}$$

Problem 8.4.1. *Show that the diagonal entries of an antisymmetric matrix vanish. Construct a 4×4 example.*

The product of two symmetric matrices need not be symmetric. This is because

$$(MN)^T = N^T M^T \tag{8.4.4}$$

just like in the case of the inverse of a product. Consequently even if $M = M^T$ and $N = N^T$, $(MN)^T \neq MN$ unless the matrices commute.

We will now verify Eqn. (8.4.4) since the manipulations involved will be useful to you later.

$$
\begin{aligned}
(MN)_{ij}^T &= (MN)_{ji} \ \text{by definition} & (8.4.5) \\
&= \sum_k M_{jk} N_{ki} & (8.4.6) \\
&= \sum_k M_{kj}^T N_{ik}^T & (8.4.7) \\
&= \sum_k N_{ik}^T M_{kj}^T & (8.4.8) \\
&= (N^T M^T)_{ij} \ \text{so that} & (8.4.9) \\
(MN)^T &= N^T M^T. & (8.4.10)
\end{aligned}
$$

We next define $M^\dagger$, the *adjoint* of a matrix as follows:

$$M_{ij}^\dagger = M_{ji}^*, \tag{8.4.11}$$

which means that the *adjoint of a matrix is obtained by transposing the given matrix and taking the complex conjugate of all the elements*. One frequently calls $M^\dagger$ as "M-dagger" and taking the adjoint of M as "taking the dagger of M."

A matrix that obeys

$$M^\dagger = \pm M \tag{8.4.12}$$

is said to be *hermitian/antihermitian*. We shall focus on the former. First note that any real symmetric matrix is automatically hermitian also since there is nothing to conjugate. Here is a complex example:

$$M = \begin{bmatrix} 1 & -i \\ i & 2 \end{bmatrix}. \tag{8.4.13}$$

This matrix is not symmetric but it is hermitian: the sign difference between the off-diagonal imaginary elements is exactly offset when we follow the transposition by complex conjugation.

Problem 8.4.2. *Prove that if M is hermitian*

$$M_{ii} = M_{ii}^* \tag{8.4.14}$$

i.e., that the diagonal elements of a hermitian matrix are real. Show that iM is antihermitian.

Problem 8.4.3. *Show that*

$$(MN)^\dagger = N^\dagger M^\dagger. \tag{8.4.15}$$

Consequently the product of two hermitian matrices is not generally hermitian unless they commute.

Next we define a *unitary matrix* as one whose adjoint equals its inverse

$$UU^\dagger = I = U^\dagger U \tag{8.4.16}$$
$$U^{-1} = U^\dagger. \tag{8.4.17}$$

For example

$$\frac{1}{\sqrt{2}} \begin{bmatrix} 1 & i \\ i & 1 \end{bmatrix} \tag{8.4.18}$$

is unitary.

Problem 8.4.4. *Verify this.*

Problem 8.4.5. *Show that the following matrix U is unitary. Argue that the determinant of a unitary matrix must be a unimodular complex number. What is it for this example?*

$$U = \begin{bmatrix} \frac{1+i\sqrt{3}}{4} & \frac{\sqrt{3}(1+i)}{2\sqrt{2}} \\ -\frac{\sqrt{3}(1+i)}{2\sqrt{2}} & \frac{i+\sqrt{3}}{4} \end{bmatrix}. \tag{8.4.19}$$

If the matrix happens to be real, there is nothing to conjugate and we define an *orthogonal matrix* as one which obeys

$$OO^T = I = O^T O \tag{8.4.20}$$

$$O^{-1} = O^T \tag{8.4.21}$$

The rotation matrix R_θ is an example.

Problem 8.4.6. *Verify that the rotation matrix R_θ is orthogonal.*

The above equation means the following: *the rows of an $N \times N$ orthogonal matrix constitute the components of N orthonormal vectors. The same goes for the columns.*

Problem 8.4.7. *Verify this assertion by writing out Eqn. (8.4.20) in some detail, using the notation for matrix elements we used in Eqn. (8.3.22).*

We will mainly discuss just unitary matrices since orthogonal matrices are a special (real) case.

The product of two unitary matrices is unitary.

The reason is as follows. Let U_1 and U_2 be unitary. Then

$$U_1 U_2 \cdot (U_1 U_2)^\dagger = U_1 U_2 \cdot (U_2^\dagger U_1^\dagger) = U_1 U_1^\dagger = I \tag{8.4.22}$$

showing that $U_1 U_2$ is unitary.

Since we have run across a whole zoo of matrices, here is a summary:

Matrix	Elements	Remarks
M	M_{ij}	Generic matrix
M^T	$M_{ij}^T = M_{ji}$	$M^T = \pm M \rightarrow$ Symmetric/Antisymmetric
		$O^T = O^{-1} \rightarrow$ Orthogonal
$M^\dagger$	$M_{ij}^\dagger = M_{ji}^*$	$M^\dagger = \pm M \rightarrow$ Hermitian/Antihermitian
		$U^\dagger = U^{-1} \rightarrow$ Unitary

Functions of matrices

Now that we know how to multiply matrices, we can take any power of a matrix and hence define functions of a matrix. The simplest example are polynomials such as

$$F(M) = M^4 + 5M^7. \tag{8.4.23}$$

More fancy examples are obtained by taking infinite series. A very common series is defined by

$$e^M = \sum_0^\infty \frac{M^n}{n!}. \qquad (8.4.24)$$

Of course the sum has no meaning if it does not converge. *The convergence of the matrix series means that each matrix element of the infinite sum of matrices converges to a limit.* The following simple example should illustrate how this works in a special case you can handle right now. We will not discuss the general theory of convergence of matrix functions in any detail.

Problem 8.4.8. *Show that if*

$$L = \begin{bmatrix} 0 & -1 \\ 1 & 0 \end{bmatrix} \; then \qquad (8.4.25)$$

$$L^2 = -I. \qquad (8.4.26)$$

Now consider

$$F(L) = e^{\theta L} \qquad (8.4.27)$$

and show by writing out the series and using $L^2 = -I$, that the series converges to a familiar matrix discussed earlier in the chapter.

The exponential of a single matrix behaves pretty much like the exponential of an ordinary number since the key difference—noncommutativity—does not show up with just one matrix, since any matrix commutes with any power of itself. Thus for example

$$e^{\theta L} e^{\phi L} = e^{(\theta+\phi)L} \qquad (8.4.28)$$

whereas in general

$$e^{\theta L} e^{\phi M} \neq e^{\theta L + \phi M} \qquad (8.4.29)$$

for noncommuting matrices.

Problem 8.4.9. *Take the L given in Problem (8.4.8.) and M a matrix with 1 and -1 along the diagonal. Show that they do not commute. Expand the exponentials in both sides of Eqn. (8.4.29) to second order in the angles θ and ϕ and verify that exponents do not simply add beyond first order. Show to second order that*

$$e^{\theta L} e^{\phi M} = e^{(\theta L + \phi M + \frac{1}{2}\theta\phi[L,M])} \qquad (8.4.30)$$

Problem 8.4.10. *Show that if H is hermitian,*

$$U = e^{iH} \qquad (8.4.31)$$

is unitary. (Write the exponential as a series and take the adjoint of each term in the sum and re-exponentiate. Use the fact that exponents can be combined if only one matrix is in the picture.)

8.4.1. Action on row vectors

We have so far treated matrices as machines that convert column vectors placed to their *right* into new column vectors as the case of the income problem

$$C = MP \qquad (8.4.32)$$

where the matrix M converts the column vector representing the parental income P to the column vector representing the children's income C.

But a matrix can also multiply a row vector R placed to its *left* and yield a new row vector R':

$$\begin{bmatrix} R'_1 & R'_2 \end{bmatrix} = \begin{bmatrix} R_1 & R_2 \end{bmatrix} \begin{bmatrix} N_{11} & N_{12} \\ N_{21} & N_{22} \end{bmatrix} \quad \text{or} \tag{8.4.33}$$

$$R' = RN. \tag{8.4.34}$$

Returning to the income problem you will agree that we could just as well store the two incomes as row vectors. The rows would be the transposes of the columns. Thus the children's income would be represented by

$$C^T = \begin{bmatrix} C_1 & C_2 \end{bmatrix} \tag{8.4.35}$$

and similarly

$$P^T = \begin{bmatrix} P_1 & P_2 \end{bmatrix} \tag{8.4.36}$$

would represent that of the parents. Question: what matrix acting to the left, would convert C^T to P^T? The answer is found by taking the transpose of both sides of Eqn. (8.4.32). *The rule that the transpose of the product is the product of the transpose in reverse holds even if some of the matrices are vectors if we remember that the transpose of a column vector is a row vector and vice versa.*

Problem 8.4.11. *Verify by explicit matrix multiplication that the rule holds for a product of a square matrix and a column vector.*

Thus we find

$$C^T = P^T M^T, \tag{8.4.37}$$

that is to say, *if M converts the parental column to the children's column by acting to the right, M^T converts the parent's row to the children's row by acting to the left.*

Problem 8.4.12. *By writing out the matrix elements explicitly verify that the above equation correctly represents the income flows.*

More generally if we associate with each column vector a row vector obtained by transposition and vice versa, any relation involving vectors and matrices will imply another one where

- *All matrices are replaced by their transposes.*

- *All factors are written in reverse order.*

For example if

$$C = MNP + V \qquad (8.4.38)$$

is a relation between matrices M, N and column vectors C, P, V, then it is also true that

$$C^T = P^T N^T M^T + V^T. \qquad (8.4.39)$$

Consider finally column vectors with complex entries. *We define their adjoints to be row vectors obtained by transposition and complex conjugation. Then we assert that any relation involving them and complex matrices implies another one where*

- *All matrices are replaced by their adjoints.*

- *All factors are written in reverse order.*

For example if

$$C = MNP + V \qquad (8.4.40)$$

is a relation between matrices M, N and column vectors C, P, V, then it is also true that

$$C^\dagger = P^\dagger N^\dagger M^\dagger + V^\dagger. \qquad (8.4.41)$$

where $C^\dagger$ and $P^\dagger$ are now vectors obtained by transposing and conjugating the columns C and P. We say that we have gone from one equation to its adjoint by "taking the adjoint."

Recall that every complex equation implies another obtained by complex conjugation of both sides. The adjoint operation is the matrix generalization of that. In fact taking the adjoint is very similar to complex conjugation as the following discussion will show.

Recall that every complex number can be written as $z = x + iy$, where x and y are real. In other words

$$z \equiv \frac{z + z^*}{2} + \frac{z - z^*}{2} \qquad (8.4.42)$$

where the first number is invariant under complex conjugation, i.e., real; and the second changes sign under complex conjugation, i.e., is pure imaginary.

Likewise for any matrix

$$M \equiv \frac{M + M^\dagger}{2} + \frac{M - M^\dagger}{2} \qquad (8.4.43)$$

where the first term in the right-hand side is hermitian (invariant under the dagger operation) and the second is *antihermitian*, i.e., changes sign under the dagger. We can also write the above equation as

$$M \equiv \frac{M + M^\dagger}{2} + i\frac{M - M^\dagger}{2i} \qquad (8.4.44)$$

so that it resembles $z = x + iy$. Now the coefficient of i is hermitian.

Continuing the analogy recall that if θ is real, $e^{i\theta}$ is unimodular, i.e., the number times its conjugate is unity. In the matrix case we saw that e^{iH} is unitary (matrix times the adjoint is the unit matrix) if H is hermitian.

Problem 8.4.13. *Show that if H is hermitian, so is $U^\dagger H U$, where U is unitary.*

Problem 8.4.14. *Show that the determinant of a unitary matrix is unimodular. Show that the determinant of an orthogonal matrix is ± 1.*

Problem 8.4.15. *Show that if $MN = 0$, one of them must have zero determinant.*

Problem 8.4.16. (Very Important). *Say hello to the* Pauli matrices

$$\sigma_x = \begin{bmatrix} 0 & 1 \\ 1 & 0 \end{bmatrix} \quad \sigma_y = \begin{bmatrix} 0 & -i \\ i & 0 \end{bmatrix} \quad \sigma_z = \begin{bmatrix} 1 & 0 \\ 0 & -1 \end{bmatrix} \tag{8.4.45}$$

which you will see on numerous occasions. (The subscripts 1, 2, and 3 are sometimes used instead of x, y, and z.)
Show that they are hermitian. Show that their square equals the unit matrix. Show that as a result of these two features they must also be unitary. Verify explicitly. Show that

$$[\sigma_x, \sigma_y] = 2i\sigma_z \text{ et cycl.} \tag{8.4.46}$$

Show that any two of them anticommute, i.e., the anticommutator

$$[M, N]_+ \equiv MN + NM \tag{8.4.47}$$

vanishes. Show that as a result

$$\sigma_x \sigma_y = i\sigma_z. \tag{8.4.48}$$

Find the determinants and the inverses of the three matrices by any method you want. If you can avoid a calculation, that is fine.

Problem 8.4.17. *Show that*

$$[\boldsymbol{\sigma} \cdot \mathbf{a}] [\boldsymbol{\sigma} \cdot \mathbf{b}] = \mathbf{a} \cdot \mathbf{b} I + i\boldsymbol{\sigma} \cdot (\mathbf{a} \times \mathbf{b}), \tag{8.4.49}$$

where $\mathbf{a}$ and $\mathbf{b}$ are ordinary three dimensional vectors and

$$\boldsymbol{\sigma} = \mathbf{i}\sigma_x + \mathbf{j}\sigma_y + \mathbf{k}\sigma_z. \tag{8.4.50}$$

Problem 8.4.18. *Using the above result for the case* $\mathbf{a} = \mathbf{b}$, *show that*

$$e^{i\mathbf{a}\cdot\boldsymbol{\sigma}} = \cos a I + i \sin a\, \hat{\mathbf{a}} \cdot \boldsymbol{\sigma} \qquad (8.4.51)$$

where a *is the length of* $\mathbf{a}$ *and* $\hat{\mathbf{a}} = \mathbf{a}/a$.

Definition 8.7. *The* trace *of a matrix* M, *denoted by* $\mathrm{Tr}\, M$ *stands for the sum of all the diagonal elements:*

$$\mathrm{Tr}\, M = \sum_i M_{ii}. \qquad (8.4.52)$$

Problem 8.4.19. *Show that*

$$\mathrm{Tr}\, MN = \mathrm{Tr}\, NM \qquad (8.4.53)$$
$$\mathrm{Tr}\, ABC = \mathrm{Tr}\, BCA = \mathrm{Tr}\, CAB. \qquad (8.4.54)$$
$$\mathrm{Tr}\, U^{\dagger}MU = \mathrm{Tr}\, M \ \textit{if}\ U \ \textit{is unitary}. \qquad (8.4.55)$$

Problem 8.4.20. *Consider four Dirac matrices that obey*

$$M_i M_j + M_j M_i = 2\delta_{ij} I \qquad (8.4.56)$$

where the Kronecker delta *symbol is defined as follows:*

$$\delta_{ij} = 1 \ \textit{if}\ i = j,\ 0 \ \textit{if}\ i \neq j. \qquad (8.4.57)$$

Thus the square of each Dirac matrix is the unit matrix and any two distinct Dirac matrices anticommute. Using the latter property show that the matrices are traceless. (Use Eqn. (8.4.54).)

8.5. Summary

The main points from this chapter are as follows.

- Know that the entry M_{ij} sits at row i and column j.

-

$$(M + N)_{ij} = M_{ij} + N_{ij}$$
$$(MN)_{ij} = \sum_k M_{ik} N_{kj}.$$

-

$$M_{ij}^T = M_{ji}$$

is the transpose, while

$$M_{ij}^\dagger = M_{ji}^*$$

is the adjoint.

- Know that the unit matrix I, has 1's along the diagonal and 0's elsewhere and obeys $IM = M$ for all M. The zero matrix has 0's everywhere.

- The commutator is defined to be

$$[M, N] = MN - NM.$$

- The matrix M_C, the cofactor, has as its ij entry, $(-1)^{i+j}$ times the determinant of the matrix obtained by deleting row i and column j from of M.

- The determinant for a 2×2 matrix is

$$|M| = M_{11}M_{22} - M_{12}M_{21}.$$

Know its extension to 3×3 and bigger matrices in terms of cofactors. Know that $|MN| = |M||N|$, $|M| = |M^T|$ and that $|M|$ changes sign when two rows or columns are exchanged and vanishes if two rows or columns are proportional.

- The inverse defined by $MM^{-1} = M^{-1}M = I$ is given by

$$M^{-1} = \frac{M_C^T}{|M|},$$

where M_C^T is the transpose of the cofactor.

- A matrix is symmetric if $M^T = M$, hermitian if $M^\dagger = M$, and unitary if $M^\dagger = M^{-1}$.

- The transpose, adjoint, or inverse of a product is the product of the transposes, inverses, or adjoints in reverse. Thus for example $(MN)^T = N^T M^T$.

- The trace of a matrix is the sum of it diagonal entries

$$\operatorname{Tr} M = \sum_i M_{ii}.$$

The trace of a product is invariant under cyclic permutation of the factors:

$$\operatorname{Tr}(ABC) = \operatorname{Tr}(BCA) = \operatorname{Tr}(CAB).$$

Note also that $\operatorname{Tr} U^\dagger M U = \operatorname{Tr} M$ for U a unitary matrix.

LINEAR VECTOR SPACES

9.1. Linear Vector Spaces: Basics

In this section you will be introduced to *linear vector spaces*. The general strategy is this. It is assumed you are familiar with the arrows from elementary physics encoding the magnitude and direction of velocity, force, displacement, torque, etc., and know how to add them and multiply them by scalars and the rules obeyed by these operations. For example you know that scalar multiplication is distributive: the multiple of a sum of two vectors is the sum of the multiples. *We want to abstract from this simple case a set of basic features or axioms and say that any set of objects obeying the same, forms a linear vector space.* The cleverness lies in deciding which of the properties to keep in the generalization. If you keep too many, there will be no other examples and if you keep too few, there will be no interesting results to develop from the axioms.

The point of this generalization is that any result we can prove starting from the axioms of a vector space apply uniformly to all cases of it. For example, we know in elementary vector analysis that any vector can be expressed as a linear combination of the unit vectors i, j, and k. You may have also heard that it is possible to express any periodic function as a sum over sines and cosines, i.e., as a Fourier series. These two ideas will be seen to be one and the same if viewed from the point of vector spaces. Or consider the rather obvious result that the dot product of any two vectors cannot exceed the product of the lengths of the vectors. In the language of vector spaces, this result is no different from the relation

$$\left[\int f(x)g(x)dx\right]^2 \leq \left[\int f(x)^2 dx\right]\left[\int g(x)^2 dx\right] \qquad (9.1.1)$$

where f and g are some real functions defined in some interval. This not-so-obvious relation is important, and is used to prove the famous Uncertainty Relations of Heisenberg.

The following is the list of properties the mathematicians have wisely chosen as requisite for a vector space. As you read them please compare them to the world of arrows and make sure that these are indeed properties possessed by these familiar vectors. (To help you along, the corresponding statements for arrows are briefly given after the axioms.)

Definition 9.1. *A linear vector space* $\mathcal{V}$ *is a collection of objects* $|1\rangle, |2\rangle, \ldots |V\rangle$ $\ldots |W\rangle \ldots$ *called vectors,* **(arrows)** *for which there exists*
(i) a definite rule for forming the vector sum, denoted $|V\rangle + |W\rangle$, **(place the tail of one on the tip of the other...)**
(ii) a definite rule for multiplication by scalars $a, b...$, *denoted* $a|V\rangle$ **(rescale or stretch the vector by the scalar factor)**
with the following features:

A-I *The result of these operations is another element of the space, a feature called* closure: $|V\rangle + |W\rangle \in \mathcal{V}$. **(sum of two arrows is an arrow)**

A-II *Scalar multiplication is* distributive in the vectors: $a(|V\rangle + |W\rangle) = a|V\rangle + a|W\rangle$. **(rescaling the sum of arrows is same as summing rescaled arrows)**

A-III *Scalar multiplication is* distributive in the scalars: $(a + b)|V\rangle = a|V\rangle + b|V\rangle$. **(obvious)**

A-IV *Scalar multiplication is* associative: $a(b|V\rangle) = ab|V\rangle$. **(obvious)**

A-V *Addition is* commutative: $|V\rangle + |W\rangle = |W\rangle + |V\rangle$. **(sum of two arrows is independent of order, i.e., we can place the tail of the second on the tip of the first or the other way around)**

A-VI *Addition is* associative: $|V\rangle + (|W\rangle + |Z\rangle) = (|V\rangle + |W\rangle) + |Z\rangle$. **(when adding three arrows, we can add the first two and then the third or add the first to the sum of the second and third)**

A-VII *There exists a* null vector $|0\rangle$ *obeying* $|V\rangle + |0\rangle = |V\rangle$. **(arrow of zero length)**

A-VIII *For every vector* $|V\rangle$ *there exists an* inverse under addition, $|-V\rangle$ *such that* $|V\rangle + |-V\rangle = |0\rangle$. **(inverse is the arrow reversed in direction)**

There is a good way to remember all of these: *do what comes naturally.*
Note that conspicuously missing above are the requirements that every vector must have a magnitude and direction, which was the first and most salient feature drilled into our heads when we first heard about them! So you might think that

in dropping this requirement the baby is thrown out with the bath water. But you will have ample time to appreciate the wisdom behind this choice as you go along and see a great unification and synthesis of diverse ideas under the heading of vector spaces. You will see examples of vector spaces that involve entities that you cannot intuitively perceive as having either a magnitude or a direction! While you should be duly impressed with all this, remember that it does not at all hurt to think of these generalizations in terms of arrows and to use your intuition to prove theorems or, at the very least, anticipate them.

Definition 9.2. *The set of numbers* $a, b, \ldots$ *used in scalar multiplication of vectors is called the* field *over which the vector space is defined.*

If the field consists of all real numbers, we have a *real vector space*, if they are complex, we have a *complex vector space*. The vectors themselves are neither real nor complex; the adjective applies only to the scalars. The usual space of arrows is a real vector space since we cannot interpret rescaling by a complex number as a stretching process.

Observe that we are using a new symbol $|V\rangle$ to denote a generic vector. This object is called *ket V* and this nomenclature is due to Dirac whose notation will be discussed at some length later. We purposely do not use the symbol **V** to denote the vectors as the first step in weaning you away from the limited concept of the vector as an arrow. You are however not discouraged from associating with $|V\rangle$ an arrow-like object until you have seen enough vectors that are not arrows and are ready to drop the crutch. When discussing the space of arrows in this notation I may refer to the unit vectors **i**, **j**, and **k** as $|1\rangle$, $|2\rangle$, and $|3\rangle$ respectively. The vector $\mathbf{V} = 3\mathbf{i} + 4\mathbf{j}$ will be referred to as $|V\rangle = 3|1\rangle + 4|2\rangle$ and so on.

The set of all arrows, which inspired the idea of a vector space, of course qualifies as a vector space. But we cannot tamper with it. For example the set of all arrows with positive z-components do not form a vector space: for one thing, there is no inverse.

We return to the fact that the axioms make no reference to magnitude or direction. The point is that while the arrows have these qualities, members of a vector space need not. Now this statement is empty unless I can give you examples. Here are two.

Consider the set of all 2×2 matrices. We know how to add them and multiply them by scalars (multiply all four matrix elements by that scalar). The corresponding rules obey closure, associativity and distributive requirements. The null matrix has all zeros in it and the inverse under *addition* of a matrix is the matrix with all elements negated. You must agree that here we have a genuine vector space consisting of things which don't have an obvious length or direction associated with them. When we want to highlight the fact that the matrix M is an element of a vector space, we may want to refer to it as, say, ket number 4 or: $|4\rangle$.

Problem 9.1.1. *Do all hermitian* 2×2 *matrices form a vector space under addition? Is there a requirement on the scalars that multiply them? How about all unitary matrices for any choice of scalars? How about* 2×2 *matrices with integer coefficients? Is there a restriction on the scalars multiplying them?*

As a second example consider all functions $f(x)$ defined in an interval $0 \leq x \leq L$. We define scalar multiplication by a simply as $af(x)$ and addition as pointwise addition: the sum of two functions f and g has the value $f(x) + g(x)$ at the point x. The null function is zero everywhere and the additive inverse of f is $-f$.

Problem 9.1.2. *Do functions that vanish at the end points* $x = 0$ *and* $x = L$ *form a vector space? How about periodic functions obeying* $f(0) = f(L)$? *How about functions that obey* $f(0) = 4$? *If the functions do not qualify, list the things that go wrong.*

Let us note that the axioms imply

- $|0\rangle$ is unique, i.e., if $|0'\rangle$ has all the properties of $|0\rangle$, then $|0\rangle = |0'\rangle$.

- $0|V\rangle = |0\rangle$.

- $|-V\rangle = -|V\rangle$.

- $|-V\rangle$ is the unique additive inverse of $|V\rangle$.

The proofs are left to the following exercise. You don't have to know the proofs, but do have to know the statements.

Problem 9.1.3. *Verify these claims. For the first consider* $|0\rangle + |0'\rangle$ *and use the advertised properties of the two null vectors in turn. For the second start with* $|0\rangle = (0+1)|V\rangle + |-V\rangle$. *For the third, begin with* $|V\rangle + (-|V\rangle) = 0|V\rangle = |0\rangle$. *For the last, let* $|W\rangle$ *also satisfy* $|V\rangle + |W\rangle = |0\rangle$. *Since* $|0\rangle$ *is unique, this means* $|V\rangle + |W\rangle = |V\rangle + |-V\rangle$. *Take it from here.*

Problem 9.1.4. *Consider the set of all entities of the form* (a, b, c) *where the entries are real numbers. Addition and scalar multiplication are defined as follows:*

$$(a, b, c) + (d, e, f) = (a + d, b + e, c + f)$$

$$\alpha(a, b, c) = (\alpha a, \alpha b, \alpha c).$$

Write down the null vector and inverse of (a, b, c). *Show that vectors of the form* $(a, b, 1)$ *do not form a vector space.*

Linear independence

The next concept in vector spaces is that of *linear independence* of a set of vectors $|1\rangle, |2\rangle \ldots |n\rangle$. Let us work our way to this through some examples involving arrows. Consider first vectors $|1\rangle, |2\rangle, \ldots$ that lie along the x-axis. It is evident that given any two of them, $|1\rangle$ and $|2\rangle$, we can write one as a (positive or negative) multiple of the other:

$$|1\rangle = a|2\rangle. \tag{9.1.2}$$

Equivalently we can form a linear combination of them which vanishes

$$|1\rangle - a|2\rangle = 0. \tag{9.1.3}$$

Notice that on the right-hand side I have used 0 in place of $|0\rangle$. This is strictly speaking incorrect since a set of vectors can only add up to a vector and not a number. It is however common to represent the null vector by 0.

If we take two or more colinear vectors, it is clear that

- we can write any one of them as a linear combination of the others,

- or equivalently, that we can form a nontrivial linear combination (with not all coefficients equal to zero) that gives the null vector.

We summarize all this by saying that a set of two or more colinear vectors is *linearly dependent*.

Next we consider arrows in a plane. Say we take any two nonparallel vectors $|1\rangle$ and $|2\rangle$. Now it is not possible to write one in terms of the other, or equivalently, to form a (nontrivial) linear combination of the two which adds up to the null vector. We say the two vectors are now *linearly independent*. Suppose we now bring in a third vector $|3\rangle$. If it is parallel to either one, it can clearly be written as a combination of the other two (namely as a multiple of the one it is parallel to) or equivalently, we can form a linear combination of the three vectors that adds up to zero. (This combination will involve just the two parallel vectors, but is still nontrivial since not all coefficients in the linear combination vanish.) Thus this triplet of vectors, in which the third is parallel to one of the first two, is linearly dependent. It turns out that the triplet of vectors is linearly dependent even if the third is not parallel to either of the first two. This is because we can write one of them, say $|3\rangle$, as a linear combination of the other two. To find the combination, draw a line from the tail of $|3\rangle$ in the direction of $|1\rangle$ as in Fig. 9.1. Next draw a line antiparallel to $|2\rangle$ from the tip of $|3\rangle$. These lines will necessarily intersect since $|1\rangle$ and $|2\rangle$ are not parallel by assumption. The intersection point P will determine how much of $|1\rangle$ and $|2\rangle$ we want: we go from the tail of $|3\rangle$ to P using the appropriate multiple of $|1\rangle$ and go from P to the tip of $|3\rangle$ using the appropriate multiple of $|2\rangle$. Thus any three vectors in a plane are linearly dependent.

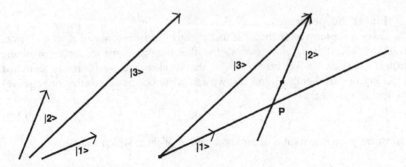

Figure 9.1. Linear dependence of three coplanar vectors.

We are now ready to extend this notion to a general vector space. Consider a linear relation of the form

$$\sum_{i=1}^{n} a_i |i\rangle = |0\rangle.$$ (9.1.4)

We may assume without loss of generality that the left-hand side does not contain any multiple of $|0\rangle$, for if it did, it could be shifted to the right, and combined with the $|0\rangle$ there to give $|0\rangle$ once more. (We are using the fact that any multiple of $|0\rangle$ equals $|0\rangle$.)

Definition 9.3. *A set of vectors is said to be* linearly independent *if the only such linear relation as Eqn. (9.1.4) is the trivial one with all $a_i = 0$. If the set of vectors is not linearly independent, we say it is* linearly dependent.

Equation (9.1.4) tells us that it is not possible to linearly combine members of a linearly independent set to get the null vector except for the trivial case where we take zero times each vector. On the other hand if the set of vectors is linearly dependent, such a nontrivial relation will exist, and it must contain at least two nonzero coefficients (why?). Let us say $a_3 \neq 0$. Then we could write

$$|3\rangle = \sum_{i=1,\, \neq 3}^{n} \frac{-a_i}{a_3} |i\rangle$$ (9.1.5)

thereby expressing $|3\rangle$ in terms of the others. *In other words, if a set of vectors is linearly dependent, we can express at least one of its members as a linear combination of the others.*

Problem 9.1.5. *Consider three elements from the vector space of real 2×2 matrices:*

$$|1\rangle = \begin{bmatrix} 0 & 1 \\ 0 & 0 \end{bmatrix} \quad |2\rangle = \begin{bmatrix} 1 & 1 \\ 0 & 1 \end{bmatrix} \quad |3\rangle = \begin{bmatrix} -2 & -1 \\ 0 & -2 \end{bmatrix}.$$

Are they linearly independent? Support your answer with details. (Notice we are calling these matrices vectors and using kets to represent them to emphasize their role as elements of a vector space.)

Problem 9.1.6. *Show that the following row vectors are linearly dependent: (1,1,0), (1,0,1), and (3,2,1). Show the opposite for (1,1,0), (1,0,1), and (0,1,1).*

Definition 9.4. *A vector space has* dimension n *if it can accommodate a maximum of n linearly independent vectors. It will be denoted by $V^n(R)$ if the field is real and by $V^n(C)$ if the field is complex.*

In view of the earlier discussions, the plane is two dimensional and the set of all arrows not limited to the plane define a three dimensional vector space. How about 2×2 matrices? They form a four-dimensional vector space. Here is a proof. The following vectors are linearly independent:

$$|1\rangle = \begin{bmatrix} 1 & 0 \\ 0 & 0 \end{bmatrix} \quad |2\rangle = \begin{bmatrix} 0 & 1 \\ 0 & 0 \end{bmatrix} \quad |3\rangle = \begin{bmatrix} 0 & 0 \\ 1 & 0 \end{bmatrix} \quad |4\rangle = \begin{bmatrix} 0 & 0 \\ 0 & 1 \end{bmatrix}$$

since it is impossible to form linear combinations of any three of them to give the fourth, since any three of them will have a zero in the one place where the fourth does not. So the space it at least four dimensional. Could it be bigger? No, since any arbitrary 2×2 matrix can be written in terms of them:

$$\begin{bmatrix} a & b \\ c & d \end{bmatrix} = a|1\rangle + b|2\rangle + c|3\rangle + d|4\rangle.$$

If the scalars a, b, c, d are real, we have a *real four-dimensional space*, if they are complex we have a *complex four-dimensional space*.

Theorem 9.1. *Any vector $|V\rangle$ in an n-dimensional space can be written as a linear combination of n linearly independent vectors $|1\rangle \dots |n\rangle$.*

Proof: If there were a vector $|V\rangle$ for which this were not possible, it would join the given set of vectors and form a set of $n + 1$ linearly independent vectors, which is not possible in an n-dimensional space by definition. ∎

Definition 9.5. *A set of n linearly independent vectors in an n-dimensional space is called a* basis.

Thus we can write, on the strength of the above, that for any $|V\rangle$ belonging to the vector space

$$|V\rangle = \sum_{i=1}^{n} v_i|i\rangle \tag{9.1.6}$$

where the vectors $|i\rangle$ form a basis.

Definition 9.6. *The coefficients of expansion* v_i *of a vector in terms of a linearly independent basis* $(|i\rangle)$ *are called the* components *of the vector in that basis.*

Theorem 9.2. *The expansion in Eqn. (9.1.6) is unique.*

Proof: Suppose the expansion is not unique. We must then have a second expansion:

$$|V\rangle = \sum_{i=1}^{n} v'_i |i\rangle. \qquad (9.1.7)$$

Subtracting Eqn. (9.1.7) from Eqn. (9.1.6) (i.e., multiplying the second by the scalar -1 and adding the two equations) we get

$$|0\rangle = \sum_i (v_i - v'_i)|i\rangle \qquad (9.1.8)$$

which implies that

$$v_i = v'_i \qquad (9.1.9)$$

since the basis vectors are linearly independent and only a trivial linear relation between them can exist. ∎

Note that given a basis, the components are unique, but if we change the basis, the components will change. We refer to $|V\rangle$ as the vector in the abstract, having an existence of its own and satisfying various relations involving other vectors. When we choose a basis the vectors assume concrete forms in terms of their components and the relation between vectors is satisfied by the components. Imagine for example three arrows in the plane, $\mathbf{A}, \mathbf{B}, \mathbf{C}$ satisfying $\mathbf{A} + \mathbf{B} = \mathbf{C}$ according to the laws for adding arrows. So far no basis has been chosen and we do not need a basis to make the statement that the vectors form a triangle. Now we choose a basis and write each vector in terms of the components. The components will satisfy $C_i = A_i + B_i$, $i = 1, 2$. If we choose a different basis, the components will change in numerical value, but the relation between them expressing the equality of $\mathbf{C}$ to the sum of the other two will still hold between the new components.

In the case of nonarrow vectors, adding them in terms of components proceeds as in the elementary case thanks to the axioms. If

$$|V\rangle \;=\; \sum_i v_i |i\rangle \text{ and} \qquad (9.1.10)$$

$$|W\rangle \;=\; \sum_i w_i |i\rangle \text{ then} \qquad (9.1.11)$$

$$|V\rangle + |W\rangle \;=\; \sum_i (v_i + w_i)|i\rangle, \qquad (9.1.12)$$

where we have used the axioms to carry out the regrouping of terms. Here is the conclusion:

To add two vectors, add their components.

There is no reference to taking the tail of one and putting it on the tip of the other, etc., since in general the vectors have no head or tail. Of course if we are dealing with arrows, we can add them either using the tail and tip routine or by simply adding their components in a basis.

In the same way, we have:

$$a|V\rangle = a \sum_i v_i |i\rangle = \sum_i a v_i |i\rangle. \tag{9.1.13}$$

In other words

To multiply a vector by a scalar, multiply all its components by the scalar.

9.2. Inner Product Spaces

The matrix and function examples must have convinced you that we can have a vector space with no preassigned definition of length or direction for the elements. However we can make up quantities that have the same properties that the lengths and angles do in the case of arrows. The first step is to define a sensible analog of the dot product. Now you might rightfully object that the dot product

$$\mathbf{V} \cdot \mathbf{W} = |V||W|\cos\theta \tag{9.2.1}$$

is itself defined in terms of lengths and angles and we seem to be going around in circles. But recall that there is another expression for it

$$\mathbf{V} \cdot \mathbf{W} = V_x W_x + V_y W_y + V_z W_z \tag{9.2.2}$$

involving just the components. Since we have already defined the notion of components of a vector for general vector space, this will be our starting point for generalization. However, we must first fully understand how one goes from Eqn. (9.2.1) to Eqn. (9.2.2).

Consider two vectors (arrows) in the plane

$$\mathbf{V} = v_1 \mathbf{1} + v_2 \mathbf{2} \tag{9.2.3}$$

$$\mathbf{W} = w_1 \mathbf{1} + v_2 \mathbf{2}, \tag{9.2.4}$$

where **1** and **2** are any two *linearly independent* basis vectors. What is their dot product in terms of their components, given just definition Eqn. (9.2.1)? We start by writing

$$\mathbf{V} \cdot \mathbf{W} = (v_1\mathbf{1} + v_2\mathbf{2}) \cdot (w_1\mathbf{1} + w_2\mathbf{2}) \tag{9.2.5}$$

and ask if we can open up the brackets naively. The answer is yes and it is because the definition, Eqn. (9.2.1), satisfies the following three axioms:

B-i $\mathbf{A} \cdot \mathbf{B} = \mathbf{B} \cdot \mathbf{A}$ (symmetry)

B-ii $\mathbf{A} \cdot \mathbf{A} \geq 0 \quad \mathbf{A} \cdot \mathbf{A} = 0$ if and only if $\mathbf{A} = 0$ (positive semidefiniteness)

B-iii $\mathbf{A} \cdot (b\mathbf{B} + c\mathbf{C}) = b\mathbf{A} \cdot \mathbf{B} + c\mathbf{A} \cdot \mathbf{C}$ (linearity)

Convince yourselves that this is indeed the case. (The middle axiom will not be invoked in this discussion and is given for completeness.)

Armed with this result we expand brackets in Eqn. (9.2.5) as follows. First, we know from the linearity axiom (B-iii) that the dot product is linear in the second vector of the dot product. Thus

$$\mathbf{V} \cdot \mathbf{W} = (v_1\mathbf{1} + v_2\mathbf{2}) \cdot w_1\mathbf{1} + (v_1\mathbf{1} + v_2\mathbf{2}) \cdot w_2\mathbf{2}. \tag{9.2.6}$$

How do we open up the bracket in the first factor of the dot product? We use axiom (B-i), which says $\mathbf{A} \cdot \mathbf{B} = \mathbf{B} \cdot \mathbf{A}$; reverse the order of factors, and *then* use linearity to obtain

$$\mathbf{V} \cdot \mathbf{W} = w_1v_1\mathbf{1} \cdot \mathbf{1} + w_1v_2\mathbf{1} \cdot \mathbf{2} + w_2v_1\mathbf{2} \cdot \mathbf{1} + w_2v_2\mathbf{2} \cdot \mathbf{2}. \tag{9.2.7}$$

At this stage we find that to go any further, we need the dot products of the basis vectors. All we know is that they are a linearly independent (non-parallel, in this case) basis. Thus the dot product $\mathbf{V} \cdot \mathbf{W}$ will involve not only the components of $\mathbf{V}$ and $\mathbf{W}$, but also four non-zero dot products of the basis vectors among themselves. We do not usually see this aspect because we use the freedom to trade the given linearly independent basis for another (by forming suitable linear combinations) in which the basis vectors $\mathbf{e}_i$, $[i = 1, 2]$ obey

$$\mathbf{e}_i \cdot \mathbf{e}_j = \begin{cases} 1 & \text{for } i = j \\ 0 & \text{for } i \neq j \end{cases} \equiv \delta_{ij}.$$

The $\mathbf{e}_i$ could be the familiar unit vectors $\mathbf{i}$ and $\mathbf{j}$ or any rotated version of these. *Assume from now on that we are working with such a basis, called an orthonormal basis.*[1] In this orthonormal basis the dot product simplifies to

$$\mathbf{V} \cdot \mathbf{W} = \sum_i v_i w_i, \tag{9.2.8}$$

---

[1] Here is a sketch of how to get an orthonormal basis out of two non-parallel vectors in a plane. We first rescale one of the vectors by its length to get a unit vector. Next, we subtract from the second vector its projection along the first, and rescale the result. We end up with two vectors whose dot products with each other is zero, and whose dot product with themselves is unity.

which is of the familiar form.

We now want to invent a generalization of the dot product to any linear vector space called the *inner product* or *scalar product*. Given any two vectors $|V\rangle$ and $|W\rangle$

$$|V\rangle = \sum_i v_i |i\rangle \tag{9.2.9}$$

$$|W\rangle = \sum_j w_j |j\rangle \tag{9.2.10}$$

we are looking for a formula in terms of their components v_i and w_j. We denote the inner product by the symbol $\langle V|W\rangle$. It is once again a number (generally complex) dependent on the two vectors. One refers to the first factor in the inner product, as the *bra* and the second factor, as the *ket*. The idea (due to Dirac) is that together they comprise the bracket $\langle V|W\rangle$.

If we follow the same route as with arrows, we will reach the stage when we have to go about expressing the dot products of linear combinations in terms of dot products of the individual terms, i.e., open up brackets. In the case of arrows we turned to the axioms B-i to B-iii, (derived from the concrete formula, Eqn. (9.2.1)) for guidance. Here we have no such thing. So we *postulate* that the inner product will obey the following axioms:

B-I $\langle V|W\rangle = \langle W|V\rangle^*$ (skew-symmetry)

B-II $\langle V|V\rangle \geq 0$,   0 if and only if $|V\rangle = |0\rangle$ (positive semidefiniteness)

B-III $\langle V|(a|W\rangle + b|Z\rangle) \equiv \langle V|aW + bZ\rangle = a\langle V|W\rangle + b\langle V|Z\rangle$ (linearity in ket).

Note that only the first axiom is different from the one for the dot product—it says the dot product depends on the order of the two factors. Its significance will become clear shortly.

What do the axioms say about opening the brackets in an expression like

$$\langle aA + bB|cC + dD\rangle, \tag{9.2.11}$$

where the upper case letters stand for vectors and lower case letters for scalars that multiply them in the linear combination? The third axiom B-III (linearity in ket) allows us to open up the ket very simply and to obtain

$$\langle aA + bB|cC + dD\rangle = c\langle aA + bB|C\rangle + d\langle aA + bB|D\rangle. \tag{9.2.12}$$

What about the linear combination in the bra? There is no axiom for that because the necessary rule follows from the first axiom: it tells us that we can swap the vectors in the bra and ket if we complex conjugate the answer. Thus

$$\langle aA + bB|C\rangle = \langle C|aA + bB\rangle^* \text{ by B-I}$$

$$\begin{aligned} &= \quad (a\langle C|A\rangle + b\langle C|B\rangle)^* \\ &= \quad a^*\langle C|A\rangle^* + b^*\langle C|B\rangle^* \\ &= \quad a^*\langle A|C\rangle + b^*\langle B|C\rangle, \end{aligned} \tag{9.2.13}$$

which expresses the *antilinearity* of the inner product with respect to the bra. In other words, the inner product of a linear superposition with another vector is the corresponding superposition of inner products if the superposition occurs in the ket, while it is the superposition with all coefficients conjugated if the superposition occurs in the bra. This asymmetry, unfamiliar in real vector spaces, is here to stay and you will get used to it as you go along. In any event we find

$$\langle aA + bB | cC + dD\rangle = a^*c\langle A|C\rangle + b^*c\langle B|C\rangle + a^*d\langle A|D\rangle + b^*d\langle B|D\rangle. \tag{9.2.14}$$

Applying this to the inner product $\langle V|W\rangle$ we obtain

$$\langle V|W\rangle = \sum_i \sum_j v_i^* w_j \langle i|j\rangle. \tag{9.2.15}$$

As for $\langle i|j\rangle$, the inner product of basis vectors, we invoke

Theorem 9.3. (Gram–Schmidt). *Given any basis, we can form linear combinations of the basis vectors to obtain an orthonormal basis, i.e., one obeying*

$$\langle i|j\rangle = \begin{cases} 1 & for\ i = j \\ 0 & for\ i \neq j \end{cases} \equiv \delta_{ij}.$$

Postponing the proof for a moment, *let us assume that the procedure has been implemented and that the current basis is orthonormal.* Upon feeding this fact into Eqn. (9.2.15), the double sum collapses to a single one due to the Kronecker delta to give the very important formula

$$\langle V|W\rangle = \sum_i v_i^* w_i. \tag{9.2.16}$$

This is the form of the inner product we will use from now on. Note that it looks just like the dot product except for the complex conjugation of the v_i's. We can now appreciate the point behind the first axiom that led to this. *But for the complex conjugation of the components of the first vector, $\langle V|V\rangle$ would not even be real, not to mention positive.* But now it is given by

$$\langle V|V\rangle = \sum_i |v_i|^2 \geq 0 \tag{9.2.17}$$

and vanishes only for the null vector. This makes it sensible to refer to $\langle V|V\rangle$ as the length or norm squared of a vector.

Let us consider an example. Let

$$|V\rangle = (3 - 4i)|1\rangle + (5 - 6i)|2\rangle \qquad (9.2.18)$$
$$|W\rangle = (1 - i)|1\rangle + (2 - 3i)|2\rangle \qquad (9.2.19)$$

be two vectors expanded in terms of an orthonormal basis $|1\rangle$ and $|2\rangle$. Then we have

$$\langle V|V\rangle = (3 + 4i)(3 - 4i) + (5 + 6i)(5 - 6i) = 86 \qquad (9.2.20)$$
$$\langle W|W\rangle = (1 + i)(1 - i) + (2 + 3i)(2 - 3i) = 15 \qquad (9.2.21)$$
$$\langle V|W\rangle = (3 + 4i)(1 - i) + (5 + 6i)(2 - 3i)$$
$$= 35 - 2i = \langle W|V\rangle^*. \qquad (9.2.22)$$

Having developed the inner product for an arbitrary linear vector space, we can introduce the notion of length or perpendicularity and use some of the same terminology from the old days of the dot product:

Definition 9.7. *We will say that two vectors are perpendicular if their inner product vanishes.*

Definition 9.8. *We will refer to* $\sqrt{\langle V|V\rangle} \equiv |V|$ *as the length of the vector. A* normalized vector *has unit norm.*

We will also frequently refer to the inner product or scalar product as the dot product.

Consider next Eqn. (9.2.16). Since the vector $|V\rangle$ is uniquely specified by its components in a given basis, we may, in this basis, associate with it a column vector:

$$|V\rangle \to \begin{bmatrix} v_1 \\ v_2 \\ \vdots \\ v_n \end{bmatrix} \text{ in this basis.} \qquad (9.2.23)$$

Likewise

$$|W\rangle \to \begin{bmatrix} w_1 \\ w_2 \\ \vdots \\ w_n \end{bmatrix} \text{ in this basis.} \qquad (9.2.24)$$

The inner product $\langle V|W\rangle$ is given by the matrix product of the transpose conjugate of the column vector representing $|V\rangle$ with the column vector representing $|W\rangle$:

$$\langle V|W\rangle = [v_1^*, v_2^*, \ldots v_n^*]\begin{bmatrix} w_1 \\ w_2 \\ \vdots \\ w_n \end{bmatrix} = V^\dagger W \qquad (9.2.25)$$

Note that $\langle V|W\rangle$, like the dot product, has the same form in any orthonormal basis: $\langle V|W\rangle = \sum_i v_i^* w_i = \sum_i v_i'^* w_i'.$

Expansion of vectors in an orthonormal basis

Suppose we wish to expand a vector $|V\rangle$ in an orthonormal basis. To find the components that go into the expansion we proceed as follows. We take the dot product of both sides of the assumed expansion with $|j\rangle$: (or $\langle j|$ if you are a purist)

$$|V\rangle = \sum_i v_i|i\rangle \qquad (9.2.26)$$

$$\langle j|V\rangle = \sum_i v_i \underbrace{\langle j|i\rangle}_{\delta_{ij}} \qquad (9.2.27)$$

$$= v_j, \qquad (9.2.28)$$

i.e., to find the j-th component of a vector we take the dot product with the j-th unit vector, exactly as with arrows. Using this result we may write

$$|V\rangle = \sum_i |i\rangle\langle i|V\rangle. \qquad (9.2.29)$$

Let us make sure the basis vectors look as they should. If we set $|V\rangle = |j\rangle$ in Eqn. (9.2.29), we find the correct answer: the i-th component of the j-th basis vector is δ_{ij}. Thus for example the column representing basis vector number 4 will have a 1 in the 4-th row and zero everywhere else. The abstract relation

$$|V\rangle = \sum_i v_i|i\rangle \qquad (9.2.30)$$

becomes in this basis

$$\begin{bmatrix} v_1 \\ v_2 \\ \vdots \\ v_n \end{bmatrix} = v_1\begin{bmatrix} 1 \\ 0 \\ \vdots \\ 0 \end{bmatrix} + v_2\begin{bmatrix} 0 \\ 1 \\ 0 \\ \vdots \\ 0 \end{bmatrix} + \cdots v_n\begin{bmatrix} 0 \\ 0 \\ \vdots \\ 1 \end{bmatrix}. \qquad (9.2.31)$$

Let us work out an example. Consider the vector $|V\rangle$ with components

$$|V\rangle = \begin{bmatrix} 1+i \\ \sqrt{3}+i \end{bmatrix}, \tag{9.2.32}$$

where I have abused notation and equated an abstract ket to its components in a basis. It is to be expanded in a new orthonormal basis $|I\rangle$, $|II\rangle$ with components

$$|I\rangle = \frac{1}{\sqrt{2}} \begin{bmatrix} 1 \\ 1 \end{bmatrix} \qquad\qquad |II\rangle = \frac{1}{\sqrt{2}} \begin{bmatrix} 1 \\ -1 \end{bmatrix}. \tag{9.2.33}$$

Let us write

$$|V\rangle = v_I|I\rangle + v_{II}|II\rangle \tag{9.2.34}$$

and determine v_I and v_{II}. The idea, as in elementary vector analysis, is to take the dot product of both sides with the corresponding normalized basis vector. Thus

$$v_I = \langle I|V\rangle = \frac{1}{\sqrt{2}}[1,1] \begin{bmatrix} 1+i \\ \sqrt{3}+i \end{bmatrix} = \frac{1}{\sqrt{2}}(1+\sqrt{3}+2i). \tag{9.2.35}$$

Likewise

$$v_{II} = \frac{1}{\sqrt{2}}(1-\sqrt{3}). \tag{9.2.36}$$

As a check on the calculation, let us recompute the norm squared of the vector and see if it equals $|1+i|^2 + |\sqrt{3}+i|^2 = 6$. We find

$$|v_I|^2 + |v_{II}|^2 = \frac{1}{2}[1+3+2\sqrt{3}+4+1+3-2\sqrt{3}] = 6. \tag{9.2.37}$$

The following exercises give you more practice. In doing them remember that if a basis vector is complex, you must conjugate its components in taking the dot product.

Problem 9.2.1. (Very important).

(i) *Repeat the above calculation, of expanding the vector in Eqn. (9.2.32), but in the following basis, after first demonstrating its orthonormality. At the end check that the norm squared of the vector comes out to be 6.*

$$|I\rangle = \frac{1}{\sqrt{2}} \begin{bmatrix} 1 \\ i \end{bmatrix} \qquad\qquad |II\rangle = \frac{1}{\sqrt{2}} \begin{bmatrix} 1 \\ -i \end{bmatrix} \tag{9.2.38}$$

(ii) *Repeat all of the above for the basis*

$$I\rangle = \begin{bmatrix} \frac{1+i\sqrt{3}}{4} \\ -\frac{\sqrt{3}(1+i)}{\sqrt{8}} \end{bmatrix} \qquad\qquad |II\rangle = \begin{bmatrix} \frac{\sqrt{3}(1+i)}{\sqrt{8}} \\ \frac{\sqrt{3}+i}{4} \end{bmatrix} \tag{9.2.39}$$

Gram–Schmidt Theorem

Let us now take up the Gram–Schmidt procedure for converting a linearly independent basis into an orthonormal one. The basic idea can be seen by a simple example. Imagine the two-dimensional space of arrows in a plane. Let us take two nonparallel vectors, $|I\rangle$ and $|II\rangle$, which qualify as a basis. To get an orthonormal basis out of these, we do the following:

- Rescale the first by its own length, so it becomes a unit vector. This will be the first basis vector.

- Subtract from the second vector its projection along the first, leaving behind only the part perpendicular to the first. (Such a part will remain since by assumption the vectors are nonparallel.)

- Rescale the left over piece by its own length. We now have the second basis vector: it is orthogonal to the first and of unit length.

This simple example tells the whole story behind this procedure, which will now be discussed in general terms in the Dirac notation.

Let $|I\rangle, |II\rangle, \dots$ be a linearly independent basis. The first vector of the orthonormal basis will be

$$|1\rangle = \frac{|I\rangle}{|I|} \quad \text{where} \quad |I| = \sqrt{\langle I|I\rangle}.$$

Clearly

$$\langle 1|1\rangle = \frac{\langle I|I\rangle}{|I|^2} = 1.$$

As for the second vector in the basis, consider

$$|2'\rangle = |II\rangle - |1\rangle\langle 1|II\rangle,$$

which is $|II\rangle$ minus the part pointing along the first unit vector. (Think of the arrow example as you read on.) Not surprisingly it is orthogonal to the latter:

$$\langle 1|2'\rangle = \langle 1|II\rangle - \langle 1|1\rangle\langle 1|II\rangle = 0.$$

We now divide $|2'\rangle$ by its norm to get $|2\rangle$, which will be orthogonal to the first and normalized to unity. Next consider

$$|3'\rangle = |III\rangle - |1\rangle\langle 1|III\rangle - |2\rangle\langle 2|III\rangle$$

which is orthogonal to both $|1\rangle$ and $|2\rangle$. Dividing by its norm we get $|3\rangle$, the third member of the orthonormal basis. There is nothing new with the generation of the rest of the basis.

Where did we use the linear independence of the original basis? What if we had started with a linearly dependent basis? Then at some point a vector like $|2'\rangle$ or $|3'\rangle$ would have vanished, putting a stop to the whole procedure. On the other hand, linear independence will assure us that such a thing will never happen since it amounts to having a nontrivial linear combination of linearly independent vectors that adds up to the null vector. (Go back to the equations for $|2'\rangle$ or $|3'\rangle$ and satisfy yourself that these are linear combinations of the old basis vectors.)

Let us work out an example, using the two linearly independent vectors $|V\rangle$ and $|W\rangle$ from Eqn(9.2.18-9.2.19). Since an orthonormal basis $|1\rangle$ and $|2\rangle$ is already present in the definition of these vectors, let us use the symbols $|G1\rangle$ and $|G2\rangle$ to denote the ones arising from the Gram–Schmidt procedure. From $|V\rangle$ let us form

$$|G1\rangle = \frac{|V\rangle}{\sqrt{\langle V|V\rangle}} = \frac{1}{\sqrt{86}}|V\rangle = \frac{1}{\sqrt{86}}[(3-4i)|1\rangle + (5-6i)|2\rangle] \qquad (9.2.40)$$

as our first normalized basis vector. Then we form the unnormalized vector

$$|G2'\rangle = |W\rangle - |G1\rangle\langle G1|W\rangle \qquad (9.2.41)$$

$$= [(1-i)|1\rangle + (2-3i)|2\rangle] - \frac{1}{86}(35-2i)|V\rangle \qquad (9.2.42)$$

$$= \frac{-11+60i}{86}|1\rangle + \frac{9-38i}{86}|2\rangle. \qquad (9.2.43)$$

Lastly we divide $|G2'\rangle$ by its norm ($\sqrt{61/86}$) to get $|G2\rangle$.

Problem 9.2.2. *Form an orthonormal basis in two dimensions starting with* $\mathbf{A} = 3\mathbf{i} + 4\mathbf{j}$ *and* $\mathbf{B} = 2\mathbf{i} - 6\mathbf{j}$. *Can you generate another orthonormal basis starting with these two vectors? If so, produce another.*

Problem 9.2.3. *Show how to go from the basis*

$$|I\rangle = \begin{bmatrix} 3 \\ 0 \\ 0 \end{bmatrix} \quad |II\rangle = \begin{bmatrix} 0 \\ 1 \\ 2 \end{bmatrix} \quad |III\rangle = \begin{bmatrix} 0 \\ 2 \\ 5 \end{bmatrix}$$

to the orthonormal basis

$$|1\rangle = \begin{bmatrix} 1 \\ 0 \\ 0 \end{bmatrix} \quad |2\rangle = \begin{bmatrix} 0 \\ 1/\sqrt{5} \\ 2/\sqrt{5} \end{bmatrix} \quad |3\rangle = \begin{bmatrix} 0 \\ -2/\sqrt{5} \\ 1/\sqrt{5} \end{bmatrix}.$$

When we first learn about dimensionality, we associate it with the number of perpendicular directions. In this chapter we defined it in terms of the maximum number of linearly independent vectors. The following theorem connects the two definitions.

Theorem 9.4. *The dimensionality of a space equals* $n_\perp$, *the maximum number of mutually orthogonal vectors in it.*

Proof: First note that any mutually orthogonal set is also linearly independent. Suppose we had a linear combination of orthogonal vectors adding up to zero. By taking the dot product of both sides with any one member and using the orthogonality we can show that the coefficient multiplying that vector has to vanish. This can clearly be done for all the coefficients, showing the linear combination is trivial. Now $n_\perp$ can only be equal to, greater than or lesser than n, the dimensionality of the space. The Gram–Schmidt procedure eliminates the last case by explicit construction, while the linear independence of the perpendicular vectors rules out the penultimate option. ■

Two powerful theorems apply to any inner product space obeying our axioms:

Theorem 9.5. (The Schwarz inequality).

$$|\langle V|W\rangle| \leq |V||W|. \tag{9.2.44}$$

Theorem 9.6. (The Triangle inequality).

$$|V + W| \leq |V| + |W|. \tag{9.2.45}$$

Proof: The proof of the first will be provided so you can get used to working with bras and kets. The second will be left as an exercise.

Before proving anything, note that the results are obviously true for arrows: the *Schwarz inequality* says that the dot product of two vectors cannot exceed the product of their lengths and the *Triangle inequality* says that the length of a sum cannot exceed the sum of the lengths. This is an example which illustrates the merits of thinking of abstract vectors as arrows and guessing what properties they might share with arrows. The proof will of course have to rely on just the axioms.

To prove the Schwarz inequality, consider axiom B-II applied to

$$|Z\rangle = |V\rangle - \frac{\langle W|V\rangle}{|W|^2}|W\rangle. \tag{9.2.46}$$

We get

$$
\begin{aligned}
\langle Z|Z\rangle &= \langle V - \frac{\langle W|V\rangle}{|W|^2}W \Big| V - \frac{\langle W|V\rangle}{|W|^2}W\rangle \qquad (9.2.47)\\
&= \langle V|V\rangle - \frac{\langle W|V\rangle\langle V|W\rangle}{|W|^2} - \frac{\langle W|V\rangle^*\langle W|V\rangle}{|W|^2}\\
&\quad + \frac{\langle W|V\rangle^*\langle W|V\rangle\langle W|W\rangle}{|W|^4}\\
&\geq 0, \qquad\qquad\qquad\qquad\qquad\qquad\qquad\quad (9.2.48)
\end{aligned}
$$

where we have used the antilinearity of the inner product with respect to the bra. Using

$$\langle W|V\rangle^* = \langle V|W\rangle$$

we find

$$\langle V|V\rangle \geq \frac{\langle W|V\rangle\langle V|W\rangle}{|W|^2}. \tag{9.2.49}$$

Cross multiplying by $|W|^2$ and taking square roots, the result follows. ∎

Problem 9.2.4. *When will this inequality be satisfied? Does this agree with your experience with arrows?*

Problem 9.2.5. *Prove the triangle inequality starting with $|V+W|^2$. You must use* Re $\langle V|W\rangle \leq |\langle V|W\rangle|$ *and the Schwarz inequality. Show that the final inequality becomes an equality only if $|V\rangle = a|W\rangle$ where a is a real positive scalar.*

Problem 9.2.6. *Verify that $|1\rangle = (1/\sqrt{2})[1, -i, 0]^T$, $|2\rangle = (1/\sqrt{2})[1, i, 0]^T$, $|3\rangle = [0, 0, 1]^T$ are orthonormal and find the coefficients in the expansion $|V\rangle = [3 - 4i, 5 - 6i, 8]^T = \alpha|1\rangle + \beta|2\rangle + \gamma|3\rangle$.*

9.3. Linear Operators

An operator Ω is a machine that takes in vectors in our vector space and spits out vectors also from the same space. If $|V\rangle$ is the input vector, we denote the output vector as $|V'\rangle = \Omega|V\rangle$. We say that *the operator Ω acts on the ket $|V\rangle$ to yield the ket $|V'\rangle$*. An operator is just like a function f which takes in an x and gives out $f(x)$; here the input is a vector and so is the output. To fully know an operator we need an infinite amount of information: a table of what happens to each vector when the operator does its job, that is, all the input-output pairs $(|V\rangle, |V'\rangle)$. This is the same as saying that if we want to specify a function $f(x)$ we need to know all the pairs $(x, f(x))$, that is, the output $f(x)$ associated with each input x, which is what the graph of f tells us.

Here are some operators that act on arrows:

- The unit operator, I whose instruction is: Leave the vector alone! We don't need a fancy table for this one.

- The rotation operator $R_z(\theta)$ whose instruction is: Turn the vector around the z-axis by θ. (The sense of rotation is such that a screw driver turned the same way would move the screw up the z-axis. When we look down the z-axis at the $x - y$ plane the points should move counterclockwise.)

- The operator SQ whose instruction is: Square the components of the vector.

We will restrict our study to *linear operators*. They satisfy:

$$\Omega(a|V\rangle + b|W\rangle) = a\Omega|V\rangle + b\Omega|W\rangle, \tag{9.3.1}$$

which means that the output corresponding to a linear combination of vectors is the corresponding linear combination of the outputs. In this case we need only n^2 pieces of information in n-dimensions to know the operator fully. *This is because every vector is a linear combination of n basis vectors and once the fate of the basis vectors is known, the fate of every vector is known.* Thus given the basis vectors $|1\rangle \cdots |n\rangle$, we need to know their destinations, $|1'\rangle \cdots |n'\rangle$ under the action of Ω. We can specify each member of the latter set by their n-components, bringing the total to n^2 pieces of information.

Consider as an example the rotation operator $R_z(\pi/2)$. I will now argue that it is linear. Let c be the sum of a and b. Thus the three vectors form a triangle. Now rotate this triangle around the z-axis by $\pi/2$. The rotated vectors will still form a triangle and each side of the triangle now equals a rotated vector. Thus we have in Dirac notation

$$|c'\rangle = |a'\rangle + |b'\rangle \quad \text{that is} \tag{9.3.2}$$
$$R_z(\pi/2)|c\rangle = R_z(\pi/2)|a\rangle + R_z(\pi/2)|b\rangle \tag{9.3.3}$$
$$R_z(\pi/2)(|a\rangle + |b\rangle) = R_z(\pi/2)|a\rangle + R_z(\pi/2)|b\rangle \tag{9.3.4}$$

showing that $R_z(\pi/2)$ is a linear operator. Its action on the basis vectors $\mathbf{i} \equiv |1\rangle$, $\mathbf{j} \equiv |2\rangle$ and $\mathbf{k} \equiv |3\rangle$ is (check this with a figure!)

$$R_z(\pi/2)|1\rangle = |2\rangle \tag{9.3.5}$$
$$R_z(\pi/2)|2\rangle = -|1\rangle \tag{9.3.6}$$
$$R_z(\pi/2)|3\rangle = |3\rangle. \tag{9.3.7}$$

Its action on a linear combination

$$|V\rangle = v_1|1\rangle + v_2|2\rangle + v_3|3\rangle \tag{9.3.8}$$

follows:

$$\begin{aligned} R_z(\pi/2)|V\rangle &= R_z(\pi/2)(v_1|1\rangle + v_2|2\rangle + v_3|3\rangle) \\ &= v_1 R_z(\pi/2)|1\rangle + v_2 R_z(\pi/2)|2\rangle + v_3 R_z(\pi/2)|3\rangle \\ &= v_1|2\rangle - v_2|1\rangle + v_3|3\rangle. \end{aligned} \tag{9.3.9}$$

Let us carry out this analysis for the general case of a linear operator Ω acting on the vectors in an n-dimensional space. Let

$$\Omega|j\rangle = |j'\rangle \tag{9.3.10}$$

be the action on the basis vectors. We assume $|j'\rangle$ is known through its components. Then

$$|V'\rangle = \Omega|V\rangle \tag{9.3.11}$$

$$= \Omega(\sum_j v_j|j\rangle) \tag{9.3.12}$$

$$= \sum_j v_j \Omega|j\rangle \tag{9.3.13}$$

$$= \sum_j v_j|j'\rangle \tag{9.3.14}$$

is fully known. Let us be more explicit. Let us write the components of the transformed or output vector $|V'\rangle$. To find v_i' we take the dot product with $\langle i|$:

$$v_i' = \sum_j v_j \langle i|j'\rangle. \tag{9.3.15}$$

To proceed further (as we did in the rotation example) we must know $|j'\rangle$. This information we assemble as a matrix with elements Ω_{ij} given by

$$\Omega_{ij} = \langle i|j'\rangle = \langle i|\Omega|j\rangle. \tag{9.3.16}$$

In terms of this matrix Eqn. (9.3.15) takes the form:

$$v_i' = \sum_j \Omega_{ij} v_j, \tag{9.3.17}$$

which is just the matrix equation:

$$
\begin{bmatrix} v_1' \\ \vdots \\ v_n' \end{bmatrix} =
\begin{bmatrix}
\Omega_{11} & \Omega_{12} & \cdots & \Omega_{1n} \\
\Omega_{21} & \Omega_{22} & \cdots & \Omega_{2n} \\
\vdots & \vdots & \vdots & \vdots \\
\Omega_{n1} & \Omega_{n2} & \cdots & \Omega_{nn}
\end{bmatrix}
\begin{bmatrix} v_1 \\ \vdots \\ v_n \end{bmatrix} \tag{9.3.18}
$$

The above matrix relation represents, in the chosen basis, the abstract linear operator Ω transforming a vector $|V\rangle$ into $|V'\rangle$. Just as abstract vectors turn into numbers (the components, assembled into a column) in a basis, so do linear operators turn into matrices. We use the same symbol to represent the matrix as we do to represent the operator in the abstract. The first column of the matrix contains the numbers $\Omega_{11}, \Omega_{21} \ldots \Omega_{n1}$, which are the components of the first basis vector after the operator has acted on it. Likewise *the j-th column of the matrix contains the components of the j-th basis vector after the transformation.* Given this *mnemonic* or Eqn. (9.3.16) we can write down the matrix corresponding to $R_z(\pi/2)$

$$
R_z(\pi/2) = \begin{bmatrix} 0 & -1 & 0 \\ 1 & 0 & 0 \\ 0 & 0 & 1 \end{bmatrix}. \tag{9.3.19}
$$

Problem 9.3.1. *Verify Eqn. (9.3.19) using the mnemonic or Eqn. (9.3.16).*

Problem 9.3.2. *Find the matrices corresponding to the projection operators P_x, P_y, P_z which preserve the components of a vector in the directions x, y, or z respectively but kill the other two. Either use the mnemonic or Eqn. (9.3.16). What do you get when you form the sum of the projection operators? Give arguments for why the result had to be so.*

Problem 9.3.3. *Is the operator SQ which squares the components of a vector a linear operator? Give some explanation.*

With the mnemonic still fresh in mind let us pause to note an important property of matrices describing rotations. Imagine some rigid rotation R that transforms the usual unit vectors $\mathbf{i}$, $\mathbf{j}$, and $\mathbf{k}$ to $\mathbf{i}'$, $\mathbf{j}'$, and $\mathbf{k}'$. By our mnemonic it is given by the following matrix:

$$R_{ij} = \begin{bmatrix} \mathbf{i}\cdot\mathbf{i}' & \mathbf{i}\cdot\mathbf{j}' & \mathbf{i}\cdot\mathbf{k}' \\ \mathbf{j}\cdot\mathbf{i}' & \mathbf{j}\cdot\mathbf{j}' & \mathbf{j}\cdot\mathbf{k}' \\ \mathbf{k}\cdot\mathbf{i}' & \mathbf{k}\cdot\mathbf{j}' & \mathbf{k}\cdot\mathbf{k}' \end{bmatrix} \tag{9.3.20}$$

that is to say, the n-th column contains the components of the n-th transformed basis vector. In the present problem, where the transformation is a rigid rotation *we know that $\mathbf{i}'$, $\mathbf{j}'$, and $\mathbf{k}'$ form an orthonormal set of vectors: the rotation does not change the lengths or angles between vectors.* Thus the sum of the squares of each column in R is unity while the dot product of two columns is zero. It turns out that this makes R an orthogonal matrix

$$R^T R = I. \tag{9.3.21}$$

To see this, evaluate $R^T R$ and remember that

$$\begin{aligned} \mathbf{A}\cdot\mathbf{B} &= A_x B_x + A_y B_y + A_z B_z \\ &= (\mathbf{A}\cdot\mathbf{i})(\mathbf{B}\cdot\mathbf{i}) + (\mathbf{A}\cdot\mathbf{j})(\mathbf{B}\cdot\mathbf{j}) \\ &+ (\mathbf{A}\cdot\mathbf{k})(\mathbf{B}\cdot\mathbf{k}). \end{aligned} \tag{9.3.22}$$

You will find that the ij entry of $R^T R$ is just $\mathbf{i}'\cdot\mathbf{j}' = \delta_{i'j'}$.

Note also that the *rows* of R are also orthonormal: this is because they contain the components of the old basis vectors in the rotated basis. Likewise, in a complex vector space, unitary matrices ($U^\dagger U = I$) convert orthonormal bases to each other.

Problem 9.3.4. *Verify that the rotation matrix is indeed orthogonal by carrying out the steps indicated above.*

Problem 9.3.5. *You have seen above the matrix R_z (Eqn. (9.3.19)) that rotates by $\pi/2$ about the z-axis. Construct a matrix that rotates by an arbitrary angle about the z-axis. Repeat for a rotation around the x-axis by some other angle. Verify that each matrix is orthogonal. Take their product and verify that it is also orthogonal. Show in general that the product of two orthogonal matrices is orthogonal. (Remember the rule for the transpose of a product).*

Problem 9.3.6. *Construct the matrix of the operator $R_x(\pi/2)$ that rotates around the x-axis by $90°$. Calculate $[R_z(\pi/2), R_x(\pi/2)]$.*

Problem 9.3.7. *Repeat this calculation with both rotation angles equal to $45°$.*

Consider the sum of two linear operators. The matrix representing the sum is the sum of the corresponding matrices. The proof is left as an exercise. How about the product?

Consider the relation

$$|V''\rangle = M(N|V\rangle) \tag{9.3.23}$$

by which we mean that N acts on $|V\rangle$ to give

$$N|V\rangle = |V'\rangle \tag{9.3.24}$$

and then M acts on $|V'\rangle$ to give $|V''\rangle$. Let us denote by MN the operator that will produce the same effect in one step. Thus

$$
\begin{aligned}
|V''\rangle &= M|V'\rangle & (9.3.25) \\
&= M(N|V\rangle) & (9.3.26) \\
&\equiv (MN)|V\rangle. & (9.3.27)
\end{aligned}
$$

What are the matrix elements of MN? We find them as follows. First we dot both sides of Eqn. (9.3.25) by $\langle i|$ to get

$$
\begin{aligned}
v_i'' &= \langle i|M|V'\rangle & (9.3.28) \\
&= \sum_j M_{ij} v_j' & (9.3.29) \\
&= \sum_j M_{ij} \sum_k N_{jk} v_k & (9.3.30) \\
&\equiv \sum_k (MN)_{ik} v_k, & (9.3.31)
\end{aligned}
$$

where in the last line we have the result of dotting both sides of Eqn. (9.3.27) with $\langle i|$. By comparing the last two equations we find

$$(MN)_{ik} = \sum_j M_{ij}N_{jk}, \tag{9.3.32}$$

which simply means that the matrix corresponding to MN is given by the product of the matrices representing M and N.

Recall that with each matrix M we can associate an adjoint matrix $M^\dagger$ obtained by transpose conjugation. *The adjoint matrix must in turn correspond to another distinct but related operator.* We call that operator the adjoint of the original operator and use the dagger in its label as we did with its matrix.

Definition 9.9. *An operator is hermitian or unitary if the corresponding matrix is. Likewise an operator is the inverse of another if the corresponding matrices are inverses.*

For later reference we derive the following result. Consider

$$\langle V|\Omega|W\rangle^* = \left[\sum_i \sum_j v_i^* \Omega_{ij} w_j\right]^* \tag{9.3.33}$$

$$= \sum_i \sum_j v_i \Omega_{ij}^* w_j^* \tag{9.3.34}$$

$$= \sum_i \sum_j w_j^* \Omega_{ji}^\dagger v_i \tag{9.3.35}$$

$$= \langle W|\Omega^\dagger|V\rangle \tag{9.3.36}$$

$$= \langle W|\Omega|V\rangle \quad \text{if hermitian} \tag{9.3.37}$$

$$= \langle W|\Omega^{-1}|V\rangle \quad \text{if unitary} \tag{9.3.38}$$

9.4. Some Advanced Topics

Digression on dual spaces

We begin with a technical point regarding the inner product, which is a number we are trying to generate from two kets $|V\rangle$ and $|W\rangle$, which are *both represented by column vectors* in some basis. Now there is no way to make a number out of two columns by direct matrix multiplication, but there is a way to make a number by matrix multiplication of a row times a column. *Our trick for producing a number out of two columns has been to associate a unique row vector with one column (its transpose conjugate) and form its matrix product with the column representing the other.* This has the feature that the answer depends on which of the two vectors we are going to convert to the row, the two choices ($\langle V|W \rangle$ and $\langle W|V \rangle$) leading to answers related by complex conjugation as per axiom B-I.

But one can also take the following alternate view. Column vectors are concrete manifestations of an abstract vector $|W \rangle$ or ket in a basis. We can also work backwards and go from the column vectors to the abstract kets. But then it is similarly possible to work backwards and associate with each *row vector* an abstract object $\langle V|$, called *bra-V*. Now we can name the bra's as we want but let us do the following. Associated with every ket $|V \rangle$ is a column vector. Let us take its *adjoint*, or transpose conjugate and form a row vector. The abstract bra associated with this will bear the same label, i.e., be called $\langle V|$. *Thus there are two vector spaces, the space of kets and a dual space of bras, with a ket for every bra and vice versa (the components being related by the adjoint operation).* Inner products are really defined only between bras and kets and hence between elements of two distinct but related vector spaces. There is a basis of vectors $|i \rangle$ for expanding kets and a similar basis $\langle i|$ for expanding bras. The basis ket $|i \rangle$ is represented in the basis we are using by a column vector with all zeros except for a 1 in the i-th row, while the basis bra $\langle i|$ is a row vector with all zeros except for a 1 in the i-th column.

All this may be summarized as follows:

$$|V \rangle \leftrightarrow \begin{bmatrix} v_1 \\ v_2 \\ \cdot \\ \cdot \\ \cdot \\ v_n \end{bmatrix} \leftrightarrow [v_1^*, v_2^*, \dots v_n^*] \leftrightarrow \langle V|. \qquad (9.4.1)$$

where $\leftrightarrow$ means "within a basis."

There is however nothing wrong with the first viewpoint of associating a scalar product with a pair of columns or kets (making no reference to another dual space) and living with the asymmetry between the first and second vector in the inner product (which one to transpose conjugate?) After all, we live with the fact that the cross product is sensitive the order of the factors.

If you found the above discussion heavy going, you can temporarily ignore it. The only thing you must remember is that in the case of a general nonarrow vector space:

- Vectors can still be assigned components in some orthonormal basis, just as with arrows, but these may be complex.

- The inner product of any two vectors is given in terms of these components by Eqn. (9.2.16). This product obeys all the axioms.

Adjoint operation

We have seen that we may pass from the column representing a ket to the row representing the corresponding bra by the adjoint operation, i.e., transpose conjugation. Let us now ask: if $\langle V|$ is the bra corresponding to the ket $|V \rangle$ what bra corresponds to $a|V \rangle$ where a is some scalar? By going to any basis it is readily found that

$$a|V \rangle \rightarrow \begin{bmatrix} av_1 \\ av_2 \\ \cdot \\ \cdot \\ \cdot \\ av_n \end{bmatrix} \rightarrow [a^*v_1^*, a^*v_2^*, \dots a^*v_n^*] \rightarrow \langle V|a^*. \qquad (9.4.2)$$

It is customary to write $a|V \rangle$ as $|aV \rangle$ and the corresponding bra as $\langle aV|$. What we have found is that

$$\langle aV| = \langle V|a^*. \qquad (9.4.3)$$

As a result, we can say that if we have an equation among kets such as

$$a|V\rangle = b|W\rangle + c|Z\rangle + \cdots ;$$ (9.4.4)

this implies another one among the corresponding bras:

$$\langle V|a^* = \langle W|b^* + \langle Z|c^* + \cdots .$$ (9.4.5)

The two equations above are said to be *adjoints of each other*. Just as any equation involving complex numbers implies another obtained by taking the complex conjugates of both sides, an equation between (bras) kets implies another one between (kets) bras. If you think in a basis, you will see that this follows simply from the fact that if two columns are equal, so are their transpose conjugates.

Here is the rule for taking the adjoint:

To take the adjoint of a linear equation relating kets (bras), replace every ket (bra) by its bra (ket) and complex conjugate all coefficients.

We can extend this rule as follows. Suppose we have an expansion for a vector:

$$|V\rangle = \sum_{i=1} v_i|i\rangle$$ (9.4.6)

in terms of basis vectors. The adjoint is

$$\langle V| = \sum_{i=1} \langle i|v_i^*$$

Recalling that $v_i = \langle i|V\rangle$ and $v_i^* = \langle V|i\rangle$, it follows that the adjoint of

$$|V\rangle = \sum_{i=1} |i\rangle\langle i|V\rangle$$ (9.4.7)

is

$$\langle V| = \sum_{i=1} \langle V|i\rangle\langle i|$$ (9.4.8)

from which comes the rule:

To take the adjoint of an equation involving bras and kets and coefficients, reverse the order of all factors, exchanging bras and kets and complex conjugating all coefficients.

Taking the adjoint in expressions with operators

Recall that every relation involving matrices, column vectors and scalars implied another one where we took the adjoint of everything in sight and reversed the order of factors. (For scalars, the adjoint is just the complex conjugate. Since scalars commute with everything, there is no real need to reverse their order, but this may be done for uniformity.) *This is the concrete representation in a basis of the statement that every relation between operators, kets and scalars has a counterpart with operators replaced by their adjoints and kets replaced by their bras and scalars replaced by their conjugates.* For example

$$|V\rangle = 5i\Omega|W\rangle + |Z\rangle$$ (9.4.9)

implies that

$$\langle V| = -5i\langle W|\Omega^\dagger + \langle Z|.$$ (9.4.10)

9.5. The Eigenvalue Problem

We now turn to a basic problem in the theory of linear operators. We first learn
what the problem is and what its solution are, and then turn to applications.

Consider some general nontrivial operator Ω. When it acts on a vector $|V\rangle$
it gives us another one related to it in some complicated way, involving rotations,
rescalings, and so on. But a nontrivial operator (which is not the identity or a
multiple of it) can have some privileged vectors, called its *eigenvectors* on which
its action is to simply multiply by a number, which is called the corresponding
eigenvalue. Take for example $R_z(\pi/2)$. While its effect on a generic vector is to
rotate it, its action on vectors along the z-axis is to leave them alone. Thus we can
say that any vector along z-axis is an eigenvector of $R_z(\pi/2)$ with eigenvalue 1.

Consider one more example, the operator M which has the following matrix
representation:

$$M = \begin{bmatrix} 0 & 1 \\ 1 & 0 \end{bmatrix}. \tag{9.5.1}$$

This operator exchanges the components of any vector:

$$\begin{bmatrix} 0 & 1 \\ 1 & 0 \end{bmatrix} \begin{bmatrix} v_1 \\ v_2 \end{bmatrix} = \begin{bmatrix} v_2 \\ v_1 \end{bmatrix}. \tag{9.5.2}$$

This will affect a generic vector in a nontrivial way. But it will have no effect
on a vector with equal components. This statement has a geometric interpretation.
Imagine the vectors as lying in the x-y plane. Exchanging the components amounts
to reflecting them on a mirror passing through the origin at $45°$ exactly between the
two axes. (Hence the name M.) Clearly any vector along the mirror is unaffected
by this reflection, as shown in Fig. 9.2. Let us choose as a representative from this
direction one with $v_1 = v_2 = 1$.

From the figure it is clear that we can get another set of special vectors:
these lie perpendicular to the mirror and get reflected to minus themselves, i.e.,
correspond to eigenvalue -1. These have equal and opposite components. Let us
choose as a representative of this direction one with $v_1 = -v_2 = 1$. Thus

$$\begin{bmatrix} 0 & 1 \\ 1 & 0 \end{bmatrix} \begin{bmatrix} 1 \\ \pm 1 \end{bmatrix} = \pm 1 \begin{bmatrix} 1 \\ \pm 1 \end{bmatrix}. \tag{9.5.3}$$

It seems apparent that there are no other eigenvectors.

We would like to ask the following questions at this point:

- How are the eigenvectors and eigenvalues to be found in the case of a general
  operator?

- Will the eigenvalues always be ± 1? Will they always be real?

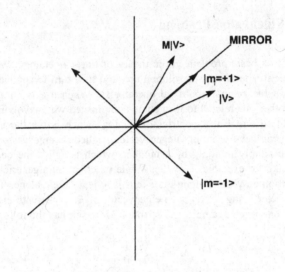

Figure 9.2. The effect of the operator M is to act as a mirror through the origin at $45°$.

- Will the eigenvectors corresponding to different eigenvalues always be perpendicular as in the case of M?

- Will it generally be true that given one eigenvector, any multiple of it is also an eigenvector? In other words, will they always be defined only in their direction (as being parallel or perpendicular to the mirror in our example)?

- Will an $n \times n$ matrix always have n eigenvectors and eigenvalues?

So we begin with the general eigenvalue equation:

$$\Omega|V\rangle = \omega|V\rangle. \tag{9.5.4}$$

Thus $|V\rangle$ is an *eigenvector of Ω with eigenvalue ω*. It is common to name the eigenvector by its eigenvalue. I will also use, as far as possible, lower case symbols for the eigenvalue corresponding to an operator denoted by the corresponding upper case symbol. Henceforth we will follow this convention and rewrite the eigenvalue equation as[2]

$$\Omega|\omega\rangle = \omega|\omega\rangle. \tag{9.5.5}$$

Suppose we wish to see if an operator has any eigenvectors and if so, what they are and what the corresponding eigenvalues are. We carry out the following ma-

---

[2]This nomenclature tells us at once that the vector $|\omega\rangle$ is an eigenvector of Ω with eigenvalue ω. This can be very useful in, say quantum physics, where we may be working with many eigenvectors of many operators. This nomenclature is not common in mathematics literature.

nipulations on the eigenvalue equation:

$$\Omega|\omega\rangle = \omega|\omega\rangle = \omega I|\omega\rangle \qquad (9.5.6)$$

$$(\Omega - \omega I)|\omega\rangle = |0\rangle \qquad (9.5.7)$$

$$|\omega\rangle = (\Omega - \omega I)^{-1}|0\rangle. \qquad (9.5.8)$$

There, we solved the equation! That is the good news. The bad news is that the result is trivial since any matrix acting on the null vector (all components zero) is going to give the null vector. We have just found that the null vectors obeys the eigenvalue equation for any eigenvalue! Of course what we really want is a nonzero eigenvector (as in the mirror example) and some way to select out its eigenvalues (± 1) from all possible numbers. Looking at the last equation we see that the only possible hope is if the inverse operator multiplying the null vector is infinite. Suppose we go to some basis where all operators have become matrices. The inverse of any matrix is the cofactor transpose divided by the determinant. The cofactor of a finite matrix is clearly finite. *Thus the determinant of $\Omega - \omega I$ must vanish if a nontrivial eigenvector is to exist.* If you imagine computing the determinant of such an $n \times n$ matrix you will see that it leads to a polynomial in ω of order n. This is called the *characteristic polynomial*. An n-th order polynomial will have n roots, i.e., vanish at n points, possibly complex. These are the allowed eigenvalues. To find the eigenvectors, we must go back to the eigenvalue equation with this eigenvalue. Let us work out the example of M. The eigenvalue equation is

$$M|m\rangle = m|m\rangle \qquad (9.5.9)$$

where we are labeling the eigenket by the eigenvalue. In concrete form (in our basis) the characteristic equation becomes

$$\begin{vmatrix} 0-m & 1 \\ 1 & 0-m \end{vmatrix} = m^2 - 1 = 0 \qquad (9.5.10)$$

with roots ± 1 as advertised earlier. Let us now find the eigenvector corresponding to $m = 1$. Its components obey

$$\begin{bmatrix} 0 & 1 \\ 1 & 0 \end{bmatrix} \begin{bmatrix} v_1 \\ v_2 \end{bmatrix} = +1 \begin{bmatrix} v_1 \\ v_2 \end{bmatrix}. \qquad (9.5.11)$$

Comparing coefficients we get two equations:

$$v_1 = v_2 \qquad (9.5.12)$$

$$v_2 = v_1. \qquad (9.5.13)$$

Notice that the two equations are redundant. But let us not complain: this is what made the determinant vanish and kept the problem alive. But how are we to find the two-dimensional vector given just one equation? The one good equation residing in the two equations above tells us the *direction* of the eigenvector: its components are equal and it points in the 45° direction, halfway between between the basis vectors. But what is its length? *The eigenvalue equation will never tell us.* We know the reason in the case of the mirror, but what is it in general?

To understand this go back to Eqn. (9.5.5). Multiply both sides by a scalar a from the left, use the linearity of the operator to slide it past the operator to the ket as follows

$$a\Omega|\omega\rangle \ = \ a\omega|\omega\rangle \tag{9.5.14}$$
$$\Omega(a|\omega\rangle) \ = \ \omega(a|\omega\rangle) \tag{9.5.15}$$

and you see that *if* $|\omega\rangle$ *is an eigenvector with eigenvalue* ω *so is* $a|\omega\rangle$. Thus the eigenvalue equation can determine the eigenvector only in its "direction" (i.e., determine the ratio of its components) but not its absolute value. One can pick any vector in this direction. We usually pick one that is normalized. In the present case that gives

$$|m = +1\rangle \to \frac{1}{\sqrt{2}}\begin{bmatrix} 1 \\ 1 \end{bmatrix}. \tag{9.5.16}$$

where I use $\to$ rather than $=$ because you can't really equate an abstract vector to its representation in a basis. Hereafter I will not be so pedantic and use the $=$ sign which really means "equal in this basis."

The eigenvector is still not unique, we can multiply all components by a unimodular phase factor without changing the norm. The usual procedure is to keep things as simple as possible and stop with the solution above. Note that eigenvectors related by rescaling are never counted as different solutions. There is however a second truly distinct eigenvector corresponding to eigenvalue -1. We can find it the same way and obtain

$$|m = -1\rangle = \frac{1}{\sqrt{2}}\begin{bmatrix} 1 \\ -1 \end{bmatrix}. \tag{9.5.17}$$

Assembling together the full solution we have:

$$|m = +1\rangle = \frac{1}{\sqrt{2}}\begin{bmatrix} 1 \\ 1 \end{bmatrix} \qquad |m = -1\rangle = \frac{1}{\sqrt{2}}\begin{bmatrix} 1 \\ -1 \end{bmatrix}. \tag{9.5.18}$$

Problem 9.5.1. *Provide the details leading to Eqn. (9.5.18).*

Let us return to $R_z(\pi/2)$ and consider its eigenvalue problem. We know that any vector along the z-axis will be *invariant* with eigenvalue 1. Are there any more eigenvectors? It is hard to imagine any other vector (not along the z-axis) that will not get changed in its direction by R. Since our intuition stops here, let us follow the machinery. The characteristic equation for the eigenvalue r is

$$\begin{vmatrix} -r & -1 & 0 \\ 1 & -r & 0 \\ 0 & 0 & 1-r \end{vmatrix} = (r-1)(r^2+1) = 0 \qquad (9.5.19)$$

with the solutions

$$r = 1, i, -i. \qquad (9.5.20)$$

We already know that the eigenvector corresponding to $r = 1$ is

$$|r = 1\rangle = \begin{bmatrix} 0 \\ 0 \\ 1 \end{bmatrix} \qquad (9.5.21)$$

as you may verify.

It is clear why our intuition did not produce the other two eigenvectors: they do not correspond to real eigenvalues. Plugging back the eigenvalue i we get the following equations for the components of the eigenvector:

$$\begin{bmatrix} 0 & -1 & 0 \\ 1 & 0 & 0 \\ 0 & 0 & 1 \end{bmatrix} \begin{bmatrix} v_1 \\ v_2 \\ v_3 \end{bmatrix} = i \begin{bmatrix} v_1 \\ v_2 \\ v_3 \end{bmatrix} \qquad (9.5.22)$$

which tell us

$$-v_2 = iv_1 \qquad (9.5.23)$$
$$v_1 = iv_2 \qquad (9.5.24)$$
$$v_3 = iv_3. \qquad (9.5.25)$$

The last equation tells us $v_3 = 0$. The first two parrot each other and tell us $v_2 = -iv_1$. So we make a choice $v_1 = 1/\sqrt{2}, v_2 = -i/\sqrt{2}, v_3 = 0$. We can similarly solve for the third vector. Here is the full set of eigenvectors labeled by the eigenvalues:

$$|r = 1\rangle = \begin{bmatrix} 0 \\ 0 \\ 1 \end{bmatrix} \quad |r = i\rangle = \frac{1}{\sqrt{2}} \begin{bmatrix} 1 \\ -i \\ 0 \end{bmatrix} \quad |r = -i\rangle = \frac{1}{\sqrt{2}} \begin{bmatrix} 1 \\ i \\ 0 \end{bmatrix} \qquad (9.5.26)$$

Problem 9.5.2. *Find the eigenvalues and normalized eigenvectors of $R_z(\theta)$.*

Problem 9.5.3. *Find the eigenvalues and normalized eigenvectors of*

$$
\begin{bmatrix} 1 & 3 & 1 \\ 0 & 2 & 0 \\ 0 & 1 & 4 \end{bmatrix}
\begin{bmatrix} 0 & 0 & 1 \\ 0 & 0 & 0 \\ 1 & 0 & 0 \end{bmatrix}
\begin{bmatrix} 1 & 1 & 0 \\ 1 & 0 & 1 \\ 0 & 1 & 1 \end{bmatrix}
$$

and

$$
\begin{bmatrix} 0 & 1 & 0 \\ 1 & 0 & 1 \\ 0 & 1 & 0 \end{bmatrix}
\begin{bmatrix} 2 & 2 & 0 \\ 2 & 2 & 2 \\ 0 & 2 & 2 \end{bmatrix}
\begin{bmatrix} 5 & 0 & \sqrt{3} \\ 0 & 3 & 0 \\ \sqrt{3} & 0 & 3 \end{bmatrix}.
$$

You are no doubt wondering what all this is good for. We turn to that after making one final set of observations. If you look at the eigenvalues of M, Eqn. (9.5.18), you will notice that its two eigenvalues are real. Perhaps this is the property of all real matrices? No, look at the eigenvalues of $R_z(\pi/2)$, Eqn. (9.5.20), which equal 1, $\pm i$. On the other hand the eigenvalues of $R_z(\pi/2)$ and in general $R_z(\theta)$ (Problem (9.5.2.)) are unimodular. Notice also that in all three cases the eigenvectors are mutually orthogonal. (For complex eigenvectors remember to complex conjugate the components of the first vector in the inner product.) All these results are covered by the following two very important theorems:

Theorem 9.7. *The eigenvalues of a hermitian operator are real and its eigenvectors corresponding to different eigenvalues are orthogonal.*

Theorem 9.8. *The eigenvalues of a unitary matrix are unimodular and the eigenvectors corresponding to different eigenvalues are orthogonal.*

Proof: Since the proof is similar for the hermitian and unitary cases in the early stages they will be treated together initially. Thus let Ω be the operator, hermitian or unitary and let $|\omega_1\rangle$ and $|\omega_2\rangle$ be two of its eigenkets:

$$
\Omega|\omega_1\rangle = \omega_1|\omega_1\rangle \tag{9.5.27}
$$

$$
\Omega|\omega_2\rangle = \omega_2|\omega_2\rangle. \tag{9.5.28}
$$

Dotting the first with $\langle\omega_2|$ and the second with $\langle\omega_1|$ we find

$$
\langle\omega_2|\Omega|\omega_1\rangle = \omega_1\langle\omega_2|\omega_1\rangle \tag{9.5.29}
$$

$$
\langle\omega_1|\Omega|\omega_2\rangle = \omega_2\langle\omega_1|\omega_2\rangle. \tag{9.5.30}
$$

Now take the complex conjugate of the lower equations and use Eqn. (9.3.36).[3] This gives

$$
\langle\omega_2|\Omega^\dagger|\omega_1\rangle = \omega_2^*\langle\omega_2|\omega_1\rangle \tag{9.5.31}
$$

---

[3] Or, if you like, take the adjoint, which means we write all factors in reverse order after changing bras to kets and vice versa, operators to their adjoints and scalars into their complex conjugates.

Now consider the hermitian case $\Omega = \Omega^\dagger$. Then the left-hand sides of Eqns. (9.5.29) and Eqn. (9.5.31) become equal. Subtracting the right-hand sides we get

$$0 = (\omega_1 - \omega_2^*)\langle\omega_2|\omega_1\rangle \qquad (9.5.32)$$

which means *one of the two factors must vanish*. If we consider the case where 1 and 2 refer to the same vector we find the eigenvalue is real:

$$\omega_1 = \omega_1^* \qquad (9.5.33)$$

since $\langle\omega_1|\omega_1\rangle \neq 0$ for a nontrivial eigenket. If we next consider $1 \neq 2$ with $\omega_1 \neq \omega_2$, we find from the same equation that it is the dot product which must vanish:

$$\langle\omega_2|\omega_1\rangle = 0, \qquad (9.5.34)$$

i.e., the eigenvectors are orthogonal.

Notice that even after saying 1 and 2 referred to two different eigenvectors, it was additionally required that $\omega_1 \neq \omega_2$. This is because the operator can be *degenerate:* eigenvalues corresponding to two distinct, nonparallel vectors can sometimes be equal. The proof fails there. We shall assume for the moment no degeneracy and return to the degenerate case later.

In the unitary case setting $\Omega^\dagger = \Omega^{-1}$ in Eqn. (9.5.31), we find

$$\langle\omega_2|\Omega^{-1}|\omega_1\rangle = \omega_2^*\langle\omega_2|\omega_1\rangle \qquad (9.5.35)$$
$$\omega_1^{-1}\langle\omega_2|\omega_1\rangle = \omega_2^*\langle\omega_2|\omega_1\rangle \qquad (9.5.36)$$
$$0 = (\omega_1^{-1} - \omega_2^*)\langle\omega_2|\omega_1\rangle, \qquad (9.5.37)$$

where we have used the fact that the eigenvector of any operator is also the eigenvector of the inverse with the inverse eigenvalue. (Prove this.)

Once again the zero must be either due to the inner product or the prefactor in brackets. If we choose $|\omega_1\rangle = |\omega_2\rangle$, we find the prefactor vanishes:

$$\omega_1^* = \omega_1^{-1}, \qquad (9.5.38)$$

i.e., the eigenvalue is unimodular. If we choose the kets to be distinct with distinct eigenvalues in Eqn. (9.5.37), the inner product vanishes:

$$\langle\omega_2|\omega_1\rangle = 0, \qquad (9.5.39)$$

i.e., the vectors with different eigenvalues are orthogonal. ∎

An important implication of these theorems is that every hermitian or unitary operator generates very naturally an orthonormal basis: the basis of its eigenvectors. Recall that just like abstract kets or bras have a life of their own, so do operators, and that they turn into numbers, namely matrices, only in a given basis. As we change the basis, the matrix representing the operator will change, as will the components of the vectors. Of course relations between operators and vectors, such as $|V'\rangle = \Omega|V\rangle$ will remain true even though the numbers representing all three keep changing with the basis. With all this mind we ask: How does an operator look in the basis of its eigenvectors? *It is diagonal with its eigenvalues on the diagonal.* Recall the mnemonic: the n-th column is the image of the n-th basis vector after the operator has acted on it. On the eigenbasis, the action is simply that of rescaling by the eigenvalue. More formally,

$$\langle \omega_j|\Omega|\omega_i\rangle = \langle \omega_j|\omega_i|\omega_i\rangle = \omega_i\delta_{ij}. \tag{9.5.40}$$

The action of a diagonal matrix is a lot simpler than that of a nondiagonal one. *Thus if a problem involving a hermitian or unitary operator is given to us, it will be often profitable to cast the problem in the eigenbasis of the operator to use its simpler form therein. While all bases are mathematically equal, the eigenbasis is more equal than others.* In the section on applications you will see this point exploited repeatedly.

Problem 9.5.4. *We saw in Problem(8.4.10.) that if H is hermitian, $U = e^{iH}$ is unitary. Show that any eigenvector of H with eigenvalue h is also an eigenvector of U with eigenvalue e^{ih}. Thus the theorem that assures us that hermitian operators have an orthonormal set of eigenvectors implies the same for unitary operators which can be written as e^{iH}. It turns out (see the next problem) that every unitary operator can be so written.*

Problem 9.5.5. *Let us specialize to a complex two-dimensional vector space. It takes four complex numbers or eight real numbers to fully specify a complex matrix acting on its vectors. Consider a unitary matrix U and hermitian matrix H. Show that hermiticity places four real conditions on H, so that it takes four real numbers to specify H. Show likewise that $U^\dagger U = 1$ reduces the eight potential real degrees of freedom to four free ones by imposing four real constraints. Thus setting any given $U = e^{iH}$, we will be able to solve for H. The same logic holds in higher dimensions.*

Problem 9.5.6. *The* Cayley–Hamilton Theorem *states that every matrix obeys its characteristic equation. In other words, if $P(\omega)$ is the characteristic polynomial for the matrix Ω, then $P(\Omega)$ vanishes as a matrix. This means that it will annihilate any vector. First prove the theorem for a hermitian Ω with nondegenerate eigenvectors.*

The Cayley–Hamilton theorem can be used to find inverses. Writing out $P(\Omega) = 0$, we must get an expression of the form

$$a_n\Omega^n + a_{n-1}\Omega^{n-1} + \cdots + a_0 = 0. \tag{9.5.41}$$

Multiplying both sides by Ω^{-1} and rearranging, we have an expression for Ω^{-1} in terms of powers of Ω going up to Ω^{n-1}. Repeat Problem 8.3.6. using this theorem.

Theorem 9.9. *If Λ and Ω are nondegenerate, hermitian and commuting, they share a common eigenbasis.*

Proof: We shall establish the theorem by showing that every eigenvector of Ω is also an eigenvector of Λ and *vice versa*. Let $|\omega\rangle$ be a (nondegenerate) eigenvector of Ω:

$$\Omega|\omega\rangle = \omega|\omega\rangle.$$

Acting from the left with Λ, and using $\Lambda\Omega = \Omega\Lambda$ we find

$$\Omega(\Lambda|\omega\rangle) = \omega(\Lambda|\omega\rangle) \tag{9.5.42}$$

which tells us that $\Lambda|\omega\rangle$ is an eigenvector of Ω with eigenvalue ω. Now, Ω being nondegenerate, has only one such eigenvector, namely $|\omega\rangle$, up to the usual latitude in length. So we conclude $\Lambda|\omega\rangle$ must be proportional to $|\omega\rangle$:

$$\Lambda|\omega\rangle = \lambda|\omega\rangle \tag{9.5.43}$$

where λ is the proportionality constant. *But the above equation tells us that $|\omega\rangle$ is an eigenvector of Λ (with eigenvalue λ).* Clearly this argument can be made for every eigenvector of Ω and also in the reverse direction to show that every eigenvector of Λ is an eigenvector of Ω. ∎

9.5.1. Degeneracy

When one or more eigenvalues are repeated in solving for the roots of the characteristic polynomial, we say the operator is degenerate. In the nondegenerate case, having found the eigenvalues, we went back to the eigenvalue equation, one eigenvalue at a time, and found the corresponding eigenvector (in direction). With an eigenvalue repeated, we will clearly have some problems. Consider the following example of an operator M represented by a matrix (also denoted by M):

$$M = \begin{bmatrix} 1 & 0 & 1 \\ 0 & 2 & 0 \\ 1 & 0 & 1 \end{bmatrix}. \tag{9.5.44}$$

The characteristic equation is

$$(m - 2)^2 m = 0 \tag{9.5.45}$$

with roots $m = 0, 2, 2$. The case $m = 0$ is solved as usual to give

$$|m = 0\rangle = \frac{1}{\sqrt{2}} \begin{bmatrix} 1 \\ 0 \\ -1 \end{bmatrix}. \qquad (9.5.46)$$

The eigenvalue $m = 2$ gives us the following equations for the eigenvector:

$$-v_1 + v_3 = 0 \qquad (9.5.47)$$
$$0 = 0 \qquad (9.5.48)$$
$$v_1 - v_3 = 0. \qquad (9.5.49)$$

Instead of the usual case where one out of three equations is useless, here two equations are useless. All we can get out of these three equations is

$$v_1 = v_3, \qquad (9.5.50)$$

which does not specify even a direction since nothing is known about v_2. *Indeed the above equation merely ensures that the eigenvector lies in the plane perpendicular to the vector with $m = 0$.* Thus any vector in this plane is an eigenvector with eigenvalue 2! We can, if we want, choose any two orthonormal vectors in this plane as our eigenvectors. One arbitrary choice is setting $v_2 = 1$ for the first member of this pair and choosing the other to be orthogonal to this one:

$$|m = 2, \text{eigenvector } 1\rangle = \frac{1}{\sqrt{3}} \begin{bmatrix} 1 \\ 1 \\ 1 \end{bmatrix} \quad |m = 2, \text{eigenvector } 2\rangle = \frac{1}{\sqrt{6}} \begin{bmatrix} 1 \\ -2 \\ 1 \end{bmatrix}.$$
$$(9.5.51)$$

Thus in the degenerate case there exist infinitely many orthonormal sets of eigenvectors.

Problem 9.5.7. *Find another set for the above case with $v_2 = 0$ for eigenvector 1.*

Problem 9.5.8. *Find the eigenvalues and normalized eigenvectors of P_x, the projection operator in the x-direction.*

Problem 9.5.9. *Find an orthonormal eigenbasis for the following matrices*

$$\begin{bmatrix} 1 & -1 & -1 \\ -1 & 1 & -1 \\ -1 & -1 & 1 \end{bmatrix} \quad \begin{bmatrix} 5 & 0 & 2 \\ 0 & 1 & 0 \\ 2 & 0 & 2 \end{bmatrix} \quad \begin{bmatrix} 3 & 1 & 0 \\ 1 & 3 & 0 \\ 0 & 0 & 2 \end{bmatrix} \qquad (9.5.52)$$

In each case find the inverse by the Cayley–Hamilton Theorem. The answers are not given at the end. If you want, verify the inverse by multiplication with the given matrix.

There are two main changes in the face of degeneracy. First, the kets cannot be fully specified by the eigenvalue. Thus $|m = 2\rangle$ does not specify an eigenvector up to its normalization, but an entire eigenplane. In general, if an eigenvalue is repeated r-times, there will be r useless equations and an r-dimensional eigenspace will result. Secondly, even though eigenvectors from different eigenspaces will still be orthogonal, we can get nonorthogonal eigenvectors *within the degenerate space*. Let us note for future reference that *if $|m, 1\rangle$ and $|m, 2\rangle$ are two eigenvectors of M with eigenvalue m, so is any linear combination*:

$$M(a_1|m, 1\rangle + a_2|m, 2\rangle) = a_1 M|m, 1\rangle + a_2 M|m, 2\rangle = m(a_1|m, 1\rangle + a_2|m, 2\rangle).$$
$$(9.5.53)$$

Problem 9.5.10. *Show that the following matrices commute and find a common eigenbasis;*

$$M = \begin{bmatrix} 1 & 0 & 1 \\ 0 & 0 & 0 \\ 1 & 0 & 1 \end{bmatrix} \quad N = \begin{bmatrix} 2 & 1 & 1 \\ 1 & 0 & -1 \\ 1 & -1 & 2 \end{bmatrix} \quad (9.5.54)$$

Problem 9.5.11. Important quantum problem. *Consider the three* spin-1 matrices:

$$S_x = \frac{1}{\sqrt{2}} \begin{bmatrix} 0 & 1 & 0 \\ 1 & 0 & 1 \\ 0 & 1 & 0 \end{bmatrix} \quad S_y = \frac{1}{\sqrt{2}} \begin{bmatrix} 0 & -i & 0 \\ i & 0 & -i \\ 0 & i & 0 \end{bmatrix} \quad S_z = \begin{bmatrix} 1 & 0 & 0 \\ 0 & 0 & 0 \\ 0 & 0 & -1 \end{bmatrix},$$
$$(9.5.55)$$

which represent the components of the internal angular momentum *of some elementary particle at rest. That is to say, the particle has some angular momentum unrelated to* $\mathbf{r} \times \mathbf{p}$. *The operator* $S^2 = S_x^2 + S_y^2 + S_z^2$ *represents the total angular momentum squared. The dynamical state of the system is given by a state vector in the complex three dimensional space on which these spin matrices act. By this we mean that all available information on the particle is stored in this vector. According to the laws of quantum mechanics*

- *A measurement of the angular momentum along any direction will give only one of the eigenvalues of the corresponding spin operator.*

- *The probability that a given eigenvalue will result is equal to the absolute value squared of the inner product of the state vector with the corresponding eigenvector. (The state vector and all eigenvectors are normalized.)*

- *The state of the system immediately following this measurement will be the corresponding eigenvector.*

(a) What are the possible values we can get if we measure spin along the z-axis?
(b) What are the possible values we can get if we measure spin along the x or

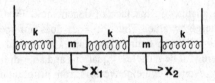

Figure 9.3. The coupled mass system.

y-axis?

(c) Say we got the largest possible value for S_x. What is the state vector immediately afterwards?

(d) If S_z is now measured what are the odds for the various outcomes? Say we got the largest value. What is the state just after the measurement? If we remeasure S_x at once, will we once again get the largest value?

(e) What are the outcomes when S^2 is measured?

(f) From the four operators S_x, S_y, S_z, S^2, what is the largest number of commuting operators we can pick at a time?

(g) A particle is in a state given by a column vector

$$|V\rangle = \begin{bmatrix} 1 \\ 2 \\ 3 \end{bmatrix}.$$

First rescale the vector to normalize it. What are the odds for getting the three possible eigenvalues of S_z? What is the statistical or weighted average of these values? Compare this to $\langle V|S_z|V\rangle$.

(h) Repeat all this for S_x.

9.6. Applications of Eigenvalue Theory

We will deal with a few examples which involve hermitian and unitary operators. In all the examples the following notion will be exploited: *every hermitian operator generates an orthonormal basis, namely the basis of its eigenvectors and in this basis it has its eigenvalues on the diagonal and zeros elsewhere.*

9.6.1. Normal modes of vibration: two coupled masses

Consider the problem depicted in Fig. 9.3.

Two masses m are resting on a frictionless table between two walls, as shown in the figure. The first mass is connected to the left wall by a spring of force constant k, and to the second mass by an identical spring; and the second mass is likewise connected to the right wall by another identical spring. The distance between the walls is such that when all the masses are at rest and in equilibrium, none of the springs is stretched or compressed. The x-axis runs from wall to wall and x_1 and x_2 refer to the displacement of each mass *from its equilibrium position.*

Let us assume that at time $t = 0$ all masses are at rest, but displaced from equilibrium by $x_1(0)$ and $x_2(0)$, respectively. The problem is to find the subsequent displacements $x_1(t), x_2(t)$.

The first step is to write down the equations of motion. If the masses are displaced by x_1 and x_2, the first experiences a leftward force of magnitude kx_1 due to the spring to its left and a rightward force $k(x_2 - x_1)$ from the (middle) spring to its right. The second mass experiences a leftward force $k(x_2 - x_1)$ due the middle spring and a leftward force kx_2 due to the right most spring. Thus from Newton's Second Law

$$\ddot{x}_1(t) = -\frac{2k}{m}x_1(t) + \frac{k}{m}x_2(t) \tag{9.6.1}$$

$$\ddot{x}_2(t) = \frac{k}{m}x_1(t) - \frac{2k}{m}x_2(t). \tag{9.6.2}$$

This problem is so important we will attack it at various levels.

First note that we are dealing with a pair of simultaneous differential equations.[4] Thus the evolutions of $x_1(t)$ and $x_2(t)$ are intertwined. (When we try to solve for $x_1(t)$, we find it depends on what $x_2(t)$ is doing and vice versa.) This in turn comes from the coupling between the masses due to the middle spring. Cut that spring out and you see at once that each mass will oscillate with frequency $\omega = \sqrt{k/m}$. But of course the middle spring is there and we must deal with the coupled equations.

Let us approach the coupled equations as we did a pair of simultaneous algebraic equations: try to form combinations that isolate one or the other unknown. (In the case of algebraic equations we were just trying to find two fixed numbers subject to two algebraic equations, while here we are trying to find two functions of time subject to two differential equations.) If we add and subtract the two equations and divide both sides by $\sqrt{2}$ (where the last operation is not necessary but convenient) we find

$$\ddot{x}_+(t) = -\frac{k}{m}x_+(t) \tag{9.6.3}$$

$$\ddot{x}_-(t) = -\frac{3k}{m}x_-(t), \quad \text{where} \tag{9.6.4}$$

[4]No prior knowledge of differential equations is needed to follow this discussion, except for the fact that a differential equation is a relation in which a function and its derivatives appear.

$$x_{\pm}(t) \;=\; \frac{x_1(t) \pm x_2(t)}{\sqrt{2}}. \tag{9.6.5}$$

This is very nice. *It means the variables $x_{\pm}$ behave like two decoupled simple harmonic oscillators of frequencies $\sqrt{k/m}$ and $\sqrt{3k/m}$, respectively.* As for the solutions describing these motions, the equation they obey tell us they are functions which when differentiated twice are proportional to themselves. These are just sines and cosines of appropriate frequency.[5] The requirement that the initial velocity equals zero means we rule out the sines and end up with

$$x_+(t) \;=\; x_+(0)\cos\sqrt{k/m}t \tag{9.6.6}$$

$$x_-(t) \;=\; x_-(0)\cos\sqrt{3k/m}t. \tag{9.6.7}$$

Notice that if we set $t = 0$, both sides reduce nicely to the initial values $x_{\pm}(0)$.

We have now solved the problem we posed ourselves, of finding $x_1(t)$ and $x_2(t)$ given $x_1(0)$ and $x_2(0)$. Here is the procedure

- From $x_1(0)$ and $x_2(0)$ form the combinations $x_{\pm}(0)$.

- Go forward in time to evaluate $x_{\pm}(t)$ which differ from the initial values simply by the cosine factor of appropriate frequency.

- Revert back to old coordinates to find $x_1(t)$ and $x_2(t)$.

Thus for example

$$
\begin{aligned}
x_1(t) \;&=\; \frac{x_+(t)+x_-(t)}{\sqrt{2}} \\
&=\; \frac{x_+(0)\cos\sqrt{k/m}t + x_-(0)\cos\sqrt{3k/m}t}{\sqrt{2}} \\
&=\; \frac{(x_1(0)+x_2(0))\cos\sqrt{k/m}t + (x_1(0)-x_2(0))\cos\sqrt{3k/m}t}{2} \\
&=\; \tfrac{1}{2}[x_1(0)(\cos\sqrt{k/m}t + \cos\sqrt{3k/m}t) \\
&\;+\; x_2(0)(\cos\sqrt{k/m}t - \cos\sqrt{3k/m}t)].
\end{aligned}
\tag{9.6.8}
$$

A similar calculation gives

$$
x_2(t) = \frac{1}{2}[x_1(0)(\cos\sqrt{k/m}t - \cos\sqrt{3k/m}t)
$$
$$
+ x_2(0)(\cos\sqrt{k/m}t + \cos\sqrt{3k/m}t)]. \tag{9.6.9}
$$

While we have solved this particular coupled mass problem with some inspired guessing, many questions remain, such as:

---

[5] They can also be exponentials, which are equivalent to sines and cosines.

- How do we know in general if such magic variables like $x_\pm$, which describe decoupled oscillators, exist?

- How are the frequencies of vibration, $\sqrt{k/m}$, $\sqrt{3k/m}$, related to the parameters in the problem?

- How do we go about finding them, especially when we have several coupled variables?

The answers to these questions lie in the domain of the eigenvalue problem which we have so far studied without any serious motivation. We will now see how a systematic procedure for solving such problems emerges from this formalism.

First we write the equations of motion (9.6.1-9.6.2) in matrix form

$$\left[\begin{array}{c} \ddot{x}_1(t) \\ \ddot{x}_2(t) \end{array} \right] = \left[\begin{array}{cc} -\frac{2k}{m} & \frac{k}{m} \\ \frac{k}{m} & -\frac{2k}{m} \end{array} \right] \left[\begin{array}{c} x_1(t) \\ x_2(t) \end{array} \right], \qquad (9.6.10)$$

or in more compact form and in obvious notation:

$$\ddot{x} = M x \qquad (9.6.11)$$

where x is a column vector, called the *state vector*, and M is a 2×2 matrix.

Next we recognize that the matrix equation (9.6.11) stands for an abstract equation

$$|\ddot{x}\rangle = M |x\rangle \qquad (9.6.12)$$

written in a basis $|1\rangle, |2\rangle$ where the basis states have the following interpretation:

$$|1\rangle \rightarrow \left[\begin{array}{c} 1 \\ 0 \end{array} \right] = \left[\begin{array}{c} \text{mass 1 displaced by unity} \\ \text{mass 2 undisplaced} \end{array} \right] \qquad (9.6.13)$$

$$|2\rangle \rightarrow \left[\begin{array}{c} 0 \\ 1 \end{array} \right] = \left[\begin{array}{c} \text{mass 1 undisplaced} \\ \text{mass 2 displaced by unity} \end{array} \right] \qquad (9.6.14)$$

A general state is written as

$$|x\rangle = x_1 |1\rangle + x_2 |2\rangle \qquad (9.6.15)$$

in the abstract, and in this basis as

$$\left[\begin{array}{c} x_1 \\ x_2 \end{array} \right] = x_1 \left[\begin{array}{c} 1 \\ 0 \end{array} \right] + x_2 \left[\begin{array}{c} 0 \\ 1 \end{array} \right]. \qquad (9.6.16)$$

The present basis is very natural physically, since the components of the state vector, x_1 and x_2, have a direct interpretation as the displacements of the masses. On the other hand, this basis is inconvenient for the solution of the problem since M is non-diagonal and couples the variables.

So now we argue as follows. Since the matrix M is a real and symmetric (due to Newton's third law), it has its own basis of eigenvectors $|m_I\rangle \equiv |I\rangle, |m_{II}\rangle \equiv |II\rangle$ with eigenvalues m_I and m_{II}. It takes very little time to find that

$$|I\rangle = \frac{1}{\sqrt{2}} \begin{bmatrix} 1 \\ 1 \end{bmatrix} \quad m_I = -\frac{k}{m} \quad |II\rangle = \frac{1}{\sqrt{2}} \begin{bmatrix} 1 \\ -1 \end{bmatrix} \quad m_{II} = -\frac{3k}{m}. \quad (9.6.17)$$

So let us expand the state vector as follows:

$$|x\rangle = x_I|I\rangle + x_{II}|II\rangle. \quad (9.6.18)$$

where $x_I = \langle I|x\rangle$ and $x_{II} = \langle II|x\rangle$. *If we go to this basis, M will become diagonal and the components of $|x\rangle$ in that basis, x_I and x_{II}, will evolve independently.* In other words, Eqn. (9.6.12) will become

$$\begin{bmatrix} \ddot{x}_I \\ \ddot{x}_{II} \end{bmatrix} = \begin{bmatrix} -\frac{k}{m} & 0 \\ 0 & -\frac{3k}{m} \end{bmatrix} \begin{bmatrix} x_I \\ x_{II} \end{bmatrix}. \quad (9.6.19)$$

Thus x_I and x_{II} obey the equations of two decoupled harmonic oscillators with frequencies $\omega_I = \sqrt{k/m}$ and $\omega_{II} = \sqrt{3k/m}$.

We now understand $x_\pm$ from the earlier analysis: they are just the projections of the state vector on the eigenvectors of M. Likewise, the frequencies of oscillations are just (square roots of minus) the eigenvalues of M.

That x_I and x_{II} obey these decoupled equations can also be seen as follows. Let us apply $|\ddot{x}(t)\rangle = M|x(t)\rangle$ to Eqn. (9.6.18). We find

$$\ddot{x}_I(t)|I\rangle + \ddot{x}_{II}(t)|II\rangle = M x_I(t)|I\rangle + M x_{II}(t)|II\rangle = m_I x_I(t)|I\rangle + m_{II} x_{II}(t)|II\rangle. \quad (9.6.20)$$

Equating the coefficients of the orthogonal vectors on both sides we find the two equations governing x_I and x_{II}. This analysis also makes clear why we choose to work with the eigenbasis of M: its action on them is just to multiply by the eigenvalue. In any event we have

$$x_I(t) = x_I(0) \cos \omega_I t \quad (9.6.21)$$
$$x_{II}(t) = x_{II}(0) \cos \omega_{II} t, \quad (9.6.22)$$

which constitutes the complete solution to the problem we set out to solve, except it is in the new basis. What the solution tells us is this. We want to know the future given some initial data at $t = 0$. We are used to giving them in terms of the initial displacements of the masses: $x_1(0)$ and $x_2(0)$. Their evolution is coupled. But suppose we give instead $x_I(0)$ and $x_{II}(0)$. Their future values are trivially determined: multiply each initial value by $\cos \omega_i t$, $i =$

I, II. As for the initial values, they are the components of the initial state vector in the eigenbasis and so

$$x_I(0) = \langle I | x(0) \rangle = \left[\begin{array}{cc} \frac{1}{\sqrt{2}} & \frac{1}{\sqrt{2}} \end{array} \right] \left[\begin{array}{c} x_1(0) \\ x_2(0) \end{array} \right] = \frac{x_1(0) + x_2(0)}{\sqrt{2}} \qquad (9.6.23)$$

$$x_{II}(0) = \langle II | x(0) \rangle = \left[\begin{array}{cc} \frac{1}{\sqrt{2}} & \frac{-1}{\sqrt{2}} \end{array} \right] \left[\begin{array}{c} x_1(0) \\ x_2(0) \end{array} \right] = \frac{x_1(0) - x_2(0)}{\sqrt{2}}. \qquad (9.6.24)$$

The preceding two equations express the new coordinates in terms of the old ones at time zero, but these relations clearly hold at all times. Inverting them we find

$$x_1(t) = \frac{x_I(t) + x_{II}(t)}{\sqrt{2}}$$

$$= \frac{x_I(0) \cos \omega_I(t) + x_{II}(0) \cos \omega_{II}(t)}{\sqrt{2}}$$

$$= \frac{(x_1(0) + x_2(0)) \cos \omega_I(t) + (x_1(0) - x_2(0)) \cos \omega_{II}(t)}{2} \qquad (9.6.25)$$

$$x_2(t) = \frac{(x_1(0) + x_2(0)) \cos \omega_I(t) - (x_1(0) - x_2(0)) \cos \omega_{II}(t)}{2}. \qquad (9.6.26)$$

which can be written in matrix form as

$$\left[\begin{array}{c} x_1(t) \\ x_2(t) \end{array} \right] = \frac{1}{2} \left[\begin{array}{cc} c_I(t) + c_{II}(t) & c_I(t) - c_{II}(t) \\ c_I(t) - c_{II}(t) & c_I(t) + c_{II}(t) \end{array} \right] \left[\begin{array}{c} x_1(0) \\ x_2(0) \end{array} \right] \qquad (9.6.27)$$

$$x(t) = U(t) x(0), \qquad (9.6.28)$$

where $c_i(t) = \cos \omega_i t$. This is the complete solution to the problem. *Given any initial state vector in the old basis, we are able to find the state vector at all future times by simply multiplying the initial state by a matrix $U(t)$.* This matrix or the abstract operator it represents is called the *propagator.*

The eigenvectors have a privileged role as initial states. *If at $t = 0$ we start in an eigenstate, we remain in a state which is that eigenvector times an overall time dependent factor.* This is clearest from the solutions to the equations of motion in the eigenbasis: if at $t = 0$, $x_I = 1$ and $x_{II} = 0$, then at later times $x_I(t) = \cos \omega_I t$ and $x_{II}(t) = 0$. Coming back to the old basis this means that under time evolution

$$\frac{1}{\sqrt{2}} \left[\begin{array}{c} 1 \\ 1 \end{array} \right] \rightarrow \frac{1}{\sqrt{2}} \left[\begin{array}{c} 1 \\ 1 \end{array} \right] \cos \sqrt{\frac{k}{m}} t. \qquad (9.6.29)$$

Let us understand this. If we start the system with both masses equally displaced, the middle spring is unaffected. In response to the end springs the masses begin to vibrate independently at frequency $\sqrt{k/m}$. Since they are always in step the distance between them stays fixed and the middle spring never comes into play. The masses continue to vibrate with equal displacements. This is called a *normal mode of vibration*. *In a normal mode, the coordinates all have a common time dependence which can be pulled out of the state vector. The coordinates oscillate in step.* Stated differently, the direction of the state vector never changes. If you look at the other eigenvector you get the other normal mode: the masses are initially displaced by equal and opposite amounts. This too will persist with time. The middle spring is distorted twice as much as the end ones now and the net restoring force on each mass is three times as due to one spring and we get a frequency $\sqrt{3k/m}$. The masses continue to have equal and opposite displacements for all times. No other ratio of initial displacements (i.e., "direction" of state vector) is preserved with time. For example if the first mass is displaced by unity and the other undisplaced, the initial ratio of displacements does not persist.

Problem 9.6.1. *Find the future state of a system with $x_1(0) = 1$ and $x_2(0) = 0$ at $t = 0$.*

Problem 9.6.2. *Find the normal modes and eigenfrequencies if the middle spring has a force constant $2k$.*

Problem 9.6.3. *Find the normal modes and frequencies for a triple mass problem where there are three equal masses $m = 1$ hooked up in series with four springs all having force constant $k = 1$.*

We have introduced the normal modes at the end of the discussion. They can also enter at the very beginning in an equivalent approach to the problem which will now be described. Let us go back to the basic Eqn. (9.6.12). Our goal is to find all solutions to the problem, for all possible initial conditions (with zero initial velocity). But we first set a more modest goal and look for solutions of special form

$$|x(t)\rangle = |x(0)\rangle f(t). \tag{9.6.30}$$

There is no guarantee that such a solution, with all components sharing a common time dependence, exists. But let us plug this *ansatz* into the equation of motion and divide both sides by $f(t)$ to find:

$$\frac{\ddot{f}(t)}{f(t)}|x(0)\rangle = M|x(0)\rangle. \tag{9.6.31}$$

Since the right-hand side has no dependence on time, neither can the left-hand side. this means that

$$\frac{\ddot{f}(t)}{f(t)} = \text{constant} = -\omega^2, \tag{9.6.32}$$

where we have chosen to name the constant $-\omega^2$ in anticipation of what is to unfold. The solution to this equation with zero initial velocity is

$$f(t) = f(0)\cos\omega t. \qquad (9.6.33)$$

We are not done yet: plugging Eqn. (9.6.32) into Eqn. (9.6.31), we find an eigenvalue equation:

$$M|x(0)\rangle = -\omega^2|x(0)\rangle. \qquad (9.6.34)$$

Thus we find that

- The *ansatz* Eqn. (9.6.30) will work if $f(t)$ is a cosine.

- The (negative squared of the) frequency of the cosine must be one of the eigenvalues of the eigenvalue Eqn. (9.6.34).

- The initial state vector must be the corresponding eigenvector.

So we have found all (two) solutions of the assumed form. What about the general solution? From the fact that the equation of motion is linear one can readily verify that given two solutions, any linear combination with constant coefficients is also a solution. Let us apply this to the normal modes $|I(t)\rangle = |I\rangle\cos\omega_I t$ and $|II(t)\rangle = |II\rangle\cos\omega_{II}t$. Thus

$$|x(t)\rangle = x_I(0)|I\rangle\cos\omega_I t + x_{II}(0)|II\rangle\cos\omega_{II}t \qquad (9.6.35)$$

is also a solution. How general is it? At $t = 0$, the initial state is

$$|x(0)\rangle = x_I(0)|I\rangle + x_{II}(0)|II\rangle. \qquad (9.6.36)$$

But the normal modes are the eigenvectors of a hermitian operator and hence form a complete orthonormal basis. Thus you can form any initial state you want. This means we can solve the initial value problem with any displacement of the masses.

This simple problem exemplifies the strategy employed to solve a whole family of problems in mechanics, quantum mechanics, and so on. For example, in quantum mechanics there is a state vector (recall the spin problem) $|\psi\rangle$ which contains all the information about the system and evolves in time according to Schrödinger's equation:

$$i\hbar\frac{\partial|\psi\rangle}{\partial t} = H|\psi\rangle \qquad (9.6.37)$$

where H is a hermitian operator called the *hamiltonian*. How do we find the state $|\psi(t)\rangle$ given $|\psi(0)\rangle$? Exactly as we did here. We will first find the eigenvectors of H, go to that basis, solve the problem trivially, and come back to the old basis in which H was given. There will also be normal modes for which the state at a later time is the initial state times a time-dependent factor. You may remember learning in some elementary course that a system like the hydrogen atom will remain stable if it is in some allowed quantum state with some quantized value of energy. The normal modes are precisely these states: the stable states of hydrogen are given by the eigenvectors of the hamiltonian and the corresponding eigenvalues are the energy levels you learned about.

9.6.2. Quadratic forms

Consider a real function of two variables expanded in a Taylor series about a stationary point which we take for convenience to be the origin of coordinates:

$$f(x,y) = f(0,0) + \frac{x^2}{2}f_{xx} + \frac{y^2}{2}f_{yy} + \frac{xy}{2}f_{xy} + \frac{yx}{2}f_{yx} + \text{higher-order terms.}$$
$$(9.6.38)$$

In this expansion we have no first order derivatives since we are expanding about a stationary point and we have purposely avoided lumping the cross derivative terms using the equality $f_{xy} = f_{yx}$.

This stationary point could be a maximum, minimum, or saddle point. With just one variable we knew for example that if $f_{xx} > 0$, then we were at a minimum since any deviation cause a positive definite increase $\frac{x^2}{2}f_{xx}$. The problem here is due to the cross terms xy which have no definite sign. Thus even if all the second derivatives were positive, we could not guarantee the function would go up in all directions. On the other hand if these cross terms were absent we could say that we have a minimum if $f_{xx} > 0$, $f_{yy} > 0$, a maximum if both were negative and a saddle point if they were of mixed sign.

To proceed further let us associate a vector $|r\rangle$ with the displacement from the stationary point. Clearly in our coordinate system

$$|r\rangle \rightarrow \begin{bmatrix} x \\ y \end{bmatrix} \qquad (9.6.39)$$

and

$$\Delta f = f(x,y) - f(0,0) = \frac{1}{2} [x\ y] \begin{bmatrix} f_{xx} & f_{xy} \\ f_{yx} & f_{yy} \end{bmatrix} \begin{bmatrix} x \\ y \end{bmatrix} \qquad (9.6.40)$$

$$= \frac{1}{2} \langle r|F|r\rangle, \qquad (9.6.41)$$

where F is the matrix of second derivatives. It is real and symmetric and has two orthonormal eigenvectors $|I\rangle$ and $|II\rangle$. Suppose we expand $|r\rangle$ in terms of these with coefficients x_I and x_{II}. Then $\langle r|F|r \rangle$ has no cross terms and

$$\Delta f = f(x,y) - f(0,0) = \frac{1}{2}(f_I x_I^2 + f_{II} x_{II}^2) \qquad (9.6.42)$$

where f_I and f_{II} are the eigenvalues of F. Now we know by inspection that if both eigenvalues are positive, we are at a minimum since no matter what our displacement from the stationary point, f goes up. Likewise if both eigenvalues are negative, we are at a maximum. In the mixed case we have a saddle point with a maximum along the eigenvector with negative eigenvalue and minimum along the other one. *In summary, to decide if a stationary point is a maximum, minimum, or saddle point, we must solve the eigenvalue problem of the matrix of second derivatives.*

Suppose our f really described the saddle on a horse and that we had chosen our axes so that they were at some oblique direction with respect to the horse. The eigenvectors would then end up pointing parallel and perpendicular to the length of the horse, i.e., aligned with the saddle point's natural axes. So if in some movie you see a dashing physicist hesitate for a split second before jumping on the horse, remember he is diagonalizing a two by two matrix.

Let us consider as an example

$$f(x,y) = xy(4+x) - (2x+y)^2. \qquad (9.6.43)$$

Since

$$f_x = 2x(y-4) \qquad f_y = x^2 - 2y, \qquad (9.6.44)$$

the stationary points are at

$$(x=0, y=0), \qquad (x = \pm\sqrt{8}, y = 4) \qquad (9.6.45)$$

The second derivatives are

$$f_{xx} = 2y - 8, \quad f_{yy} = -2, \quad f_{xy} = 2x. \qquad (9.6.46)$$

At the origin the matrix of second derivatives is diagonal with both entries negative, i.e, the origin is a maximum. At the other two points it becomes

$$\begin{bmatrix} 0 & \pm\sqrt{32} \\ \pm\sqrt{32} & -2 \end{bmatrix}. \qquad (9.6.47)$$

The eigenvalues are found to be $-1 \mp \sqrt{33}$, so that we are at a saddle point. The corresponding eigenvectors are $[-\frac{\sqrt{32}}{1+\sqrt{33}}, 1]^T$ and $[-\frac{\sqrt{32}}{1-\sqrt{33}}, 1]^T$.

All this is depicted in Fig. 9.4.

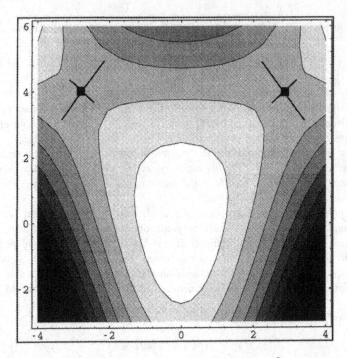

Figure 9.4. The contour plot of $f(x,y) = xy(4+x) - (2x+y)^2$. The bright areas are higher. Besides the maximum at the origin, note the two saddle points with the eigendirections indicated. As we move along the short direction, the function starts going up (lighter shading) while along the other, it goes down.

Problem 9.6.4. *Consider* $f(xy) = x + y - xy - y^2/2$. *Find its stationary point and see if it is a maximum, minimum, or saddle point. Locate the eigendirections.*

Problem 9.6.5. *By solving once and for all the* 2×2 *eigenvalue equation for a general hermitian matrix verify that the sum and product of the eigenvalues equal the trace and determinant, respectively. Thus if the determinant is negative we surely have a saddle point. If positive, both eigenvalues have the same sign, but what is that sign?*

Problem 9.6.6. *Using the results from the previous problem show that* $f(x,y) = (x-1)^3 - 3x - 4y^2$ *has a maximum at the origin and a saddle point at* $(2,0)$. *Show likewise that* $f(x,y) = x^4 + y^4 - 2x^2(1-y^2) + 2y^2$ *has a saddle point at the origin and minima at* $(\pm 1, 0)$.

Problem 9.6.7. *Find and classify the stationary points of* $f(x, y) = y(4 + x) - (2x + y)^2$.

The expression $f = \langle r | F | r \rangle$, is an example of a *quadratic form* with an obvious generalization to complex vector spaces. (See the following problem.)

Problem 9.6.8. Important *Consider Eqn. (9.2.15) expressing* $\langle V | W \rangle$ *in terms of the components of the vectors and the inner products* $\langle i | j \rangle$ *among the basis vectors. Since we know nothing about the basis other than its linear independence, we can pick* $\langle i | j \rangle$ *freely, as long as the inner product in Eqn. (9.2.15) obeys Axioms B-I to B-III. Starting with* $\langle V | W \rangle = V^\dagger S W$, *where* V *and* W *are column vectors containing the components of* $|V\rangle$ *and* $|W\rangle$, *and* $S_{ij} = \langle i | j \rangle$ *is the matrix of inner products, show that B-III is automatically satisfied, B-I requires that* S *must be hermitian, and B-II demands that it have positive eigenvalues. The orthonormal basis with* $S = I$, *is a special case that satisfies all this.*

9.7. Function Spaces

We have seen that functions can be elements of a vector space. We return to this theme for a more detailed analysis. We defined addition of two functions to be the pointwise addition: if $f + g = h$, then $h(x) = f(x) + g(x)$. On the other hand we know that to add two vectors, we must add their components. *Thus the function f has an infinite number of components, namely the values it takes in the continuum of points labeled by the real variable x.* Since we have been working so far in finite dimensional spaces, let us back off a bit and see if we can start with what we know. Let us say we want our functions to be defined in the interval $0 \leq x \leq L$. Let us assume that they are smooth and differentiable. A good example is the function describing the displacement $f(x)$ of a string clamped at $x = 0$ and $x = L$. Say we want to tell someone on another planet what our string is doing at some time. One way is to fax the snapshot. Let us say that we can only send some digital information by Morse code. Then we can proceed as follows. We divide the interval of length L into N equal parts and measure the displacements $f(x_i) \equiv f_i$ at N points x_i $|i = 1, 2 \cdots N$. If N is large enough the person at the other hand can get a fairly good idea of what the string is doing by connecting the dots. If he is not satisfied, we can double or triple N. (Even the fax only scans the picture at a countable number of points!) At fixed N, the functions are elements of a finite N-dimensional vector space. We can assemble the values of the function into a column vector. We can add and scalar multiply the column vectors and do all the standard vector manipulations of a finite dimensional linear vector space.

Let us associate a ket $|f\rangle$ with each such discretized function. The equation

$$\begin{bmatrix} f_1 \\ f_2 \\ \vdots \\ f_N \end{bmatrix} = f_1 \begin{bmatrix} 1 \\ 0 \\ \vdots \\ 0 \end{bmatrix} + f_2 \begin{bmatrix} 0 \\ 1 \\ \vdots \\ 0 \end{bmatrix} + \cdots f_N \begin{bmatrix} 0 \\ 0 \\ \vdots \\ 1 \end{bmatrix} \tag{9.7.1}$$

is the concrete realization of the abstract vector equation

$$|f\rangle = f_1|1\rangle + f_2|2\rangle + \cdots + f_N|N\rangle. \tag{9.7.2}$$

The basis vector $|j\rangle$ describes a situation where the displacement is unity at the discrete location j and zero elsewhere.

Now consider an inner product. The obvious choice is

$$\langle f|g\rangle = \sum_{i=1}^{N} f_i g_i. \tag{9.7.3}$$

It obeys all the axioms. Since we are still speaking of a string, the space is real and there is no need to conjugate anything.

Now we let $N \to \infty$ by increasing the number of points without limit. There is no problem with addition and scalar multiplication. The scalar product however diverges as we subdivide further and further. The way out is to modify the definition by a positive prefactor $\Delta = L/N$ which does not violate any of the axioms for the inner product. But now

$$\langle f|g\rangle = \lim_{\Delta \to 0} \sum_{i=1}^{N} f_i g_i \Delta \tag{9.7.4}$$

$$\to \int_0^L f(x)g(x)dx \tag{9.7.5}$$

by the usual definition of an integral. *Thus the inner product of two functions is the integral of their product.* We will say that two functions are orthogonal if this inner product vanishes, and say a function is normalized if the integral of its square equals unity. The space of square integrable functions is an example of a *Hilbert space.* Thus one can speak of an orthonormal set of functions in a Hilbert space just as in finite dimensions. Here is an example of such a set of functions defined in the interval $0 \le x \le L$ and vanishing at the end points:

$$|m\rangle \to m(x) = \sqrt{\frac{2}{L}} \sin\frac{m\pi x}{L} \qquad m = 1, 2, \ldots \infty \tag{9.7.6}$$

$$\langle m|n\rangle = \frac{2}{L} \int_0^L \sin\frac{m\pi x}{L} \sin\frac{n\pi x}{L} = \delta_{mn}. \tag{9.7.7}$$

Problem 9.7.1. *Verify the normalization for the case $m = n$ and the orthogonality for $m, n \leq 2$.*

Consider another example. Let us forget the string and consider complex functions in the same interval. We no longer demand that they vanish at the end points, but only that they be *periodic*

$$f(0) = f(L). \tag{9.7.8}$$

The correct inner product is of course modified as follows

$$\langle f|g \rangle = \int_0^L f^*(x)g(x)dx. \tag{9.7.9}$$

An orthonormal set for this case is

$$|m\rangle \rightarrow \frac{1}{\sqrt{L}}e^{2\pi imx/L} \quad m = 0, \pm 1, \pm 2, \ldots . \tag{9.7.10}$$

Problem 9.7.2. *Verify the orthonormality of these functions.*

9.7.1. Generation of orthonormal bases

Where did we get these orthonormal functions? By solving the eigenvalue equation of some hermitian operator of course! *The operator in question is related to D, the derivative operator. It acts on an input function and spits out its derivative:*

$$D|f\rangle = |\frac{df}{dx}\rangle. \tag{9.7.11}$$

Since both the input and the output are viewed as vectors, D is no different from any other operator we have seen before, except for the fact that it acts on infinite-dimensional vectors. Let us first note that D is a linear operator since the act of differentiation is a linear operation that does not involve, say, the square of the function being differentiated. However D is not hermitian. Here is why. Consider first the finite-dimensional case where a matrix obeying:

$$M_{ij} = M_{ji}^* \tag{9.7.12}$$

is called hermitian. We have seen that this implies (see Eqn. (9.3.37))

$$\langle V|M|W \rangle^* = \langle W|M|V \rangle. \tag{9.7.13}$$

Let us subject D to this test. Let us ask if

$$\langle f|D|g \rangle = \langle g|D|f \rangle^*?$$

This is the same as asking if

$$\int_0^L f^*(x) \frac{dg}{dx} dx$$

equals

$$\left(\int_0^L g^*(x) \frac{df}{dx} dx \right)^* = \int_0^L g(x) \frac{df^*}{dx} dx \tag{9.7.14}$$

$$= gf^* \Big|_0^L - \int_0^L f^*(x) \frac{dg}{dx} dx \tag{9.7.15}$$

where in reaching the last line, we have used the rule for integral of a product, or done "integration by parts". We see that hermiticity of D is lost on two counts. First we have the surface term coming from the end points. Next the integral in the last equation has the wrong sign. Both get fixed if we do the following:

- Use the operator $-iD$. The extra i will change sign under conjugation and kill the minus sign in front of the integral, as you must check.

- Restrict the functions to those that are *periodic*: $f(0) = f(L)$. (A special case of this is $f(0) = f(L) = 0$.) This amounts to joining the two ends of the interval into a circle of circumference L and demanding that the functions are continuous on this circle. This kills the surface term.

To summarize $-iD$ is a hermitian operator on periodic functions. (We could just as easily use $76iD$. The key feature is the introduction of the i.)

Let us now find its eigenfunctions. They obey

$$-i\frac{df}{dx} = \lambda f, \tag{9.7.16}$$

where λ is the eigenvalue. This equation asks us to find a function which reproduces itself under differentiation. The differential operator is a nontrivial operator which generally changes function to other functions—for example, x^2 goes into $2x$. On the other hand we know that on the exponential functions its action is simply that of rescaling. Thus

$$f(x) = Ae^{i\lambda x} \tag{9.7.17}$$

where A is the arbitrary scale of the *eigenfunction*. We solved an infinite-dimensional eigenvalue equation by inspection! There was no characteristic equation to solve, nothing! (Naively you would think that in infinite dimensions you will have to find the roots of an infinite-order characteristic polynomial. We got away easily because the infinite-dimensional vectors we are looking for have components $f(x)$ that vary smoothly with the index x.) Notice something remarkable:

the eigenvalue can be any number you want! In fact λ can even be complex. How about the notion that hermitian operators will have real eigenvalues only? It is alive and well, only we have failed to weed out functions that do not meet the periodicity requirement which in turn ensures $-iD$ is hermitian. By demanding

$$e^{i\lambda L} = e^{i\lambda 0} = 1 \tag{9.7.18}$$

we find the following restriction on the eigenvalues:

$$\lambda = \frac{2\pi m}{L} \qquad m = 0, \pm 1, \pm 2 \ldots \tag{9.7.19}$$

The normalization condition is easily found to be $A = 1/\sqrt{L}$ so that the final set of orthonormal eigenvectors is given by

$$|m\rangle \rightarrow f_m(x) = \frac{1}{\sqrt{L}} \exp\left[\frac{2\pi i m x}{L}\right]. \tag{9.7.20}$$

Thus we regain the exponential basis functions quoted earlier. Their orthogonality is assured by repeating the finite-dimensional proof almost verbatim. (The proof is literally the same in the Dirac notation which does not change its form as we go to infinite dimensions. The key ingredient—that surface terms vanish—is hidden in that notation, where we simply assume $\langle f|(-iD)|g\rangle = \langle g|(-iD)|f\rangle^*$.) There is however one subtlety with infinite dimensions. True, we have found an infinite number of orthonormal functions. But do they form a complete basis? In the finite-dimensional case if we have N orthonormal eigenvectors in N dimensions we are done. The corresponding statement for infinite dimensions is more complicated. For example if we drop the functions with $m = \pm 67$, we still have an infinite number of orthonormal functions, which are clearly not a complete basis: there is no way to write $e^{\pm 2\pi i \cdot 67 \cdot x/L}$ in terms of the other functions. We have to prove something called *completeness* to know we have all of them. This point will be discussed briefly at the end of the following subsection on the vibrating string. For the present we ignore this point and proceed to write for any periodic function of period L

$$f(x) = \sum_m f_m \frac{1}{\sqrt{L}} \exp\left[\frac{2\pi i m x}{L}\right] \tag{9.7.21}$$

which is the concrete version of

$$|f\rangle = \sum_m f_m |m\rangle. \tag{9.7.22}$$

The expansion in Eqn. (9.7.21) is called the *exponential Fourier series* and the coefficients f_m are called *Fourier coefficients*. (Note that we take both signs of the integer m since $\exp[\frac{2\pi i m x}{L}]$ and $\exp[-\frac{2\pi i m x}{L}]$ are linearly independent and orthogonal functions.) The expansion of a periodic function in terms of these exponentials is analogous to the expansion of an arrow in terms of **i, j,** and **k.** To find the coefficient of the expansion we simply dot both sides with the basis vector:

$$f_m = \langle m|f \rangle = \int_0^L \frac{1}{\sqrt{L}} \exp\left[-\frac{2\pi i m x}{L}\right] f(x) dx. \qquad (9.7.23)$$

As a concrete example consider the function defined as follows inside the interval $[-1 \le x \le 1]$

$$f(x) = x^2 \qquad [-1 \le x \le 1] \qquad (9.7.24)$$

and which is periodically repeated outside this interval with period $L = 2$. We can expand it in terms of the function $\frac{1}{\sqrt{2}} \exp[2\pi i m x/2] = \frac{1}{\sqrt{2}} \exp[i\pi m x]$. The coefficient f_m is

$$
\begin{aligned}
f_m &= \frac{1}{\sqrt{2}} \int_{-1}^1 x^2 e^{-im\pi x} dx \\
&= \frac{1}{\sqrt{2}} \left[\frac{2\sin m\pi}{m\pi} + \frac{4\cos m\pi}{m^2\pi^2} - \frac{4\sin m\pi}{m^3\pi^3} \right] \\
&= \frac{1}{\sqrt{2}} \frac{(-1)^m \cdot 4}{m^2\pi^2}, \qquad (9.7.25)
\end{aligned}
$$

where we have exploited the fact that the function is even in x so that $e^{-im\pi x}$ can be replaced by its even part $\cos m\pi x$, and where in the last step, we have assumed $m \ne 0$. As for the case $m = 0$ you can have fun taking the limit of the above expression using the Taylor series for the sines and cosines, or more easily, compute the integral of $\frac{1}{\sqrt{2}} x^2$. In any event, we find

$$f(x) = \frac{1}{3} + \sum_{m=\pm 1, \pm 2..} \frac{2(-1)^m}{m^2\pi^2} e^{i\pi m x} \qquad (9.7.26)$$

Notice that the $1/3$ up front is just the mean value of f over the period. As a check on the calculation, consider $f(1)$ which equals 1. Thus it must be true that

$$1 = \frac{1}{3} + \sum_{m=\pm 1, \pm 2..} \frac{2(-1)^m}{m^2\pi^2} e^{i\pi m} = \frac{1}{3} + \sum_{m=\pm 1, \pm 2..} \frac{2}{m^2\pi^2} \qquad (9.7.27)$$

which reduces to

$$\sum_1^\infty \frac{1}{m^2} = \frac{\pi^2}{6},$$

which is true, as per any standard Table of Sums.[6]

To get a feeling for how the series with a finite number of terms approximates the real answer, consider Fig. 9.5.

---

[6]By comparing at $x = 0$, you should find that $1 - 1/4 + 1/9 - 1/16 + \ldots = \pi^2/12$.

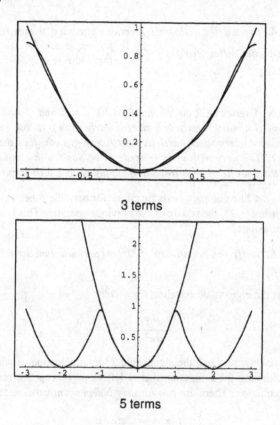

3 terms

5 terms

Figure 9.5. The Fourier series for $f(x) = x^2$ for $-1 \leq x \leq 1$ and a few more periods. In the second case, the nonperiodic function x^2 is shown for comparison.

At the top is the approximation going up to $m = 3$. Notice how well the series fits the function x^2, also shown alongside. At the bottom is the fit with $m = 5$. Now I show three periods and the non-periodic version of x^2 as well as the version periodicized through the series. The weakest point is at $|x| = 1$, where the periodicized function has a cusp and the Fourier series needs more terms to approximate it satisfactorily there than anywhere else.

Problem 9.7.3. *(a): Expand $f(x) = 2xh/L$ for $0 \leq x \leq L/2$ and $f(x) = 2h(L - x)/L$ for $L/2 \leq x \leq L$ in an exponential Fourier series. (b) What do you think is the value of $\sum_{odd} \frac{1}{m^2}$ where $\sum_{odd}$ stands for the sum over all positive odd integers?*

Problem 9.7.4. *Expand the following functions specified within one period, in terms of exponential series:* (a) $f(x) = \begin{cases} 1 & \text{for } -\pi < x < 0 \\ 0 & \text{for } 0 < x < \pi \end{cases}$

(b) $f(x) = e^{-x} \quad -1 \le x \le 1$.

Problem 9.7.5. *A series LCR circuit has* $R = 60$, $L = 1$ *and* $C = 2 \cdot 10^{-6}$. *If the applied voltage is a wave form which rises linearly from 0 to 200 in .01 seconds, instantly plunges to zero, and starts its next linear growth, find the steady state current. (Hint: The term with* $m = 0$ *needs to be dealt with carefully. You may have to take the limit as* $m \to 0$ *instead of simply setting it to zero.)*

If you do not like complex basis functions for periodic functions we can get a real basis by using $-D^2$, the square of the previous operator. This is also hermitian on periodic functions.

Problem 9.7.6. *Verify the hermiticity of* D^2 *on periodic functions by integrating by parts twice.*

Let us look at the eigenvalue equation for $-D^2$:

$$-\frac{d^2 f}{dx^2} = \lambda^2 f, \tag{9.7.28}$$

where we have chosen to call the eigenvalue λ^2 in anticipation of what is coming. We are now looking for a function that is left alone (except for rescaling) upon double differentiation. There are *two linearly independent* candidates

$$f(x) = \sin \lambda x \tag{9.7.29}$$
$$f(x) = \cos \lambda x \tag{9.7.30}$$

for each eigenvalue. *Thus we have a degeneracy and we can take any linear combination of these.* In particular we can form $e^{\pm i\lambda x}$ from these. That is to say, the $-iD$ operator can help $-D^2$ make up its mind on choice of basis within the degenerate space. Conversely, consider linear combinations of two distinct eigenfunctions of $-iD$: $\exp[\pm \frac{2\pi i m x}{L}]$. These will no longer be eigenfunctions of $-iD$. However $-D^2$ is blind to the sign of the eigenvalue of $-iD$ and accepts any linear combination as an eigenfunction. If we demand that the combinations be real functions, we get the sines and cosines.

Proceeding, we will once again find that periodicity will require that $\lambda = 2\pi m/L$. We do not however consider negative m since we do not get new functions that way: the cosine is invariant under $m \to -m$ and the sine goes into minus itself. Stated differently, since exponentials with $+m$ and $-m$ were combined to give the sines and cosines; nothing new will come from starting with $-m$ and $+m$. Once more with feeling: in the eigenvalue problem of $-D^2$

- There is one basis vector for every integer, positive or negative.

- We can associate an exponential function with each.

- We can also pair the nonzero integers of opposite signs and form sines and cosines for each positive integer. The vector with $m = 0$ stands by itself: it is real as such.

The normalization of the trigonometric functions follows from that of the exponentials readily. If $|m\rangle$ and $|-m\rangle$ are normalized exponentials, so are the two combinations labeled by a degeneracy index α

$$|m, \alpha = 1\rangle \quad = \quad \frac{|m\rangle + |-m\rangle}{\sqrt{2}} \tag{9.7.31}$$

$$|m, \alpha = 2\rangle \quad = \quad \frac{|m\rangle - |-m\rangle}{\sqrt{2}i}. \tag{9.7.32}$$

Multiplying and dividing the right-hand sides by $\sqrt{2}$ to bring in the sines and cosines it is clear that the normalized trigonometric functions are

$$|m, \alpha = 1\rangle \quad \rightarrow \quad \sqrt{\frac{2}{L}} \cos \frac{2\pi m x}{L} \tag{9.7.34}$$

$$|m, \alpha = 2\rangle \quad \rightarrow \quad \sqrt{\frac{2}{L}} \sin \frac{2\pi m x}{L}. \tag{9.7.34}$$

The expansion of a periodic function will now take the form

$$f(x) = \frac{a_0}{\sqrt{L}} + \sum_{m=1}^{\infty} a_m \sqrt{\frac{2}{L}} \cos \frac{2m\pi x}{L} + \sum_{m=1}^{\infty} b_m \sqrt{\frac{2}{L}} \sin \frac{2m\pi x}{L}. \tag{9.7.35}$$

which is the concrete version of

$$|f\rangle = \sum_{m=0}^{\infty} \sum_{\alpha=1}^{2} f_{m,\alpha} |m\alpha\rangle, \tag{9.7.36}$$

where α is a label within the degenerate space at each m that tells us if it is a sine or cosine. (In the sector $m = 0$, there is no sine.) To find any coefficient of expansions we dot both sides with the corresponding eigenfunctions. For example

$$b_{17} = \int_0^L \sqrt{\frac{2}{L}} \sin \left[\frac{2\pi \cdot 17 \cdot x}{L} \right] f(x) dx. \tag{9.7.37}$$

Problem 9.7.7. *Repeat problem 9.7.3. in terms of sines and cosines. Compare to the answer you got with the exponential Fourier series.*

Problem 9.7.8. *(i) Obtain the series in terms of sines and cosines for $f(x) = e^{-|x|}$ in the interval $-1 \le x \le 1$. (ii) Repeat for the case $f(x) = \cosh x$.*
Show that

$$f(x) = \frac{\sinh \pi}{\pi} \left[1 + 2 \sum_{1}^{\infty} \frac{(-1)^n}{1+n^2} (\cos nx - n \sin nx) \right]$$

represents the function e^x in the interval $-\pi \le x \le \pi$ (and its periodicized version outside.) Show how you can get the series for $\sinh x$ and $\cosh x$ from the above.

Convergence questions

We will not discuss the question of convergence of the Fourier series in any depth. First note that since the sines, cosines, and exponential functions are bounded by unity, we may replace the Fourier series with one where all these functions are replaced by unity and test for convergence. For example

$$f(x) = \sum_{1}^{\infty} \frac{\cos n\pi x}{n^2} \tag{9.7.38}$$

converges since the sum over $1/n^2$ does. But the sum need not always reproduce the function even if it converges: if the function has a jump, the series will converge to the average of the two values at the jump. In this sense the Fourier series will exist for any function which has a finite number of jumps per period, has a finite number of maxima and minima and $\int |f(x)| dx$ over a period exists.

Unlike in the case of power series, which either converge or do not, for infinite series of sines and cosines we must distinguish between *uniform convergence* and *absolute convergence*. If we pick a point x where the sum converges, it means that given an $\varepsilon > 0$, we can find an $N(\varepsilon, x)$ such that $|f(x) - f_{N(\varepsilon,x)}(x)| < \varepsilon$ where f_N is the partial sum up to N terms. But $N(\varepsilon, x)$ can vary from point to point. In other words, as we approach certain points, it may require more and more terms to come within ε of the limit and stay there.

Definition 9.10. *We say the sum converges uniformly in some interval $[a \le x \le b]$ if for a given ε we can find an $N(\varepsilon)$ common to all x.*

One can find series which converge absolutely but not uniformly and vice versa. We do not get into these questions. All you must note at this point is that given a Fourier series, there is in general no guarantee that its term-by-term derivative will exist or reproduce the derivative of the function in question. On the other hand it is safe to integrate the convergent series term-by-term and the result will be a series that is better behaved. The reason is that each derivative brings an extra factor n to the numerator of the Fourier coefficient, while each integration brings an extra n to the denominator.

9.7.2. Fourier integrals

We want to extend the notion of a Fourier series for periodic functions to nonperiodic functions by viewing them as limits of periodic functions whose periods L have gone to infinity.

Let us write the familiar case as

$$f(x) = \sum_m f_m \frac{e^{2\pi imx/L}}{L} \tag{9.7.39}$$

$$f_m = \int_0^L e^{-2\pi imx/L} f(x) dx. \tag{9.7.40}$$

Notice that the normalization of the eigenfunctions has been changed: instead of having a factor $1/\sqrt{L}$ in the Fourier series and in the inverse (which is the second formula above, which gives the coefficients in terms of the function) we have just a factor $1/L$ in the series. You should check that the above equations still define a consistent pair of Fourier and inverse Fourier transforms. Let us now introduce a variable

$$k = m\frac{2\pi}{L} \equiv m\Delta. \tag{9.7.41}$$

Note that when m goes up by unity, k goes up by $dk = \frac{2\pi}{L} \equiv \Delta$. In terms of k the two equations above become

$$f(x) = \frac{1}{2\pi} \sum_m f_m \frac{2\pi}{L} e^{ikx} \tag{9.7.42}$$

$$f_m = \int_0^L e^{-ikx} f(x) dx, \tag{9.7.43}$$

where we have multiplied and divided by 2π in the first equation. As $L \to \infty$, and $\Delta \to 0$, the k becomes a continuous variable, f_m becomes $f(k)$ and the sum over m becomes an integral over k:

$$\frac{2\pi}{L} \sum_m \to \int dk \tag{9.7.44}$$

so that in the limit of infinite L, we end up with

$$f(x) = \int_{-\infty}^{\infty} e^{ikx} f(k) \frac{dk}{2\pi} \tag{9.7.45}$$

$$f(k) = \int_{-\infty}^{\infty} e^{-ikx} f(x) dx \tag{9.7.46}$$

which are referred to as the *Fourier Transform* and *Inverse Fourier Transform*, respectively. Let us consider one example, the Fourier transform of the *Gaussian*:

$$f(x) = e^{-x^2}. \tag{9.7.47}$$

For this,

$$f(k) = \int_{-\infty}^{\infty} e^{-ikx} e^{-x^2} dx \tag{9.7.48}$$

$$= \int_{-\infty}^{\infty} e^{-(x-ik/2)^2} e^{-k^2/4} dx \tag{9.7.49}$$

$$= e^{-k^2/4} \int_{-\infty-ik/2}^{\infty-ik/2} e^{-z^2} dz. \tag{9.7.50}$$

In going to the second line we have completed squares in the exponent by adding and subtracting $-k^2/4$. In the next we have introduced a complex variable $z = x - ik/2$ and written the limits on the integral in terms of z. Notice that the z-integral now runs parallel to the real axis, but a distance ik below it. We now argue that since the integrand is free of singularities in the region between the real axis and this line, we may shift the contour up to the real axis in the z-integral. Recalling that this integral equals $\sqrt{\pi}$, we find

$$f(k) = \sqrt{\pi}e^{-k^2/4} \tag{9.7.51}$$

so that the transform of a gaussian is a gaussian also. Had we chosen our original function to be

$$f(x) = e^{-x^2/a^2} \tag{9.7.52}$$

so that this bell-shaped curve had a width of order a, we would have found that its transform $f(k)$ had a width of order $1/a$.

Problem 9.7.9. *Verify the above claim. It follows that a gaussian in k of width a transforms back to a gaussian in x of width $1/a$.*

In quantum theory we deal with a function $f(x)$ whose height measures the likelihood of finding the particle at the point x and whose transform $f(k)$ measures the likelihood of finding it with momentum $p = \hbar k$, where $\hbar$ is called *Planck's constant*. The above example shows that if the particle is localized in a region of width a its momentum has a spread of order $\hbar/a$. This is the origin of the uncertainty principle $\Delta x \Delta p \simeq \hbar$.

Problem 9.7.10. *Use contour integration to show that the transform of*

$$f(x) = \frac{1}{x^2 + a^2} \tag{9.7.53}$$

is

$$f(k) = \frac{\pi}{a}e^{-|k|a}. \tag{9.7.54}$$

Invert the transform to get back $f(x)$. Do not use contour integration since $f(k)$ is not an analytic function. Just do the simple exponential integrals involved.

If we combine the Fourier transform and its inverse we get

$$f(x) = \int_{-\infty}^{\infty} \frac{dk}{2\pi} e^{ikx} \int_{-\infty}^{\infty} dy e^{-iky} f(y) \tag{9.7.55}$$

$$= \int_{-\infty}^{\infty} f(y) R(x-y) dy \text{ where} \tag{9.7.56}$$

$$R(x-y) = \int_{-\infty}^{\infty} \frac{dk}{2\pi} e^{ik(x-y)}. \tag{9.7.57}$$

Earlier we had introduced the Dirac delta function $\delta(x - y)$ with the following properties:

- $\delta(x - y) = 0$ if $x \neq y$.
- $\delta(x - y)$ blows up at $x = y$ in such a way that the area under it equals unity:

$$\int \delta(x - y) dy = 1 \tag{9.7.58}$$

if the region of integration contains the point $y = x$.

- As a result of the above if $f(y)$ is any smooth function

$$\int \delta(x-y)f(y)dy \quad = \quad \int_{x-\varepsilon}^{x+\varepsilon} f(y)\delta(x-y)dy \quad \varepsilon \to 0 \tag{9.7.59}$$

$$= \quad f(x)\int_{x-\varepsilon}^{x+\varepsilon} \delta(x-y)dy \tag{9.7.60}$$

$$= \quad f(x). \tag{9.7.61}$$

In short the integral of any function with the delta function pulls out the value of the function at $y = x$.

It follows from all of the above and Eqn. (9.7.56) that

$$R(x-y) = \delta(x-y) = \int_{-\infty}^{\infty} \frac{dk}{2\pi} e^{ik(x-y)}. \tag{9.7.62}$$

This is a very useful formula for the Dirac delta function and will be recalled towards the end of the next chapter.

9.7.3. Normal modes: vibrating string

Consider a string of length L with points on it labeled by x running between 0 and L. Let $\psi(x,t)$ be its displacement from equilibrium at time t at position x. Our job is to find $\psi(x,t)$ given $\psi(x,0)$. Once again we will assume for convenience that the initial velocity is zero at all points.

This is an extension of the coupled mass problem. There we had only two coordinates to follow, here we have an infinite number: the displacement at each point x. (Note that unlike in the coupled mass problem, where x was the dynamical variable, here it is a label for the vibrating degrees of freedom $\psi(x)$.)

To proceed we need the equation of motion for the string. This is the wave equation:

$$\frac{1}{v^2}\frac{\partial^2 \psi}{\partial t^2} = \frac{\partial^2 \psi}{\partial x^2}, \tag{9.7.63}$$

where v is the velocity of propagation of waves in the string.

At any given time the string is described by a function of x. We may therefore associate with that function a ket $|\psi(t)\rangle$. The equation obeyed by the ket is then

$$\frac{1}{v^2}|\ddot{\psi}(t)\rangle = D^2|\psi(t)\rangle \tag{9.7.64}$$

which is mathematically the same as the coupled mass problem (compare with Eqns. (9.6.30 - 9.6.34) except for the fact that the two dimensional vector space has been replaced by an infinite-dimensional Hilbert space. The strategy is still the same. We first look for normal modes:

$$|\psi(t)\rangle = |\psi(0)\rangle f(t). \qquad (9.7.65)$$

Once again f will be a cosine and more precisely if

$$\frac{\ddot{f}(t)}{f(t)} = \text{constant} = -\omega^2 \qquad (9.7.66)$$

then

$$f(t) = f(0)\cos\omega t \qquad (9.7.67)$$

and

$$D^2|\psi(0)\rangle = -\frac{\omega^2}{v^2}|\psi(0)\rangle \equiv -k^2|\psi(0)\rangle. \qquad (9.7.68)$$

So what we find is that the ansatz we made, Eqn. (9.7.65), is viable only if $f(t)$ is a cosine and $\psi(0)\rangle$ is an eigenket of D^2.

Let us note for future use that

$$\omega = kv. \qquad (9.7.69)$$

Going back from the ket to the function it represents, Eqn. (9.7.68) becomes:

$$\frac{d^2\psi(x,0)}{dx^2} + k^2\psi(x,0) = 0 \qquad (9.7.70)$$

which is readily solved to give

$$\psi(x,0) = A\cos kx + B\sin kx. \qquad (9.7.71)$$

But the displacement must vanish at $x = 0$. This kills the coefficient of the cosine. As for the sine, the condition that the displacement vanish at the other end gives either $B = 0$, which renders the solution trivial, or

$$k = \frac{m\pi}{L}, \qquad (9.7.72)$$

which is the case we are interested in. The normalized eigenfunctions are then

$$|m\rangle \rightarrow \sqrt{\frac{2}{L}}\sin\frac{m\pi x}{L} \qquad (9.7.73)$$

and the normal modes of vibration are

$$|m(t)\rangle \rightarrow \sqrt{\frac{2}{L}}\sin\frac{m\pi x}{L}\cos\frac{m\pi vt}{L}. \qquad (9.7.74)$$

In the case of coupled masses, we saw that we could excite a normal mode if we pulled the two masses by equal, or equal and opposite, amounts. To excite a string into a pure normal mode takes more than two hands: you must initially deform it into a perfect sine wave $\sin(m\pi x/L)$ and let it go. If you just pluck it, you will excite all the modes that enter the Fourier expansion of the initial deformation.

By mimicking what we did with the coupled masses, we can write the general solution as

$$\psi(x,t) = \sum_{m=1}^{\infty} \psi_m \sqrt{\frac{2}{L}} \sin\frac{m\pi x}{L} \cos\frac{m\pi vt}{L}. \tag{9.7.75}$$

To find ψ_m use

$$\psi_m = \langle m|\psi(0)\rangle = \int_0^L \sqrt{\frac{2}{L}} \sin\frac{m\pi x}{L}\psi(x,0)dx. \tag{9.7.76}$$

Consider a specific example where

$$\psi(x,0) = x(L-x) \tag{9.7.77}$$

in which case

$$\psi_m = \sqrt{\frac{2}{L}}\int_0^L \sin\frac{m\pi x}{L}x(L-x)dx = \sqrt{\frac{2}{L}}\frac{4L^3}{m^3\pi^3} \quad m \text{ odd}. \tag{9.7.78}$$

Problem 9.7.11. *Work out the integral and derive the above result.*

Depicted in Fig. 9.6 is the subsequent behavior of the string

$$\psi(x,t) = \sum_{\text{odd}} \frac{8L^2}{m^3\pi^3} \sin\frac{m\pi x}{L} \cos\frac{m\pi vt}{L} \tag{9.7.79}$$

for the case $L = v = 1$ over a period $0 \le t \le 2$.

Problem 9.7.12. *A string of length L that supports waves at velocity v is displaced at $t = 0$ to have a triangular profile, symmetric about its midpoint, where the displacement is h. Find the displacement at all future times. (First write down a function that describes the initial displacement.)*

Problem 9.7.13. *A string of length π has the following initial displacement: rises linearly between 0 and $\pi/4$ to reach unit height, drops linearly to height -1 by $3\pi/4$, rises linearly back to zero at the other end. What is its subsequent evolution, assuming velocity $v = 1$?*

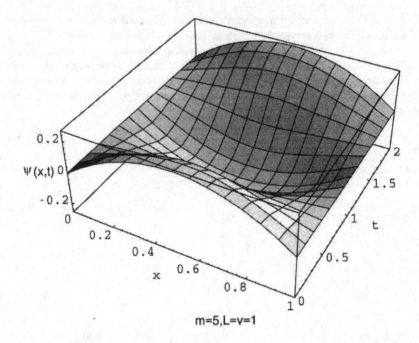

Figure 9.6. The motion of the string with $L = v = 1$ and $\psi(x,0) = x(L - x)$ for one full period $0 \leq t \leq 2$. (We truncate the series at $m = 5$.)

Sometimes you will be asked to solve the string problem for the case where the initial displacement is h for the first half of the string and zero for the second. This configuration violates the condition that the string displacement vanish at the ends. This initial condition should be interpreted to mean that the displacement grows from zero to h in a very small distance from the left end and likewise drop to zero in a very tiny distance around the midpoint. If you expand this initial function in terms of the normal modes, they will sum to the right answer everywhere except at $x = 0$ and $x = L/2$, where the series will sum to a $h/2$, the average of the two values associated with the discontinuity.

Problem 9.7.14. *Do the time evolution for the problem where the string of length L and wave velocity v has an initial displacement h for the left half and zero for the right half.*

We will now discuss the question of completeness within the limited context of the vibrating string. The question is this: we have an infinite number of basis functions, but are they complete in the sense that any displacement of the string can be written as a sum of these? Take the case of the string displaced into a triangular shape with a height of h at the midpoint. Suppose we expand it as

$$f(x) = \sum_{1}^{\infty} f_n \sqrt{\frac{2}{L}} \sin \frac{n\pi x}{L} \tag{9.7.80}$$

where

$$f_n = \frac{8h}{\pi^2 n^2} \sqrt{\frac{L}{2}} \sin \frac{\pi n}{2} \tag{9.7.81}$$

and want to see if we missed some components along the way. Suppose f were a finite-dimensional vector with components f_n in a basis known to be complete and we had expanded it in another orthonormal basis with components f_n'. In the new basis we would simply project the vectors along the new basis vectors to find the new components. If all were going well, we would find that even though the components change with the basis, the length squared of the vector would come out the same as before:

$$\sum f_n^2 = \sum f_n'^2. \tag{9.7.82}$$

If we dropped a basis vector from the new basis, it would still be orthonormal, but its lack of completeness would be betrayed by the fact that the right-hand side would come out smaller. (It can happen that the vector we picked did not have a component in the direction omitted. But some other vector will reveal the problem.) Let us now transcribe this logic to the case of the string. The original description of the string in terms of $f(x)$ is obviously complete since $f(x)$ tells us everything about the string everywhere. The new description in terms of f_n should therefore satisfy *Parseval's Theorem*

$$\int_0^L f^2(x)dx = \sum_{1}^{\infty} f_n^2. \tag{9.7.83}$$

(On the right-hand side, we use a sum, as is appropriate to a countable basis, in finding the norm squared.) The verification of this for the special initial condition is left as an exercise.

Problem 9.7.15. *Show that Parseval's theorem is satisfied by the coefficients in Eqn. (9.7.81) using*

$$\sum_{k=1}^{\infty} \frac{1}{(2k-1)^4} = \frac{\pi^4}{96}.$$

As mentioned earlier, lack of completeness can be exposed by even one failure of Parseval's theorem while no amount of successful trials can assure us the basis is complete since it can happen that none of the examples we took had components in the missing direction(s). Is there a test that will establish once and for all the completeness of the basis? There is, but it goes beyond the scope of this book.

9.8. Some Terminology

Suppose $|1\rangle, |2\rangle \ldots |i\rangle \ldots |n\rangle$ is an orthonormal basis. Let us generate a new set of vectors

$$|1'\rangle \;=\; U|1\rangle \tag{9.8.1}$$
$$|i'\rangle \;=\; U|i\rangle \tag{9.8.2}$$
$$|n'\rangle \;=\; U|n\rangle, \tag{9.8.3}$$

where U is unitary. *Then the primed vectors form another orthonormal basis.* The proof is simple. Take the adjoint of these equations to get

$$\langle i'| = \langle i|U^\dagger \tag{9.8.4}$$

Now dot this with $|j'\rangle = U|j\rangle$ to get

$$\langle i'|j'\rangle = \langle i|U^\dagger U|j\rangle = \langle i|j\rangle = \delta_{ij} \tag{9.8.5}$$

Thus a unitary transformation converts an orthonormal basis to another. In other words, unitary transformations are the generalizations of rotations to complex vector spaces. The converse is also true: given any two orthonormal bases, we can find a unitary transformation that connects them. The proof is left to you. In view of the above, it is clear that the rows and columns of U are orthonormal.

By this argument, the passage from some given orthonormal basis to the eigenbasis $|I\rangle, |II\rangle \ldots$ of a hermitian operator H must be through some U. That is

$$|I\rangle = U|1\rangle \tag{9.8.6}$$

and so on. Let us look at the matrix elements of this U. By the mnemonic, its first column contains the components of the basis vector after the action of U. But the effect of U is to convert $|1\rangle$ to the first normalized eigenvector $|I\rangle$. Thus the first column of U contains the first eigenvector components and likewise for the other columns. You can verify that with these entries, the matrix will be unitary.

Consider next the statement that a hermitian operator is diagonal in its eigenbasis with eigenvalues on the diagonal. This means that if $|I\rangle$, $|J\rangle$ are the eigenvectors of H:

$$\langle I|H|J\rangle = \langle i|U^\dagger H U|j\rangle = \text{diagonal matrix}. \tag{9.8.7}$$

Thus $U^\dagger H U$ is a diagonal matrix in the *original* basis. *We say that H has been diagonalized by the unitary transformation U.* Thus for every hermitian matrix we can find a unitary matrix U (built out of its eigenvectors) such that $U^\dagger H U$ is diagonal.

Not only does the unitary operator preserve the orthonormality of the basis vectors, it preserves the inner product between any two vectors. Thus if $|V'\rangle = U|V\rangle$ and $|W'\rangle = U|W\rangle$ then

$$\langle V'|W'\rangle = \langle UV|UW\rangle = \langle V|U^\dagger U|W\rangle = \langle V|W\rangle. \tag{9.8.8}$$

Problem 9.8.1. *Consider the coupled mass problem. Explicitly write out U, the matrix of normalized eigenvectors of M. Show that $U^\dagger M U$ is diagonal.*

9.9. Tensors: An Introduction

Let us begin by recalling the distinction between scalars and vectors. Let $\phi(\mathbf{r})$ be a scalar field and $\mathbf{W}(\mathbf{r})$ a vector field. Under a rotation of axes, the points in space get renamed. The point

(x_1, x_2, x_3) turns into

$$x_i' = \sum_{j=1}^{3} R_{ij} x_j \qquad (9.9.1)$$

where R is a rotation matrix, parametrized by some rotation angles, obeying

$$R^T = R^{-1}. \qquad (9.9.2)$$

Note that we have switched from (x, y, z) to (x_1, x_2, x_3). The advantage of this notation will become quite apparent.

If ϕ is a scalar field, its numerical value at a point is unaltered by the change of coordinates. Thus if it stands for temperature, a person in a rotated coordinate system will ascribe to a given point the same value of temperature, even though he calls the point by a different name. We express this by saying

$$\phi'(x') = \phi(x) \qquad (9.9.3)$$

where x stands for all three coordinates. *Thus a scalar is invariant under a rotation of axes.* (It may also be invariant under other transformations, such as reflections, but we restrict our discussion to rotations.)

By contrast, if $\mathbf{W}$ is a vector field, it will appear to have different components at one and the same point to the two observers. To be precise,

$$W_i'(x') = \sum_{j=1}^{3} R_{ij} W_j(x). \qquad (9.9.4)$$

Note that the components of $\mathbf{W}$ transform the same way as the components of $\mathbf{r}$, as is clear from Eqns. (9.9.1-9.9.4). *Indeed by definition, a vector in three dimensions is any object described by three components which respond to coordinate transformations the way $\mathbf{r}$ does.*

We are now ready to discuss tensors. Consider first a situation where one vector, the cause, produces another, the effect. For example, according to Newton a force $\mathbf{F}$ produces an acceleration $\mathbf{a}$. The relation between the two is

$$\mathbf{F} = m\mathbf{a}. \qquad (9.9.5)$$

Likewise in a wire, the applied electric field $\mathbf{E}$ causes a current density $\mathbf{j}$ given by

$$\mathbf{j} = \sigma \mathbf{E} \qquad (9.9.6)$$

which is just Ohm's law written in terms of the field and current density instead of voltage and current. The proportionality constant σ is the *conductivity*.

Note that in both cases the effect is linear in the cause and parallel to it, allowing us to relate the two by a scalar, m or σ. But what if the cause and effect are still linearly related but not parallel? For example if one takes electrons in the $x - y$ plane, and applies a magnetic field perpendicular to it, one finds that an electric field in the plane will produce a current in the plane, linear in the field, but no longer parallel to the field. The linear relationship takes the form

$$j_x = \sigma_{xx} E_x + \sigma_{xy} E_y \qquad (9.9.7)$$

$$j_y = \sigma_{yx} E_x + \sigma_{yy} E_y \qquad (9.9.8)$$

$$j_i = \sum_{j=1}^{2} \sigma_{ij} E_j. \qquad (9.9.9)$$

You will recognize this as a matrix relation

$$j = \sigma E \qquad (9.9.10)$$

in which the conductivity matrix σ_{ij} converts the applied field to the current by matrix multiplication. But it is also a *tensor* equation and σ_{ij} are the four components of a *second-rank tensor*. This name implies something very definite about how things look in a rotated frame. So let us see how the above equation appears in a rotated frame. First, since $\mathbf{j}$ and $\mathbf{E}$ are vectors, the old components are related to the new ones by

$$j' = Rj \tag{9.9.11}$$

$$E' = RE, \tag{9.9.12}$$

where R is the rotation matrix. Thus Eqn. (9.9.10) can be written as

$$R^T j' = \sigma R^T E' \tag{9.9.13}$$

$$j' = R\sigma R^T E' \tag{9.9.14}$$

$$j'_i = \sum_{a=1}^{2} \sum_{b=1}^{2} R_{ia} R_{lb} \sigma_{ab} E'_l. \tag{9.9.15}$$

Now, the observer in the primed frame will relate the current density and field using a conductivity tensor σ'_{il} defined by

$$j'_i = \sum_{l=1}^{2} \sigma'_{il} E'_l. \tag{9.9.16}$$

Upon comparing the two preceding equations we find

$$\sigma'_{il} = \sum_{a=1}^{2} \sum_{b=1}^{2} R_{ia} R_{lb} \sigma_{ab}. \tag{9.9.17}$$

When we compare this transformation law to the one for vectors, *we notice that the tensor has two indices and each gets rotated by its own R matrix.* The rank is two because the tensor has two indices.

In the same fashion, a tensor of rank three will have components T_{ijk} which transform as follows

$$T'_{ijk} = \sum_{a=1}^{2} \sum_{b=1}^{2} \sum_{c=1}^{2} R_{ia} R_{jb} R_{kc} T_{abc}. \tag{9.9.18}$$

It is not possible to view this tensor T as a matrix.

In general a rank N tensor will have N indices and need a product of N rotation matrices to rotate each of its indices. In this terminology a scalar has rank 0, and a vector has rank 1. This definition works in all dimensions.

Einstein proposed that in dealing with tensors we adopt the convention that any repeated index is to be summed over. In the *Einstein convention* the above equation is written as

$$T'_{ijk} = R_{ia} R_{jb} R_{kc} T_{abc}. \tag{9.9.19}$$

We will follow it from now on.

The process of summing over a common index is called *contraction* or the *inner product* and it reduces the overall rank by two. For example in

$$j_i = \sigma_{ij} E_j \tag{9.9.20}$$

we have on the left a rank one object and on the right a sum over products of a rank two object and rank one object. However the whole right-hand side transforms like a rank $3 - 2 = 1$ object due to the sum or contraction over the index j.

How does contraction get rid of two indices? Consider a contraction you already know: the dot product of two vectors

$$\phi = V_i W_i = \mathbf{V} \cdot \mathbf{W} \tag{9.9.21}$$

Potentially this has two indices. Let us look at its transformation law:

$$\phi' = V_i' W_i' = R_{ia} R_{ib} V_a W_b = R_{ai}^T R_{ib} V_a W_b = (R^T R)_{ab} V_a W_b = \delta_{ab} V_a W_b = V_a W_a = \phi.$$
$$(9.9.22)$$

Thus ϕ is a scalar that goes into itself, i.e., has rank zero. Note how the two $R's$ neutralized each other because they were inverses. This is generally what happens in a contraction.

Given two vectors V and W, one can form a second-rank tensor T with elements:

$$T_{ij} = V_i W_j. \tag{9.9.23}$$

This is called an *outer product.* (In contrast to the inner product or contraction, the outer product of two tensors gives a tensor of higher rank.) The rotation properties of this object are transparent and assure us of its rank two tensorial character. In d dimensions it will have d^2 components. However it is possible to form rank two objects with fewer components. Consider in three dimensions the tensor

$$A_{ij} = V_i W_j - W_i V_j. \tag{9.9.24}$$

Notice that:

$$A_{ij} = -A_{ji} \tag{9.9.25}$$

i.e., the tensor is *antisymmetric* in its indices. This means in particular that $A_{ii} = 0$. The six nonzero components are not fully independent due to antisymmetry: given A_{12}, A_{23}, A_{31}, the other three follow. *Thus an antisymmetric tensor in three dimensions has by coincidence the same number of independent components as a vector and can be treated as such. You will no doubt know what this vector is called: it is the cross product of* **V** *and* **W**. (This coincidence is what makes it possible to uniquely define the cross product of two vectors as a vector perpendicular to both.)

If you have been following this discussion carefully you will have noted the following: while it is true that the antisymmetric tensor A has six nonzero components, it remains to be shown that the components in the rotated frame will be related to those in the old frame before we can call it a second-rank tensor. Let us verify this:

$$A_{ij}' = V_i' W_j' - W_i' V_j' \tag{9.9.26}$$
$$= R_{ia} R_{jb} V_a W_b - R_{ia} R_{jb} W_a V_b \tag{9.9.27}$$
$$= R_{ia} R_{jb}(V_a W_b - W_a V_b) \tag{9.9.28}$$
$$= R_{ia} R_{jb} A_{ab}. \tag{9.9.29}$$

The point to note is that the transformed antisymmetric tensor is expressed in terms of the old one and nothing else.

Sometimes a tensor can have the property that its components are the same in all coordinate systems. Consider the tensor with components

$$T_{ij} = \delta_{ij}. \tag{9.9.30}$$

In a rotated system the components will be

$$T_{ij}' = R_{ia} R_{jb} \delta_{ab} \tag{9.9.31}$$
$$= R_{ia} R_{ja} = R_{ia} R_{aj}^T = \delta_{ij}. \tag{9.9.32}$$

Notice that the invariance is with respect to rotations and not some arbitrary transformations.

Since

$$x_i' = R_{ij} x_j \tag{9.9.33}$$

we may write

$$R_{ij} = \frac{\partial x_i'}{\partial x_j} \tag{9.9.34}$$

and define a vector as one which transforms as

$$V_i' = \frac{\partial x_i'}{\partial x_j} V_j \tag{9.9.35}$$

and tensors of higher rank in a similar fashion. The power of this definition is that even if the primed and unprimed coordinates are not linearly related, (as they were in the case of rotations), the definition holds. I mention this for your future use and will not discuss it further.

Before continuing with more tensor analysis, let us consider a few more examples of tensors.

If you spin an ellipsoid along its long axis going through its center of mass, (call it the y-axis) with some angular velocity, it will have a certain angular momentum in that direction. If you spin it around its short axis (call it the x-axis) with the same angular velocity it will have a different (larger) angular momentum parallel to the angular velocity. It follows that if you spin it along an axis half way between the two, i.e., such that the angular velocity has equal x and y components, the angular momentum will not have equal x and y components, i.e., will not be in the same direction as the angular velocity Thus in general, we will have

$$L_i = I_{ij}\omega_j, \tag{9.9.36}$$

where $\mathbf{L}$ and $\boldsymbol{\omega}$ are the angular momentum and velocity and I_{ij} is called the *moment of inertia tensor*. (We have been using I to denote the unit matrix. But I is also a commonly used symbol for the moment of inertia. To prevent confusion, in this section we will use I to stand for the moment of inertia only.)

If an object is made of point masses m^α with coordinates $\mathbf{r}^\alpha$, I is given by

$$I_{ij} = \sum_\alpha m^\alpha (\delta_{ij}(r^\alpha)^2 - r_i^\alpha r_j^\alpha) \equiv \sum_\alpha I_{ij}^\alpha. \tag{9.9.37}$$

Thus the moment of inertia tensor I of a collection of objects is the sum over the moment of inertia tensor I_{ij} of each object. This is a nice example of tensor addition. *As with vectors, the addition of tensors is possible only between tensors of same rank (same number of indices) and dimension (the indices run over the same range) and done component by component.*

Let us consider an example. A rigid body is made of masses placed at the edges of a unit cube in the first octant, with one corner at the origin and its edges parallel to the axes. At each corner the mass is m times the sum of the coordinates. For example at $(1, 1, 1)$ we have a mass $3m$. The moment of inertia of this object is

$$I = m \begin{bmatrix} 16 & -5 & -5 \\ -5 & 16 & -5 \\ -5 & -5 & 16 \end{bmatrix}. \tag{9.9.38}$$

Problem 9.9.1. *Derive this expression for the inertia tensor I.*

By writing out the components explicitly, or by inspecting Eqn. (9.9.37), we notice that I is a real symmetric (i.e., hermitian) matrix. This means that it will have three orthogonal eigenvectors. If $\boldsymbol{\omega}$ is along one of them, it follows from Eqn. (9.9.36) that $\mathbf{L}$ will be parallel to $\boldsymbol{\omega}$. These directions are called the *principal axes* of the object. Any object, even a (rigid) elephant, will have three such axes, a fact that is not so obvious physically (but well-known to elephant twirlers) but follows from the theorem about hermitian matrices.

Problem 9.9.2. *(a): An object is made of eight equal masses m placed at the edges of a unit cube centered at the origin. Find its moment of inertia tensor for rotations about the origin. Find its eigenvectors and eigenvalues. Note that due to degeneracy, any axis is an eigen or principal axis. Now you would have expected this for a sphere but not a cube. The result however follows from the mathematics done earlier in this chapter.*

(b): Find the moment of inertia tensor if only two of the masses, at coordinates $\pm(1/2, 1/2, 1/2)$ are present. What are the principal axes and moments of inertia? Try to guess at least some parts of the answer before doing the calculation. (Hint: Can you line up the masses along some axis? What will be the inertia for this case? Given this axes, where are the other two axes?)

If the object has a continuous distribution of mass, I can be written as an integral. It will still be hermitian and all of the above results will hold.

Consider next a crystal to which we apply an electric field. This will cause the cloud of electrons in each atom to shift away from the nucleus, producing a dipole moment, defined as the product of the electron charge and the separation from the nucleus. The dipole moment per unit volume is called $\mathbf{P}$, the *polarization*. Once again if the crystal has different polarizability in different directions, the polarization and electric field vectors will not generally be parallel. The two will be related by

$$P_i = \alpha_{ij} E_j, \tag{9.9.39}$$

where α_{ij} is called the *polarizability tensor.*

After these examples, we return to the general study of tensors. Now, we have defined a vector by the transformation rule Eqn. (9.9.35) repeated below:

$$V_i' = \frac{\partial x_i'}{\partial x_j} V_j.$$

Now I will point out another vector you know which does not obey this rule! Consider $\nabla \phi$, the gradient of a scalar. Its components in the primed frame will be

$$V_i' \;=\; \frac{\partial \phi}{\partial x_i'} \tag{9.9.40}$$

$$=\; \frac{\partial \phi}{\partial x_j} \frac{\partial x_j}{\partial x_i'} \tag{9.9.41}$$

$$=\; \frac{\partial x_j}{\partial x_i'} V_j. \tag{9.9.42}$$

In general

$$\frac{\partial x_j}{\partial x_i'} \neq \frac{\partial x_i'}{\partial x_j} \tag{9.9.43}$$

and one must make a distinction between the two vectors: those that transform like $\mathbf{r}$ and are called *contravariant vectors* and those that transform like the gradient and are called *covariant vectors.* (Tensors of higher rank will be similarly classified with respect to each of their indices.) The reason we did not make the distinction is that in the case of rotations

$$\frac{\partial x_j}{\partial x_i'} \;=\; R_{ji}^T \tag{9.9.44}$$

$$=\; R_{ij} \tag{9.9.45}$$

$$=\; \frac{\partial x_i'}{\partial x_j}. \tag{9.9.46}$$

For most purposes the above treatment of tensors will do. The notable exception is relativity. You will learn about the unavoidable distinction between covariant and contravariant tensors in special relativity. You will also learn that the electric and magnetic fields are really the six independent components of an antisymmetric tensor $f_{\mu\nu}$ where the indices run over four values (three for space and one for time). As for general relativity, the bad news is that you will need to know a lot more tensor analysis than we have covered. The good news is that it is usually taught as part of the course.

9.10. Summary

Here are the key ideas from this chapter.

- You must know that there are axioms that define a linear vector space. Preferably you will know them by heart.

- Definition of linear independence, basis, dimensionality, components, uniqueness of expansion.

- Know how to add abstract vectors (component by component) and multiply them by scalars.

- Know the concrete formula for the inner product in an orthonormal basis:

$$\langle V|W \rangle = \sum_i v_i^* w_i$$

and that it is linear in the ket, antilinear in the bra, and that $\langle V|V \rangle$ vanishes only when the vector does.

- Know how to expand a vector $|V\rangle$ in terms of a given orthonormal basis. In other words, if

$$|V\rangle = \sum_i v_i |i\rangle$$

then the coefficients of expansion are

$$v_i = \langle i|V \rangle.$$

- Know that there exists a Gram–Schmidt procedure and the key idea behind it.

- The Schwarz and triangle inequalities:

$$|\langle V|W \rangle| < |V|\,|W|$$

$$|V + W| < |V| + |W|.$$

- Definition of a linear operator

$$\Omega(a|V\rangle + b|W\rangle) = a\Omega|V\rangle + b\Omega|W\rangle.$$

and the mnemonic for finding the matrix elements

$$\Omega_{ij} = \langle i|j' \rangle = \langle i|\Omega|j \rangle.$$

- The eigenvalue problem

$$\Omega|\omega\rangle = \omega|\omega\rangle$$

is solved by (i) writing all the entities in a basis, (ii) finding the roots of the characteristic polynomial $P(\omega) = |\Omega - \omega I|$, (iii) and plugging in the eigenvalues one at time to get the eigenvectors (up to a scale). Be aware that if there is degeneracy, special treatment is needed.

- Know that the eigenvectors of a hermitian operator are orthogonal and eigenvalues are real. These eigenvectors form a basis.

- Good knowledge of coupled mass problem, normal modes and solution for propagator in terms of the former. In other words, given an equation

$$|\ddot{x}\rangle = M|x\rangle$$

(i) find the eigenvectors of M (which form an orthonormal basis), (ii) expand the initial state in terms of them by taking the appropriate inner products, (iii) attach to the expansion coefficients the simple cosine time dependence (for the case of vanishing initial velocity) to get the state at future times.

- Know that a stationary point of $f(x, y)$ is a maximum, minimum, or saddle point according as the eigenvalues of the matrix of second derivatives are both negative, positive, or of opposite sign. Know the generalization to functions of more variables.

- Be familiar with the idea of functions as vectors; their inner product

$$\langle f|g\rangle = \int f^*(x)g(x)dx$$

and the notion of functions being normalized or orthogonal on the basis of this inner product.

- Know that $-iD$ is hermitian with respect to periodic functions and generate orthonormal basis functions $\frac{1}{\sqrt{L}}\exp[\frac{2\pi imx}{L}]$ obeying

$$\langle m|m'\rangle = \int_0^L \frac{1}{\sqrt{L}}\exp\left[-\frac{2\pi imx}{L}\right]\frac{1}{\sqrt{L}}\exp\left[\frac{2\pi im'x}{L}\right] = \delta_{mm'}.$$

In terms of these we obtain the exponential Fourier series:

$$|f\rangle = \sum_{m=0,\pm1,..} f_m|m\rangle \qquad\qquad f_m = \langle m|f\rangle$$

while in concrete form

$$f(x) = \sum_m f_m \frac{1}{\sqrt{L}} \exp\left[\frac{2\pi i m x}{L}\right]$$

$$f_m = \int_0^L \frac{1}{\sqrt{L}} \exp\left[-\frac{2\pi i m x}{L}\right] f(x) dx.$$

Similarly $-D^2$ has sines and cosines as its eigenfunctions and permits the expansion in Fourier trigonometric series:

$$f(x) = \frac{a_0}{\sqrt{L}} + \sum_{m=1}^{\infty} a_m \sqrt{\frac{2}{L}} \cos\frac{2m\pi x}{L} + \sum_{m=1}^{\infty} b_m \sqrt{\frac{2}{L}} \sin\frac{2m\pi x}{L}$$

which is the concrete version of

$$|f\rangle = \sum_{m=0}^{\infty} \sum_{\alpha=1}^{2} f_{m,\alpha} |m\alpha\rangle,$$

where α is a label within the degenerate space at each m that tells us if it is a sine or cosine. (In the sector $m = 0$, there is no sine.) To find any coefficient of expansions we dot both sides with the corresponding eigenfunctions. For example

$$b_{17} = \int_0^L \sqrt{\frac{2}{L}} \sin\frac{2\pi \cdot 17 \cdot x}{L} f(x) dx.$$

- For nonperiodic functions we resort to Fourier integrals:

$$f(x) = \int_{-\infty}^{\infty} e^{ikx} f(k) \frac{dk}{2\pi} \qquad f(k) = \int_{-\infty}^{\infty} e^{-ikx} f(x) dx.$$

- The Dirac delta function $\delta(x - x_0)$ is nonzero only at $x = x_0$ where it is infinitely-tall in such a way as to enclose unit area. As a result

$$\int_{a<x_0}^{b>x_0} f(x)\delta(x - x_0) dx = f(x_0).$$

The delta function has the following expansion:

$$\delta(x - y) = \int_{-\infty}^{\infty} \frac{dk}{2\pi} e^{ik(x-y)}.$$

- The equation of the vibrating string

$$\frac{1}{v^2}\frac{\partial^2 \psi}{\partial t^2} = \frac{\partial^2 \psi}{\partial x^2}$$

(for vanishing initial velocity) is dealt with as follows. We expand $\psi(x, 0)$ in terms of the eigenfunctions of D^2 (vanishing at the ends $x = 0, L$) and append a cosine time dependence for future times:

$$\psi(x, t) = \sum_{m=1}^{\infty} \psi_m \sqrt{\frac{2}{L}} \sin \frac{m\pi x}{L} \cos \frac{m\pi vt}{L}$$

$$\psi_m = \langle m | \psi(0) \rangle = \int_0^L \sqrt{\frac{2}{L}} \sin \frac{m\pi x}{L} \psi(x, 0) dx.$$

DIFFERENTIAL EQUATIONS

10.1. Introduction

The quadratic equation

$$ax^2 + bx + c = 0 \tag{10.1.1}$$

is an example of an *algebraic equation*, one in which the unknown and powers of the unknown appear, and our task is to determine the unknown given this information. A *differential equation* on the other hand, expresses a relation between a function and its derivatives, using which we are to deduce the function. We have encountered a few even before we got to this chapter. Here is a familiar example:

$$\frac{d^2x}{dt^2} + \omega^2 x = 0 \tag{10.1.2}$$

which describes simple harmonic motion. Another example is the wave equation for the vibrating string:

$$\frac{1}{v^2}\frac{\partial^2\psi}{\partial t^2} - \frac{\partial^2\psi}{\partial x^2} = 0. \tag{10.1.3}$$

Another famous example, not discussed in this book is the sine-Gordon equation:

$$\frac{1}{v^2}\frac{\partial^2\psi}{\partial t^2} - \frac{\partial^2\psi}{\partial x^2} + \sin\psi = 0. \tag{10.1.4}$$

In each example, we are to find some unknown function given some information relating the function and its derivatives. When we do this, we will have solved the differential equation. We will learn now to classify equations and to solve a few of the simpler ones. There are many tricks of the trade but nothing changes the following rule of the game: it doesn't matter how you get the solution: if you feed it into the equation and it satisfies it, you win.

A differential equation is said to be *a partial differential equation or PDE* if partial derivatives enter the equation. The wave equation and sine-Gordon equation are such equations. If only ordinary derivatives enter, as in the harmonic motion example, we have an *ordinary differential equation or ODE*. A differential equation is *linear* if the function and its derivatives appear linearly in the equation; if not, it is a *nonlinear equation*. The sine-Gordon equation is nonlinear and the other two are linear. A differential equation can have constant coefficients (such as the ω^2 in the harmonic example) or variable coefficients as in the following example:

$$\frac{d^2 x}{dt^2} + t^2 x = 0 \qquad (10.1.5)$$

where the coefficient of x depends on the independent variable t. Finally the *order of the equation* stands for the highest derivative occurring in the equation. All three equations discussed above are second order.

In dealing with these equations, you must remember that x can sometimes be the dependent variable and sometimes the independent variable. For example in the harmonic oscillator equation, it is the dependent variable depending on the independent variable t, whereas in the wave equation it is the independent variable (along with t) on which ψ depends.

We can of course combine adjectives: the harmonic oscillator example is an ordinary, linear, second-order equation with constant coefficients, while the sine-Gordon equation is a second-order nonlinear partial differential equation with constant coefficients.

A property of linear equations that will be repeatedly invoked is the principle of superposition:

Any linear combination of the solutions of a linear equation is also a solution.

This principle is so important, we will go over it again despite having done so earlier in the book. We will work with a concrete example leaving its obvious generalization to you.

Let us be given that

$$\frac{d^2 x_1}{dt^2} + t^2 \frac{dx_1}{dt} + 5t^3 x_1 \;\; = \;\; 0 \qquad (10.1.6)$$

$$\frac{d^2 x_2}{dt^2} + t^2 \frac{dx_2}{dt} + 5t^3 x_2 \;\; = \;\; 0 \qquad (10.1.7)$$

so that x_1 and x_2 are solutions to the same linear equation. If you multiply the first equation by a constant c_1 and the second by constant c_2 and add them, you can readily show that the linear combination $c_1 x_1 + c_2 x_2$ satisfies the equation as well. All you need is the fact that taking derivatives is a linear operation: *the derivative of the linear combination is the same linear combination of the derivatives*. Notice that even though there are non-linear terms in the *dependent variable t*, all that matters is that x appears only linearly.

The above equation is linear and *homogeneous:* every term in the equation goes as the first power of x. The superposition principle can be generalized to the case where there are terms that go as the zeroth power of x. Thus if

$$\frac{d^2x_1}{dt^2} + t^2\frac{dx_1}{dt} + 5t^3x_1 = f_1(t) \tag{10.1.8}$$

$$\frac{d^2x_2}{dt^2} + t^2\frac{dx_2}{dt} + 5t^3x_2 = f_2(t) \tag{10.1.9}$$

then the same manipulations will tell us that $c_1x_1 + c_2x_2$ will satisfy the equation with $c_1f_1 + c_2f_2$ on the right-hand side. In the harmonic oscillator case such a nonzero right-hand side will stand for any external force acting on the mass in addition to the spring and friction. *The principle of superposition tells us that the response to a sum of two forces is the sum of the responses.*

Of all the adjectives, the most daunting is the word "nonlinear," it implies we cannot form linear combinations of solutions to get new solutions, as we can in the case of linear equations. We will hardly discuss these. As for the rest, the only cases we will discuss are those that you will encounter in the near future. No attempt will be made to get deep into this topic since books can be written on any one of the cases discussed above. The aim of this chapter is to ensure that you will recognize a differential equation when you see one, have an idea of how complicated it is, and in the simpler cases have a notion of how to go about solving it.

10.2. ODEs with Constant Coefficients

Let us begin with a simple example

$$\frac{dx}{dt} = a. \tag{10.2.1}$$

The equations tells us that the function grows at a steady rate a. For example $x(t)$ could be the position of a particle traveling at steady speed a or the money in my bank with a the rate at which I make deposits. (In the latter case, the deposits are made not continuously but say once a week. Thus x grows in fits. The differential equation approximates this into a steady growth. On a time scale much larger than one week, the equation will be a good description of reality.) The solution to the equation in any event is

$$x = at + c \tag{10.2.2}$$

where c in an *integration constant.* How did we find this solution? We can use any one of the tricks of the trade, but here we use the one alluded to earlier: simply guess it! The constant c enters because the equation only tells us the rate of change of the function, and this is not enough to nail it down. For example if a is the speed of a car and x its position, we cannot find x for all t given just the speed; we have to know where it was at the outset, $t = 0$. If this information, $x(0)$, is given to us we can feed that into the equation

$$x(0) = a \cdot 0 + c \qquad (10.2.3)$$

and infer that $c = x(0)$. Thus the solution obeying the given *initial condition* is

$$x(t) = x(0) + at. \qquad (10.2.4)$$

But you do not need to be given the position at $t = 0$, you can be given $x(4)$, the position at time $t = 4$. We then proceed as follows:

$$
\begin{aligned}
x(4) &= a \cdot 4 + c & (10.2.5) \\
c &= x(4) - 4a & (10.2.6) \\
x(t) &= x(4) + a(t - 4). & (10.2.7)
\end{aligned}
$$

At a formal level, the constant of integration enters because any constant gets annihilated by the derivative operation, and hence may be added with impunity.

Suppose the equation is

$$\frac{d^2 y}{dt^2} = g. \qquad (10.2.8)$$

Here we want a function whose second derivative is a constant g. Since each derivative knocks off a power of t we know we must begin with t^2. The coefficient is then adjusted to be $g/2$ to ensure we end up with g on taking two derivatives. To this answer we can add anything that gets annihilated by the two derivatives: namely a function linear in time and a constant. Putting all this together, we obtain

$$y(t) = \frac{1}{2}gt^2 + vt + c. \qquad (10.2.9)$$

Now we have two free parameters whose formal origin was just explained. More physically, if g is the acceleration of a falling rock, and y its vertical coordinate, these parameters reflect the fact that its location cannot be determined unless we are given two extra pieces of information which will nail down v and c. This stands to reason: Eqn. (10.2.8) describes *every* rock that falls under the earth's pull. If you want to know what happened to your pet rock you must add for example what its initial height $y(0)$ and velocity $\dot{y}(0)$ were. Then you will proceed as follows:

$$
\begin{aligned}
y(0) &= \frac{1}{2}g \cdot 0^2 + v \cdot 0 + c = c & (10.2.10) \\
\dot{y}(0) &= g \cdot 0 + v = v \text{ so that} & (10.2.11) \\
y(t) &= \frac{1}{2}gt^2 + \dot{y}(0)t + y(0). & (10.2.12)
\end{aligned}
$$

The trend you are seeing is real: an n-th order equation will have n free parameters in the solution. We shall get a better understanding of this fact as we go along. (The parameters can enter the solution in various ways, additive constants, multiplicative constants, etc.) The general solution can then be tailor made to fit a given situation by imposing n conditions that fix the parameters. A simple choice is to impose the value of the function and $n - 1$ derivatives at the initial time. In some situations other choices suggest themselves. Consider a mass m attached to a spring of force constant k executing simple harmonic motion at frequency $\omega = \sqrt{k/m}$. The motion is described by Equation (10.1.2). By inspection we see that the solution is

$$x(t) = A \cos \omega t + B \sin \omega t. \tag{10.2.13}$$

We can of course nail down A and B in terms of the initial position and velocity. But consider the following problem from Lagrangian mechanics where we need to compute S, the *action* defined as

$$S[x_1, t_1; x_2, t_2] = \int_{t_1}^{t_2} \left[\frac{m}{2} \dot{x}^2 - \frac{m\omega^2}{2} x^2 \right] dt, \tag{10.2.14}$$

where $x(t)$ is a trajectory that runs through point x_1 at t_1 and x_2 at t_2. Thus we need to tune A and B such that the trajectory runs through the space–time points (x_1, t_1) and (x_2, t_2) and then evaluate S, as a function of these coordinates.

Problem 10.2.1. (Difficult). *Show that for the oscillator*

$$S[x_1, t_1; x_2, t_2] = \frac{m\omega}{2 \sin \omega t} \left[(x_1^2 + x_2^2) \cos \omega t - 2x_1 x_2 \right].$$

Problem 10.2.2. *Consider the equation $\frac{dx}{dt} + Ax + B = 0$. Show that if $x(t)$ is a solution, so is $x(t) + Ce^{-At}$, where C is an arbitrary constant.*

Let us take a second look at

$$\frac{d^2 x}{dt^2} + \omega^2 x = 0 \tag{10.2.15}$$

which we solved by inspection in terms of sines and cosines, and learn how to solve it by more systematic means. Let us rewrite this as

$$(D^2 + \omega^2)x = 0 \tag{10.2.16}$$

where D is the derivative operator. Let us assume a solution of the form

$$x = A e^{\alpha t}. \tag{10.2.17}$$

Feeding this into the differential equation we find (upon remembering that the exponential function is an eigenfunction of D)

$$(\alpha^2 + \omega^2) A e^{\alpha t} = 0, \tag{10.2.18}$$

which tells us that if $A \neq 0$, then α must obey

$$\alpha^2 + \omega^2 = 0 \tag{10.2.19}$$

or

$$\alpha = \pm i\omega. \tag{10.2.20}$$

Thus we find two solutions:

$$x_\pm = A_\pm e^{\pm i\omega t}, \tag{10.2.21}$$

where the parameters $A_\pm$ are arbitrary.[1] The most general solution is an arbitrary linear combination

$$x(t) = A_+ e^{i\omega t} + A_- e^{-i\omega t}. \tag{10.2.22}$$

If we want, at this stage we can rewrite this in terms of sines and cosines as

$$x(t) = C \cos \omega t + D \sin \omega t \tag{10.2.23}$$

which in turn could be written as

$$x(t) = E \cos(\omega t - \phi). \tag{10.2.24}$$

Problem 10.2.3. *Find E and ϕ in terms of C and D above.*

Thus we see that *the differential equation with constant coefficients can be reduced to an algebraic equation by looking for solutions of the exponential form.* Once again I remind you that this is possible because the action of D on exponential functions, which are its eigenfunctions, is multiplicative. More generally if

$$(a_n D^n + a_{n-1} D^{n-1} + \cdots + a_0)x = 0 \tag{10.2.25}$$

the answer is a sum of exponentials

$$x = \sum_{i=1}^n A_i e^{i\alpha_i t}, \tag{10.2.26}$$

where α_i are the roots of the algebraic equation

$$(a_n \alpha^n + a_{n-1} \alpha^{n-1} + \cdots + a_0) = 0. \tag{10.2.27}$$

Problem 10.2.4. *Find the general solution to $(D^2 - \omega^2)x = 0$. Find the parameters that ensure that $x(0) = 2$, $\dot{x}(0) = 0$.*

Problem 10.2.5. *Find the general solution of $(D^3 - D^2 + D - 1)x = 0$. (Find one root by inspection and then proceed.)*

---

[1] If we want to describe a problem where $x(t)$ is real, we must choose these constants to be complex conjugates.

There is clearly a problem when two of the α's become equal: we can only obtain one solution from our procedure. We proceed as follows. We argue that if α_1 and α_2 are roots then

$$\psi = \frac{e^{\alpha_1 t} - e^{\alpha_2 t}}{\alpha_1 - \alpha_2} \tag{10.2.28}$$

is also a solution by linearity. In the limit of equal roots, this solution is finite and reduces to

$$\psi = \frac{de^{\alpha t}}{d\alpha} = te^{\alpha t}. \tag{10.2.29}$$

Thus the general solution when both roots equal α, is the linear combination

$$x(t) = (A + Bt)e^{\alpha t}. \tag{10.2.30}$$

Problem 10.2.6. *Verify that Eqn. (10.2.30) is indeed a solution to $(D - \alpha)^2 x = 0$.*

There is another way to deal with the degenerate roots case which I mention because the same strategy works in many places when faced with a degeneracy. If α is a repeated root, so that $x_1 = e^{\alpha t}$ is a solution, we look for another solution of the form

$$x_2(t) = x_1(t)f(t), \tag{10.2.31}$$

where $f(t)$ is to be determined by feeding this ansatz into the differential equation. The details are left to the following exercise.

Problem 10.2.7. *First show that $(D - \alpha)e^{\alpha t} f(t) = e^{\alpha t} Df$. Then show that $(D - \alpha)^2 e^{\alpha t} f(t) = e^{\alpha t} D^2 f$. Argue on the basis of the above that $(D - \alpha)^2 x = 0$ has a solution of the form $e^{\alpha t} f(t)$, where $D^2 f = 0$. Make contact with Eqn. (10.2.30).*

Problem 10.2.8. *Find the solutions to*

$$\begin{aligned}
&(i) && (D^2 + 2D + 1)x(t) = 0 && \text{with } x(0) = 1,\ \dot{x}(0) = 0 \\
&(ii) && (D^4 + 1)x(t) = 0 \\
&(iii) && (D^3 - 3D^2 - 9D - 5)x(t) = 0 && (5 \text{ is a root}) \\
&(iv) && (D + 1)^2(D^4 - 256)x(t) = 0
\end{aligned}$$

Let us study a problem of some physical interest in some detail. Consider the usual mass spring system with an additional frictional force $-2m\gamma\dot{x}$ so that the equation of motion is

$$(D^2 + 2\gamma D + \omega^2)x = 0. \tag{10.2.32}$$

We will (i) find its general solution, (ii) analyze the different domains of parameter values.

The roots now obey

$$\alpha^2 + 2\gamma\alpha + \omega^2 = 0 \qquad (10.2.33)$$

and are hence given by

$$\alpha = -\gamma \pm \sqrt{\gamma^2 - \omega^2}. \qquad (10.2.34)$$

There are three classes of solutions.

- $\omega > \gamma$ underdamped

- $\omega = \gamma$ critically damped

- $\omega < \gamma$ overdamped

In the underdamped case the solutions are of the form

$$
\begin{aligned}
x(t) &= e^{-\gamma t}[Ae^{i\omega' t} + Be^{-i\omega' t}] & (10.2.35) \\
&= e^{-\gamma t}[C\cos\omega' t + D\sin\omega' t], & (10.2.36)
\end{aligned}
$$

where A, B, C, D are free constants, $\omega' = \sqrt{\omega^2 - \gamma^2}$, and I have written the answer both in terms of exponential and trigonometric functions.

In any event, this solution corresponds to oscillations that get damped with time and become negligible for $\gamma t >> 1$.

In the case of critical damping, the roots are equal (to $-\gamma$) and the solution is

$$x(t) = (A + Bt)e^{-\gamma t}. \qquad (10.2.37)$$

There are no oscillations.

Finally, in the overdamped case, we have two falling exponentials:

$$x(t) = Ae^{-\gamma_+ t} + Be^{-\gamma_- t}, \qquad (10.2.38)$$

where

$$\gamma_\pm = \gamma \mp \sqrt{\gamma^2 - \omega^2}. \qquad (10.2.39)$$

Consider now an extension of Eqn. (10.2.25) to the inhomogeneous version:

$$(a_n D^n + a_{n-1} D^{n-1} + \cdots + a_0)x = f(t). \qquad (10.2.40)$$

In the mass spring case, this would correspond to turning on an external force f. Let us begin with a special case

$$f(t) = Fe^{i\Omega t} \qquad (10.2.41)$$

with the usual understanding that the real part will be taken at the end. In this case we assume that $x(t)$ has the form

$$x(t) = x_0 e^{i\Omega t} \qquad (10.2.42)$$

and solve for the prefactor x_0. It is readily seen that

$$x = \frac{F e^{i\Omega t}}{(a_n (i\Omega)^n + a_{n-1}(i\Omega)^{n-1} + \cdots + a_0)}. \tag{10.2.43}$$

Problem 10.2.9. *Provide the missing steps in the above result.*

What happened to the n free parameters expected in an n-th order equation? *As explained towards the end of Chapter 5, they are still lurking around, since we can add to this solution the solution when we had no $f(t)$: the linearity of the equation allows us to add the responses when we add the forces and we adding the response to the forces $f(t)$ and 0.* The final answer is

$$x = x_p(t) + x_c(t), \tag{10.2.44}$$

where x_p, the *particular solution* is given by the function in Eqn. (10.2.43) and is dependent on the applied "force" $f(t)$, the *complementary function* x_c is specified by Eqns. (10.2.26,10.2.27) and is present even if f is absent. In the mass spring case, the latter corresponds to what happens when we give the oscillator a kick and let it vibrate on its own. In the circuit case an example of a complementary function (or transient) occurs when we charge up a capacitor in an LCR circuit and close the switch.

If the right-hand side is a sum of exponentials, we can again use linearity and add the corresponding particular solutions to get the net particular solution.

Here is a simple example. Suppose we are to solve

$$\ddot{x} + 3\dot{x} + 2x = 2\cosh 3t \tag{10.2.45}$$

The roots are $\alpha = -2, -1$ for the complementary part. As for the particular solution, let us use $2\cosh t = e^{3t} + e^{-3t}$ and deal with one exponential at a time. The particular solution to the problem

$$\ddot{x} + 3\dot{x} + 2x = e^{3t} \tag{10.2.46}$$

is (upon making the ansatz $x(t) = Ae^{3t}$),

$$x_p(t) = \frac{e^{3t}}{(3^2 + 3 \cdot 3 + 2)}. \tag{10.2.47}$$

The solution for the other exponential is

$$x_p(t) = \frac{e^{-3t}}{((-3)^2 + 3 \cdot (-3) + 2)}. \tag{10.2.48}$$

Thus the general solution to the original problem is

$$x(t) = Be^{-t} + Ce^{-2t} + \frac{e^{3t}}{(3^2 + 3 \cdot 3 + 2)} + \frac{e^{-3t}}{((-3)^2 + 3 \cdot (-3) + 2)}. \tag{10.2.49}$$

If we want to impose some initial conditions, such as $x(0) = 5$, $\dot{x}(0) = 0$, we may do so by imposing the conditions

$$5 = B + C + \frac{1}{(3^2 + 3 \cdot 3 + 2)} + \frac{1}{((-3)^2 + 3 \cdot (-3) + 2)} \qquad (10.2.50)$$

$$0 = -B - 2C + \frac{3}{(3^2 + 3 \cdot 3 + 2)} - \frac{3}{((-3)^2 + 3 \cdot (-3) + 2)}. \qquad (10.2.51)$$

There is some problem obtaining the particular solution if the exponential on the right side of the equation is one of the complementary exponentials. Thus let us say we are looking at

$$(D - \alpha_1) \ldots (D - \alpha_n)x(t) = e^{\alpha t}. \qquad (10.2.52)$$

If $\alpha \neq \alpha_i$, we can assume $x(t) = Ae^{\alpha t}$, where A is a constant and solve for A as usual:

$$A = \frac{1}{(\alpha - \alpha_1) \ldots (\alpha - \alpha_n)} \qquad (10.2.53)$$

What are we to do if $\alpha = \alpha_i$ for some i? Let us say it happens for $i = n$. Now we make an ansatz $x(t) = A(t)e^{\alpha t}$. Note that A now depends on t. We find

$$(D - \alpha_1) \ldots (D - \alpha)A(t)e^{\alpha t} = e^{\alpha t}. \qquad (10.2.54)$$

Now note that

$$(D - \alpha)A(t)e^{\alpha t} = e^{\alpha t}[DA(t)] \qquad (10.2.55)$$

(as you should verify by using the rule for the derivative of a product or recalling Problem (10.2.7.)). So we are down to

$$(D - \alpha_1) \ldots (D - \alpha_{n-1})e^{\alpha t}[DA(t)] = e^{\alpha t}. \qquad (10.2.56)$$

We will now try to see if we can make the ansatz that $DA(t)$ is a constant C. In this case the D's on the left act only on the exponential and turn into α's giving us

$$(\alpha - \alpha_1) \ldots (\alpha - \alpha_{n-1})Ce^{\alpha t} = e^{\alpha t}. \qquad (10.2.57)$$

Cancelling the exponentials on both sides, we can solve for C and obtain

$$C = \frac{1}{(\alpha - \alpha_1) \ldots (\alpha - \alpha_{n-1})}. \qquad (10.2.58)$$

Given $DA(t) = C$, it means $A = Ct + B$ and

$$x(t) = \frac{te^{\alpha t}}{(\alpha - \alpha_1) \ldots (\alpha - \alpha_{n-1})} + x_c(t) \qquad (10.2.59)$$

Note that I have not shown the part proportional to B: it can be absorbed in $x_c(t)$ which contains the function $e^{\alpha t}$.

Problem 10.2.10. *Consider a damped oscillator subject to a cosine force:*

$$(D^2 + 2\gamma D + \omega_0^2)x = F\cos\Omega t \qquad (10.2.60)$$

Show that the frequency at which the steady state solution has maximum amplitude, i.e., resonance, is given by $\Omega^2 = \omega_0^2 - 2\gamma^2$. Let $F = 25, \omega_0 = 1,\ 2\gamma = \Omega = 2\omega_0$. (i) What can you say about the transient (complementary) part? (ii) Write down the solution that has initial displacement and velocity zero. (iii) Repeat for the case where the initial position is unity and velocity is zero and (iv) vice versa.

Since any periodic function can be written in terms of exponentials, we can find the response to any periodic force. It follows that in any LCR circuit we can find the charge (or current) for any applied voltage since Q and x obey the same equations except for the change of notation ($Q \to x,\ I \to v$ etc.).

Problem 10.2.11. *Solve the following subject to $y(0) = 1,\ \dot{y}(0) = 0$*

$$\begin{aligned}
&\text{(i)} & \ddot{y} - \dot{y} - 2y &= e^{2x} \\
&\text{(ii)} & (D^2 - 2D + 1)y &= 2\cos x \\
&\text{(iii)} & y'' + 16y &= 16\cos 4x \\
&\text{(iv)} & y'' - y &= \cosh x
\end{aligned}$$

10.3. ODEs with Variable Coefficients: First Order

A common case has the form:

$$\frac{dy}{dx} + p(x)y = 0. \qquad (10.3.1)$$

We solve this as follows:

$$\frac{dy}{y} = -p(x)dx \qquad (10.3.2)$$

$$\int_{y_1}^{y_2} \frac{dy}{y} = -\int_{x_1}^{x_2} p(x)dx \qquad (10.3.3)$$

$$\ln\frac{y_2}{y_1} = P(x_1) - P(x_2) \qquad (10.3.4)$$

$$y_2 \equiv y(x_2) = y(x_1)e^{P(x_1)}e^{-P(x_2)}. \qquad (10.3.5)$$

In the above steps, $P(x)$ is the indefinite integral of $p(x)$. Notice that the constant of integration drops out in the difference $P(x_1) - P(x_2)$. Let us now call the upper limit x_2 simply as x and write the solution as

$$y(x) = y(x_1)e^{P(x_1)}e^{-P(x)} \equiv Ae^{-P(x)}, \tag{10.3.6}$$

where we have lumped all the prefactors into a constant A. This is done to show that the first-order equation has one free parameter in the solution as it should. Alternately we could go back to the earlier form which relates the free constant to $y(x_1)$ which we are free to choose at will.

Next consider the extension to the inhomogeneous case:

$$\frac{dy}{dx} + p(x)y = q(x). \tag{10.3.7}$$

Now we look for a solution of the form:

$$y = e^{-P(x)}v(x) \tag{10.3.8}$$

and determine v from the equation. Notice that this is getting to be a habit: whenever we are trying to solve a problem related to an earlier one, we make an ansatz in which the solution is the product of the old solution and an undetermined function. There is no loss of generality here since v is unspecified. Indeed you can look for a solution of the form $V(x)\sin x$ and solve for V. But the hope is that my v would be simpler than your V since my v had some relevant information (the solution to the homogeneous equation) built in.

Feeding this ansatz in we find that v obeys

$$\frac{dv}{dx} = q(x)e^{P(x)} \tag{10.3.9}$$

which is readily solved to give

$$v(x) = \int^x q(x')e^{P(x')}dx' + c \tag{10.3.10}$$

$$y(x) = e^{-P(x)}\left[\int^x q(x')e^{P(x')}dx' + c\right] \tag{10.3.11}$$

Note that the solution has a free parameter c as is expected of a first-order equation. What about the freedom to add a constant to $P(x)$, the indefinite integral of p? You may check that it merely redefines c. What about the lower limit in the integral over x'? When we change it, this too changes c. Thus our only real freedom is in the choice of c. Let us make this very explicit. Let $y = y_0$ at $x = x_0$. This tells us what c is, and feeding that back we get

$$y(x) = e^{-P(x)}\left[\int_{x_0}^x q(x')e^{P(x')}dx' + y_0e^{P(x_0)}\right]. \tag{10.3.12}$$

Problem 10.3.1. *Solve for c and obtain this result. Verify that when $x = x_0$, y indeed reduces to y_0. Verify also that adding a constant to P makes no difference.*

As an example of this approach let us consider

$$y' + \frac{1+x}{x}y = \frac{e^{-x}}{x} \qquad (10.3.13)$$

for which

$$p = 1 + \frac{1}{x} \qquad (10.3.14)$$
$$P = x + \ln x \qquad (10.3.15)$$
$$y = e^{-x-\ln x}\left[\int^x e^{x'}x'\frac{e^{-x'}}{x'}dx' + C\right] \qquad (10.3.16)$$
$$= \frac{e^{-x}}{x}(x+C) \qquad (10.3.17)$$

Suppose we want the boundary condition

$$y(1) = 0. \qquad (10.3.18)$$

This is achieved by demanding

$$0 = \frac{1}{e}(1+C) \qquad (10.3.19)$$

from which we obtain finally

$$y(x) = e^{-x}\left(1 - \frac{1}{x}\right). \qquad (10.3.20)$$

Problem 10.3.2. *Show that the solution to*

$$\frac{dy}{dx} - y = e^{2x}$$

is

$$y = ce^x + e^{2x}.$$

Problem 10.3.3. *Solve $x^2y' + 2xy - x + 1 = 0$ (where prime denotes differentiation) with $y(1) = 0$.*

Problem 10.3.4. *Solve $y' + y = (x+1)^2$ with $y(0) = 0$.*

Problem 10.3.5. *Solve* $x^2y' + 2xy = \sinh x$ *with* $y(1) = 2$.

Problem 10.3.6. *Solve* $y' + \frac{y}{1-x} + 2x - x^2 = 0$.

Problem 10.3.7. *Solve* $y' + \frac{y}{1-x} + x - x^2 = 0$

Problem 10.3.8. *Solve* $(1 + x^2)y' = 1 + xy$

Problem 10.3.9. Bernoulli's equation. *We will make an exception and handle the following nonlinear equation since it can be made linear by a trick:*

$$y' + p(x)y = q(x)y^m.$$

Show that if we set $v(x) = y^{1-m}$, v *obeys a linear equation*

$$v' + (1 - m)p(x)v = (1 - m)q(x).$$

Use this trick to solve

$$a : y' + xy = xy^2$$

$$b : 3xy' + y + x^2y^4 = 0$$

10.4. ODEs with Variable Coefficients: Second Order and Homogeneous

Many problems in physics reduce to the following differential equation:

$$y'' + p(x)y' + q(x)y = 0. \tag{10.4.1}$$

First consider the case where p and q are analytic at the origin $x = 0$. The strategy is to follow *the method of Frobenius*: assume the answer has a power series expansion about the origin and determine all the coefficients of the series using the equation. This will give us the answer within the radius of convergence of the series. To illustrate the method, let us consider a problem where we know the answer:

$$y'' + \omega^2 y = 0. \tag{10.4.2}$$

Assume

$$y = \sum_{0}^{\infty} c_n x^n \tag{10.4.3}$$

and feed that into the equation to obtain:

$$\sum_{0}^{\infty} \left[n(n-1)x^{n-2}c_n + \omega^2 c_n x^n \right] = 0. \tag{10.4.4}$$

On the left we have sum over different powers of x and on the right a zero. For this to work, each power must have zero net coefficient: this is because the functions x^n are linearly independent (you cannot write one power of x as a sum of other powers) and the only linear combination that adds up to zero is the trivial one. Let us now look at the various powers. The first term in the brackets seems to start out with x^{-2} corresponding to $n = 0$. However due to the factor $n(n-1)$, this one and the next one, x^{-1} are both absent. Thus both series begin with x^0. Equating the coefficients of successive powers of x to zero, we find the following infinite set of relations:

$$x^0: \quad (2(2-1)c_2 + \omega^2 c_0) \quad = \quad 0 \tag{10.4.5}$$
$$x^1: \quad (3(3-1)c_3 + \omega^2 c_1) \quad = \quad 0 \tag{10.4.6}$$
$$\vdots \quad = \quad 0$$
$$x^{n-2}: \quad (n(n-1)c_n + \omega^2 c_{n-2}) \quad = \quad 0 \tag{10.4.7}$$
$$\vdots$$

The first of these equations relates c_2 to c_0:

$$c_2 = -\frac{\omega^2}{2}c_0. \tag{10.4.8}$$

while the second relates c_3 to c_1:

$$c_3 = -\frac{\omega^2}{6}c_1. \tag{10.4.9}$$

Thereafter every equation relates c_n to c_{n-2} by a *recursion relation*:

$$c_n = -\frac{\omega^2}{n(n-1)}c_{n-2} = \frac{\omega^4}{n(n-1)(n-2)(n-3)}c_{n-4}\cdots \tag{10.4.10}$$

all the way back to c_0 or c_1 depending on whether n is even or odd. We are completely free to pick c_0 and c_1; once this is done, the rest of the coefficients are fixed: c_0 fixes c_2, which in turn fixes c_4 and so on, while the odd coefficients are similarly slaved to c_1. The general solution is then

$$y = c_0(1 - \frac{\omega^2 x^2}{2} + \frac{\omega^4 x^4}{4!} + \cdots + (-1)^n \frac{\omega^{2n} x^{2n}}{(2n)!} + \cdots) + c_1(x - \frac{\omega^3 x^3}{3!} + \cdots +) \tag{10.4.11}$$

which we recognize to be an arbitrary linear combination of $\cos \omega x$ and $\sin \omega x$. Having found the solution this way, we must find the radius of convergence of the series. In the present case it is infinite. Of course it will be a long way from there to the other properties of these functions, such as their periodicity, functional identities ($\sin^2 + \cos^2 = 1$) and so on.

Problem 10.4.1. *Solve the equation $y' + ay = 0$ using this method.*

We will now consider a problem that arises in quantum mechanics. (Stop shaking now! You do not have to know any quantum mechanics to follow this discussion. The example is chosen to bring some drama into your life and to give a feeling for how real problems are solved by a combination of intuition and mathematical machinery.) Imagine a tiny particle like an electron that is attracted to some center by a potential $V = \frac{1}{2}kx^2$. In classical mechanics we know it will oscillate with an angular frequency $\omega^2 = k/m$. As for its energy, it can have any value, depending on x_0, its amplitude of oscillation: $E = \frac{1}{2}kx_0^2 = \frac{1}{2}m\omega^2 x_0^2$. The lowest energy is zero, in which case the particle sits at the origin. In quantum theory the whole description is different. We are told that only certain energies are allowed and these are given by the solution to the following eigenvalue equation:

$$\psi''(y) - y^2\psi(y) = -2\varepsilon\psi(y) \qquad (10.4.12)$$

In the above, ε is the energy E measured in some units appropriate to the problem:

$$\varepsilon = \frac{E}{\hbar\omega}, \qquad (10.4.13)$$

where $\hbar$ is called *Planck's constant*, and likewise y is x measured in some new units. Thus the allowed energies are (up to a factor -2) the eigenvalues of the operator $D^2 - y^2$ whose action on any function is to take its second derivative and subtract from the result y^2 times the function. In general this sequence of operations will alter the input function in a serious way. But we are looking for eigenfunctions on which the effect will be simply rescaling. If the eigenvalue is the energy, what of the corresponding eigenfunction? Quantum theory tells us that the absolute value squared of the eigenfunction gives at each point $P(y)$, the probability density of finding the particle there. In other words $P(y)dy$ is the absolute probability of finding the particle between y and $y + dy$. Since the overall scale of the eigenfunction is not determined by the eigenvalue equation, it will be chosen so that the total probability of finding the particle anywhere adds up to unity:

$$1 = \int_{-\infty}^{\infty} P(y)dy = \int_{-\infty}^{\infty} \psi^2(y)dy. \qquad (10.4.14)$$

If we try to solve Eqn. (10.4.12) by the power series method, we will run into the following problem. If we feed the ansatz

$$\psi = \sum_{0}^{\infty} c_n y^n \qquad (10.4.15)$$

into the differential equation we will find that a given power of y, say y^n has the
following coefficient for generic n:

$$((n+2)(n+1)c_{n+2} - c_{n-2} + 2\varepsilon c_n) \qquad (10.4.16)$$

which must vanish. This gives a *three term recursion relation* relating three coefficients at a time. Since there are no negative powers of y in the problem, the
equations for $n = 0, 1$ will contain only two coefficients. For example when $n = 0$
we will get a relation $\varepsilon c_0 + c_2 = 0$. If we choose c_0 arbitrarily, we can solve for
c_2 in terms of it. Armed with this, we can go the $n = 2$ equation and solve for c_4
in terms c_0, c_2 and finally reduce it to an expression involving just c_0. Likewise
we can build up all the odd coefficients in terms of c_1. But if you carry out the
calculation, you will find that it gets very unwieldy in contrast to the case of the
two term recursion relation where each c_n is a factor times c_{n-2}, which in turn is
a factor times c_{n-4} and so on.

Problem 10.4.2. *Show that*

$$c_2 = -\varepsilon c_0 \qquad (10.4.17)$$

$$c_4 = \frac{c_0 - 2\varepsilon c_2}{12} = \frac{1 + 2\varepsilon^2}{12} c_0 \qquad (10.4.18)$$

Thus we must find a way to convert this to an equation with a two term
recursion relation. (That Eqn. (10.4.12) would give a three term recursion relation
is clear on inspection: in the first term on the left, taking ψ'', we lose two powers
of y, in the second term we put in two extra powers of y, while in the right-hand
side we simply multiply by a constant. It follows that a given power of y then
drags in three different c's.)

Now there is no simple trick for converting this problem to one with a two
term recursion relation. However the following steps, which are pretty standard,
lead to the desired goal. First we ask how the solution behaves at small and large
y. At small y, we ignore the y^2 term in the equation and see that the answer is
given by a superposition of sine and cosine. Neglecting terms of order y^2 and
higher in the solution (as we did in the equation), we find the solution goes as
$a + by$. At very large y, we ignore the right-hand side and are left with

$$\psi'' = y^2 \psi \qquad (10.4.19)$$

which is solved in this limit by

$$\psi \simeq y^m e^{\pm y^2/2} \ (m \text{ finite}) \qquad (10.4.20)$$

since

$$\psi'' = \lim_{y \to \infty} \left(y^{m+2} e^{\pm y^2/2} \left[1 \pm \frac{2m+1}{y^2} + \frac{m(m-1)}{y^4} \right] \right) \qquad (10.4.21)$$

$$\to y^2 \psi \qquad (10.4.22)$$

We will reject the growing exponential since we wish to normalize the square integral of ψ to unity and this is not possible if it blows up at infinity. We therefore write

$$\psi = u(y)e^{-y^2/2}, \tag{10.4.23}$$

where u will go to a constant as we near the origin and grow as y^m for large y. It is our hope that having fed all the extra information at various limits into the ansatz, the solution of u will prove a lot easier. The equation for u is easily seen to be

$$u'' - 2yu' + (2\varepsilon - 1)u = 0. \tag{10.4.24}$$

This does not look any easier, but if you have been following the earlier discussion you will see that it admits a two term recursion relation: the first term knocks out two powers of y, while the remaining two do not change the power – the last one because it is simply multiplicative and the middle one because it reduces a power by differentiation and puts it right back when multiplying by y. The equation tells us that if

$$u = \sum_0^\infty a_n y^n \tag{10.4.25}$$

then

$$\sum_0^\infty \left[y^{n-2}(a_n n(n-1)) + y^n a_n(2\varepsilon - 1 - 2n) \right] = 0. \tag{10.4.26}$$

Due to the factor $n(n-1)$, the first sum (involving y^{n-2}) really begins with $n = 2$. Defining a new label $m = n-2$, which runs from 0 upwards the first sum becomes

$$\sum_0^\infty y^m (a_{m+2}(m+2)(m+1)). \tag{10.4.27}$$

Let us now change the dummy index m into n and combine the two sums to

$$\sum_0^\infty y^n \left[(a_{n+2}(n+2)(n+1) + a_n(2\varepsilon - 1 - 2n) \right] = 0 \tag{10.4.28}$$

from which we deduce the two term recursion relation valid for all n:

$$a_{n+2} = \frac{1 + 2n - 2\varepsilon}{(n+2)(n+1)} a_n. \tag{10.4.29}$$

As in the case of the sines and cosines, we can once again choose a_0 and a_1 arbitrarily and the rest will be determined by the above relation.

This is fine as far as the equation goes, but the physics isn't quite right. You might have learned in elementary physics courses that the energy of a quantum oscillator is limited to some special values. In our solution we saw no restriction—indeed, ε did not even have to be real!

Now the inventors of quantum mechanics could not quote physics text books yet to be written! What did they do at this point? They responded to an internal alarm that goes off at this point. Suppose we continue with the process and try to choose the overall scale of the eigenfunction so that

$$1 = \int_{-\infty}^{\infty} P(y)dy \qquad (10.4.30)$$

$$= \int_{-\infty}^{\infty} \psi^2(y)dy \qquad (10.4.31)$$

$$= \int_{-\infty}^{\infty} e^{-y^2} u^2(y)dy. \qquad (10.4.32)$$

It is clear that no choice of overall prefactor will allow us to normalize the eigenfunction to unity if the integral of ψ^2 is divergent. Now the function e^{-y^2} can overcome any finite power of y at large y and render the integral convergent. However our power series solution goes on forever. You may check for yourselves that the ratio of successive coefficients in both the even and odd series both series obey

$$\frac{a_{n+2}}{a_n} \to \frac{2}{n} \text{ for large } n \qquad (10.4.33)$$

which is exactly how the series for e^{y^2} behaves. Thus ψ is going like $e^{-y^2/2}u \simeq e^{-y^2/2}e^{y^2} = e^{y^2/2}$ which makes it impossible to normalize the probability. Now we have gone from riches to rags: from having two solutions at any energy, we are reduced to having none! Before reading on you should think what you would do if you had invented quantum mechanics and had come this far.

The answer lies in the recursion relations Eqn. (10.4.29). Observe that *if the energy ε just happened to be a half integer, i.e., $2\varepsilon = 2m + 1$, where m is an integer, then a_{m+2} would vanish as would all subsequent coefficients that relied on it.* For example if $\varepsilon = 7/2$, then $a_5 = a_7 = \cdots = 0$ while if $\varepsilon = 9/2$, we would have $a_6 = a_8 = \cdots = 0$. The strategy is clear: if the even powers stop after m terms, we kill the odd series by choosing $a_1 = 0$, while if the odd series is a polynomial, the even series is killed manually. In either case, we have quantization of the energy

$$E = \varepsilon \hbar \omega = (m + 1/2)\hbar \omega \qquad (10.4.34)$$

and a corresponding set of normalizable eigenfunctions. The polynomials we get as solutions for v are called *Hermite polynomials* and these are denoted by $H_n(y)$.

Problem 10.4.3. *Show that the first four Hermite polynomials are*

$$H_0 = 1 \qquad (10.4.35)$$
$$H_1 = 2y \qquad (10.4.36)$$

$$H_2 = -2(1 - 2y^2) \qquad (10.4.37)$$

$$H_3 = -12(y - \frac{2}{3}y^3) \qquad (10.4.38)$$

where the overall normalization (choice of a_0 or a_1) is as per some convention we need not get into. To compare your answers to the above, choose the starting coefficients to agree with the above. Show that

$$\int_{-\infty}^{\infty} e^{-y^2} H_n(y) H_m(y) dy = \delta_{nm}(\sqrt{\pi} 2^n n!) \qquad (10.4.39)$$

for the cases $m, n \leq 2$. Notice that the Hermite polynomials are not themselves orthogonal or even normalizable, we need the weight function e^{-y^2} in the integration measure. We understand this as follows: the exponential factor converts u's to ψ's, which are the eigenfunctions of a hermitian operator (hermitian with respect to normalizable functions that vanished at infinity) and hence orthogonal for different eigenvalues.

Problem 10.4.4. Consider the Legendre Equation

$$(1 - x^2)y'' - 2xy' + l(l + 1)y = 0 \qquad (10.4.40)$$

Argue that the power series method will lead to a two term recursion relation and find the latter. Show that if l is an even (odd) integer, the even(odd) series will reduce to polynomials, called P_l, the Legendre polynomials of order l. Show that

$$P_0 = 1 \qquad (10.4.41)$$

$$P_1 = x \qquad (10.4.42)$$

$$P_2 = \frac{1}{2}(3x^2 - 1) \qquad (10.4.43)$$

$$P_3 = \frac{1}{2}(5x^3 - 3x) \qquad (10.4.44)$$

(The overall scale of these functions is not defined by the equation, but by convention as above.) Pick any two of the above and show that they are orthogonal over the interval $-1 \leq x \leq 1$.

Problem 10.4.5. The functions $1, x, x^2, \cdots$ are linearly independent—there is no way, for example, to express x^3 in terms of sums of other powers. Use the Gram–Schmidt procedure to extract from this set the first four Legendre polynomials (up to normalization) known to be orthonormal in the interval $-1 \leq x \leq 1$.

10.4.1. Frobenius method for singular coefficients

Consider now the differential equation

$$x^2 y'' + xy' + (4x^2 - 3)y = 0. \tag{10.4.45}$$

The ansatz

$$y = \sum_0^\infty c_n x^n \tag{10.4.46}$$

leads to the conclusion that $c_0 = c_1 = 0$ upon considering the coefficients of x^0 and x. Since the recursion relation relates c_n to c_{n+2}, it follows that only a trivial solution (zero) of this form exists. On the other hand, the equation does have some nontrivial solutions. The point is that these are not analytic at the origin as was assumed in the ansatz. This becomes more manifest if we divide the whole equation by x^2 so that the coefficients of y' and y diverge as $1/x$ and $1/x^2$ at the origin. Frobenius's idea is to assume that a solution in the form of *a generalized power series:*

$$y = x^s \sum_0^\infty c_n x^n \tag{10.4.47}$$

exists for some suitable s; that is to say, apart from one prefactor, the solution has a Taylor series. (Note that by definition $c_0 \neq 0$, since if you say the series begins with c_1, you are merely shifting s to $s + 1$. Thus given that the smallest power in the series is s, we must have $c_0 \neq 0$.) Proceeding as usual, we find:

$$\sum_0^\infty [(n + s)(n + s - 1) + (n + s) - 3] x^{n+s} c_n + 4x^{n+s+2} c_n = 0, \tag{10.4.48}$$

Setting the coefficients of x^s and x^{s+1} to zero we find

$$(s^2 - 3)c_0 = 0 \tag{10.4.49}$$
$$[(1 + s)^2 - 3] c_1 = 0 \tag{10.4.50}$$

from which we see that

$$s = \pm\sqrt{3}, \tag{10.4.51}$$
$$c_1 = 0. \tag{10.4.52}$$

Equation (10.4.49) which determines the value of s is called the *indicial equation.* Equating the coefficients of x^{s+n}, $n \geq 2$, we find the recursion relation

$$c_{n+2} = -\frac{4}{(n + 2 + s)^2 - 3} c_n \tag{10.4.53}$$

and we get two solutions corresponding to the two choices of s.

Problem 10.4.6. *Write the general solution out to c_5.*

I will briefly discuss the general situation for equations of the form

$$x^2 y'' + x p(x) y' + q(x) y = 0 \qquad (10.4.54)$$

where $p(x), q(x)$ have a Taylor series at the origin.[2] Here is a list of possibilities when the indicial equation is solved.

- *The indicial equation has* distinct *roots, s_1 and s_2.* There will be two linearly independent solutions of the form x^{s_1} or x^{s_2} times a power series starting with the zeroth power of x. The coefficients of the series are to be found from the recursion relation. The starting coefficients will be arbitrary in each solution and constitute the two free parameters in the solution of this second-order equation.

- *The roots are degenerate, $s_1 = s_2 = s$.* One solution y_1 will be of the assumed form, the generalized series, starting with x^s. The second solution will have the form

$$y_2 = y_1 \ln x + x^s \sum_Q^\infty b_n x^n, \qquad (10.4.55)$$

where the coefficients can be determined by feeding in this ansatz. We can see why the solution has this form and also how the coefficients of this solution are related to the general solution of the recursion relation as follows.

Let us view the degenerate case as the limit of s_1 and s_2 that approach each other. Then the linear combination

$$y = \frac{x^{s_1} \sum_0^\infty a_n(s_1) x^n - x^{s_2} \sum_0^\infty a_n(s_2) x^n}{s_1 - s_2} \qquad (10.4.56)$$

is also a solution, where the coefficients $a_n(s_{1,2})$ are the solutions to the recursion relation for the cases $s_{1,2}$ with $a_0(s_1) = a_0(s_2)$. *The latter is done so the coefficients in the two series approach each other as the roots do.* With this choice we see that the solution above is simply a derivative with respect to s:

$$y_2 = \frac{d}{ds}\left[x^s \sum_0^\infty a_n(s) x^n \right] \qquad (10.4.57)$$

$$= \ln x \, x^s \sum_0^\infty a_n(s) x^n + x^s \sum_0^\infty \frac{da_n(s)}{ds} x^n \qquad (10.4.58)$$

---

[2] The series could be centered at any other point. We shift our coordinates until it becomes the origin.

which has the form stated in Eqn. (10.4.55). We are of course free to multiply the whole solution by any overall factor and also add to it any multiple of y_1.

- *The roots differ by an integer.* What can be the problem now? Can we not use the recursion relations and get two solutions each beginning with different powers, and obviously linearly independent? Not always! This will now be illustrated via *Bessel's equation*

$$x^2 y'' + xy' + (x^2 - \nu^2)y = 0. \tag{10.4.59}$$

Feeding in the series we find

$$\sum_0 a_n[(n+s)^2 - \nu^2]x^{n+s} + \sum_n a_n x^{n+s+2} = 0. \tag{10.4.60}$$

Equating the coefficient of x^s and x^{s+1} we find from the indicial equation

$$s = \pm\nu, \quad a_1 = 0. \tag{10.4.61}$$

If ν is not an integer, we get two linearly independent solutions, called $J_{\pm\nu}(x)$, the *Bessel functions of the first kind and of order* $\pm\nu$.

Let us look at the general recursion relation

$$a_n = -\frac{a_{n-2}}{(n+s)^2 - \nu^2} \tag{10.4.62}$$

for the case ν an integer and the choice $s = s_2 = -\nu$, the smaller root. *Notice that when* $n = 2\nu$, *(which is the difference between the two roots)*, a_n *blows up!* To avoid this we must choose all the preceding coefficients $a_0, a_2 \ldots a_{n-2}$ to be equal to zero. It follows that the solution y_2 actually begins with $x^{-\nu+2\nu} = x^\nu$ and coincides with y_1.

If one chooses the coefficients continuously in ν one finds in fact that

$$J_n(x) = (-1)^n J_{-n}(x). \tag{10.4.63}$$

To find the second solution for $\nu = n$, called the *Neumann Function* we must form the linear combination

$$N_n(x) = \lim_{\nu \to n} \frac{\cos(\pi\nu)J_\nu - J_{-\nu}}{\sin(\pi\nu)}. \tag{10.4.64}$$

These functions are singular at the origin. When you learn about Bessel functions later, you will surely delve deeper into these points.

Let us return to the general case of $s_1 = s_2 + m$, focusing on the solution y_2 corresponding to the smaller root. You can show (see the following exercise) that if we proceed naively a_m will blow up, (just as in the case of the Bessel functions), since the formula for it will involve dividing by the left-hand-side of the indicial equation with the replacement $s_2 \rightarrow s_2 + m = s_1$. Unless a compensating zero occurs miraculously in the numerator (in which case we can proceed naively) the following modification is needed: one must assume

$$y_2(x) = A y_1(x) \ln(x) + x^{s_2} \sum_0^\infty a_n x^n$$

and solve for a_n and A. The latter will vanish in the miraculous case alluded to, as in part c of Problem (10.4.8.). Also, in some cases the coefficients in a_n will not be totally determined. Do not worry when this happens (as in parts g and h of Problem (10.4.8.)): the indeterminate parts will define a function proportional to y_1, which can always be added to any valid y_2. (In the answers given at the end, this part is ignored, or if you want, lumped with y_1.)

Problem 10.4.7. *Consider the case where* $p(x) = \sum_n p_n x^n$, $q(x) = \sum_n q_n x^n$ *in Eqn. (10.4.54). Solve for the indicial equation and the recursion relation for* a_m *for the smaller root and verify the claims made in the previous discussion.*

Problem 10.4.8. *Solve the following equations by the method of indicial equations. Go up to x^4 or to all orders if a pattern is found.*

$$
\begin{aligned}
a: \quad & x(x+1)^2 y'' + (1-x^2)y' + (x-1)y &= 0 \\
b: \quad & x(1-x)y'' + 2(1-2x)y' - 2y &= 0 \\
c: \quad & x^2 y'' + xy' - 9y &= 0 \\
d: \quad & xy'' + \frac{1}{2}y' + 2y &= 0 \\
e: \quad & x^2 y'' - xy' + y &= 0 \\
f: \quad & 2xy'' - y' + 2y &= 0 \\
g: \quad & xy'' + xy' - 2y &= 0 \\
h: \quad & x(x-1)^2 y'' - 2y &= 0
\end{aligned}
$$

Problem 10.4.9. *Solve the following equations both ways: using Eqn. (10.3.11) as well as generalized power series:*

$$(i): \quad y' - \frac{2y}{x} - x^2 = 0,$$

$$(ii): \quad y' + \frac{2y}{x} - x^3 = 0.$$

Problem 10.4.10. *Solve* Laguerre's Equation *which enters the solution of the hydrogen atom problem in quantum mechanics (after premultiplying both sides by x)*

$$xy'' + (1 - x)y' + my = 0 \tag{10.4.65}$$

by the power series method. Show that there is a repeated root and focus on the solution which is regular at the origin. Show that this reduces to a polynomial when m is an integer. These are the Laguerre polynomials L_m. *Find the first four polynomials choosing $c_0 = 1$. Show that L_1 and L_2 are orthogonal in the interval $0 \le x \le \infty$ with a weight function e^{-x}. (Recall the gamma function.)*

10.5. Partial Differential Equations

In this section we will discuss just a few of the linear partial differential equations that you will encounter, once again with the view that you are not trying to learn all the equations you will ever encounter (there will be enough time for that) but just a few so as to get a feeling for the subject.

10.5.1. The wave equation in one and two space dimensions

Let us begin with the wave equation which we have discussed at some length in the chapter on vector spaces. We will approach it in a different way here. The wave equation and boundary conditions for a string of length L are given by

$$\frac{1}{v^2} \frac{\partial^2 \psi}{\partial t^2} = \frac{\partial^2 \psi}{\partial x^2} \tag{10.5.1}$$

$$\psi(x = 0, t) = \psi(x = L, t) = 0, \tag{10.5.2}$$

where v is the velocity of propagation of waves in the string. Now we use the trick called *separation of variables* which lies at the heart of most of what we shall do with PDEs. We look for a solution of the form

$$\psi(x, t) = X(x)T(t). \tag{10.5.3}$$

While this is not the most general solution, let us see how far we can go with this. Feeding in this *ansatz* we find from the wave equation:

$$X(x)\frac{1}{v^2}\frac{d^2 T}{dt^2} = T(t)\frac{d^2 X}{dx^2}. \tag{10.5.4}$$

Note that partial derivatives have disappeared and ordinary derivatives have taken their place since the functions in question depend on just one variable. Dividing both sides by XT (since it is not identically zero) we find

$$\frac{1}{T}\frac{1}{v^2}\frac{d^2T}{dt^2} = \frac{1}{X}\frac{d^2X}{dx^2}. \tag{10.5.5}$$

Next we argue that since the left-hand side is a function only of t, and this equals the right-hand side, which has no idea of t, the left-hand side is likewise t-independent. By the same logic the right-hand side cannot depend on x. Thus both must equal some constant, which we call with some foresight, $-k^2$. Thus we have a pair of ordinary differential equations in place of the original PDE:

$$\frac{1}{T}\frac{1}{v^2}\frac{d^2T}{dt^2} = -k^2 \tag{10.5.6}$$

$$\frac{1}{X}\frac{d^2X}{dx^2} = -k^2. \tag{10.5.7}$$

The solution to the first, with zero initial velocity (a convenient choice, but not a necessity for this approach to work) is

$$T(t) = A\cos vkt. \tag{10.5.8}$$

The solution to the second, with the condition that X vanish at $x = 0, L$ is

$$X(x) = A\sin kx \quad k = \frac{n\pi}{L} \quad n = 1, 2, 3, \dots . \tag{10.5.9}$$

Thus we have found an infinite number of solutions of the product form:

$$\psi_n(x, t) = A_n \sin\frac{n\pi x}{L}\cos\frac{nv\pi t}{L}. \tag{10.5.10}$$

By linearity any linear combination

$$\psi(x, t) = \sum_1^\infty A_n \sin\frac{n\pi x}{L}\cos\frac{nv\pi t}{L}. \tag{10.5.11}$$

is also a solution. Consider now some initial value data $\psi(x, 0)$ that is given. This allows us to solve for the A_n's:

$$\psi(x, 0) = \sum_1^\infty A_n \sin\frac{n\pi x}{L} \tag{10.5.12}$$

Since the sines form a complete basis for functions that vanish at both ends, we can fit any given initial profile with some set of A_n's. Given this the future state of the string follows uniquely. We will not discuss this case further since you have seen it in Chapter 9.

Let us instead consider wave propagation in a membrane such as a drumhead. First consider a square drum of sides L, the type you will find in any physics party. The wave equation and boundary conditions for the displacement $\psi(x, y, t)$ are

$$\frac{1}{v^2} \frac{\partial^2 \psi}{\partial t^2} = \frac{\partial^2 \psi}{\partial x^2} + \frac{\partial^2 \psi}{\partial y^2} \equiv \nabla^2 \psi \qquad (10.5.13)$$

$$\psi(x = 0, y, t) = \psi(x = L, y, t) = \psi(x, y = 0, t) = \psi(x, y = L, t) = 0. \qquad (10.5.14)$$

(The drum head has one corner at the origin and lies in the first quadrant.) Now we look for solutions of the form

$$\psi(x, y, t) = X(x)Y(y)T(t). \qquad (10.5.15)$$

Feeding this in and dividing by XYT, we find

$$\frac{1}{T} \frac{1}{v^2} \frac{d^2 T}{dt^2} = \frac{1}{X} \frac{d^2 X}{dx^2} + \frac{1}{Y} \frac{d^2 Y}{dy^2} \qquad (10.5.16)$$

Once again we argue that the functions of t, x, y that appear above cannot depend on their arguments since they are equal to sums (or differences) of two other functions that do not depend on them. Thus we write

$$\underbrace{\frac{1}{T} \frac{1}{v^2} \frac{d^2 T}{dt^2}}_{=-k_x^2 - k_y^2} = \underbrace{\frac{1}{X} \frac{d^2 X}{dx^2}}_{=-k_x^2} + \underbrace{\frac{1}{Y} \frac{d^2 Y}{dy^2}}_{-k_y^2}. \qquad (10.5.17)$$

We are thus left with three ODEs in place of the original PDE:

$$\begin{aligned} T'' + \omega^2 T &= 0 \quad \omega^2 = (k_x^2 + k_y^2)v^2 & (10.5.18) \\ X'' + k_x^2 X &= 0 & (10.5.19) \\ Y'' + k_y^2 Y &= 0. & (10.5.20) \end{aligned}$$

Solving the equations subject to the boundary conditions we find that the most general solution is

$$\psi(x, y, t) = \sum_{n=1}^{\infty} \sum_{m=1}^{\infty} A_{nm} \sin \frac{n\pi x}{L} \sin \frac{m\pi y}{L} \cos(v\sqrt{(n\pi/L)^2 + (m\pi/L)^2}\, t).$$

$$(10.5.21)$$

Problem 10.5.1. *Provide the steps leading to the above equation.*

The initial value data will be expanded as follows:

$$\psi(x, y, 0) = \sum_{n=1}^{\infty} \sum_{m=1}^{\infty} A_{nm} \sin \frac{n\pi x}{L} \sin \frac{m\pi y}{L}. \tag{10.5.22}$$

The double sine series can be used to expand any function of x and y vanishing at the boundary according to the following argument. First hold y fixed and expand the function of x using the sines in x with y-dependent coefficients:

$$\psi(x, y, 0) = \sum_{n=1}^{\infty} A_n(y) \sin \frac{n\pi x}{L}. \tag{10.5.23}$$

Since the functions $A_n(y)$ themselves vanish at $y = 0, L$ they may be expanded as

$$A_n(y) = \sum_{m=1}^{\infty} A_{nm} \sin \frac{m\pi y}{L}. \tag{10.5.24}$$

Combining the preceding two equations, the double series follows. To find the coefficient A_{nm} we multiply both sides by $\sin \frac{n\pi x}{L} \sin \frac{m\pi y}{L}$ and integrate, remembering that these functions are not normalized.

For example, let

$$\psi(x, y, 0) = x(L - x)y(L - y). \tag{10.5.25}$$

Then

$$\begin{aligned}
A_{nm} \left[\frac{L}{2}\right]^2 &= \int_0^L \int_0^L dx \, dy \, x(L - x)y(L - y) \sin \frac{n\pi x}{L} \sin \frac{m\pi y}{L} \\
&= \left[\int_0^L dx \, x(L - x) \sin \frac{n\pi x}{L}\right] \cdot [x \rightarrow y, n \rightarrow m] \\
&= \frac{4L^3}{n^3 \pi^3} \frac{4L^3}{m^3 \pi^3} \quad [n, m, \text{odd}]
\end{aligned} \tag{10.5.26}$$

leading to the solution

$$\psi(x, y, t) = \frac{64L^4}{\pi^6} \sum_{n=\text{odd}} \sum_{m=\text{odd}} \frac{1}{n^3 m^3} \sin \frac{n\pi x}{L} \sin \frac{m\pi y}{L} \cos \left[\frac{\pi vt}{L} \sqrt{m^2 + n^2}\right]. \tag{10.5.27}$$

Figure (10.1) shows the vibrations of the membrane. Unlike in the case of the string, the frequencies in the sum are not all multiples of a fundamental frequency. Hence the motion will generally not be periodic.

Problem 10.5.2. *Show that A_{nm} has the value quoted in Eqn. (10.5.26) by doing the integrals.*

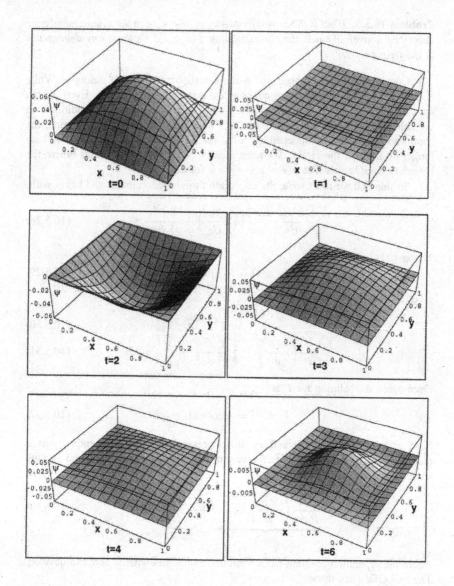

Figure 10.1. The vibrating membrane with $L = v = 1$ and initial condition $\psi(x, y, 0) = x(L - x)y(L - y)$ for selected times. Note the change of scale in the last plot.

Problem 10.5.3. *What is the lowest frequency of vibration in the square membrane described above? What is the corresponding $\psi(x, y, t)$? Is there any degeneracy in the frequencies?*

Consider now a drumhead that is more realistic: a circle of radius a. While it is permissible to use cartesian coordinates, it is not wise to do so. Even though the equations separate into three ODEs as before, the boundary conditions on the solutions will mix the coordinates up. Whereas in the square membrane, the condition on $X(x)$ was independent of y (and likewise for $Y(y)$), this will no longer be true for the circle. We must clearly go to polar coordinates where the boundary condition is that $\psi(r = a, \phi, t) = 0$ for all ϕ.

To this end we must write the Laplacian in polar coordinates and begin with

$$\frac{1}{v^2}\frac{\partial^2 \psi}{\partial t^2} = \frac{1}{r}\frac{\partial}{\partial r}\left(r\frac{\partial \psi}{\partial r}\right) + \frac{1}{r^2}\frac{\partial^2 \psi}{\partial \phi^2}. \tag{10.5.28}$$

The ansatz

$$\psi(r, \phi, t) = R(r)\Phi(\phi)T(t) \tag{10.5.29}$$

gives as usual

$$T'' + v^2 k^2 T = 0 \tag{10.5.30}$$

$$\frac{1}{R}\frac{1}{r}\frac{d}{dr}\left(r\frac{dR}{dr}\right) + \frac{1}{\Phi}\frac{1}{r^2}\frac{d^2\Phi}{d\phi^2} + k^2 = 0. \tag{10.5.31}$$

Once again the solution for T is

$$T(t) = A\cos vkt. \tag{10.5.32}$$

Look at the equation for R and Φ. In its present form the ϕ dependent part is contaminated by the factor $1/r^2$. So we multiply both sides by r^2 and end up with something that separates as follows:

$$\underbrace{\frac{r}{R}\frac{d}{dr}\left(r\frac{dR}{dr}\right) + k^2 r^2}_{=m^2} + \underbrace{\frac{1}{\Phi}\frac{d^2\Phi}{d\phi^2}}_{=-m^2} = 0, \tag{10.5.33}$$

where the separation constant has a form that anticipates what is about to develop. The two ODEs are then

$$r\frac{d}{dr}\left(r\frac{dR}{dr}\right) + (k^2 r^2 - m^2)R = 0 \tag{10.5.34}$$

$$\frac{d^2\Phi}{d\phi^2} + m^2\Phi = 0. \tag{10.5.35}$$

The solution to the ϕ equation is

$$\Phi_m = A_m e^{im\phi} + B_m e^{-im\phi}, \tag{10.5.36}$$

where m must be an integer for the solution to be unaffected by the change $\phi \to \phi + 2\pi$ which should make no physical difference. (If this condition is not satisfied we will not be assigning a unique value for the displacement of the membrane at each point on its surface.) We are then left with the radial equation

$$r^2 R'' + r R' + (k^2 r^2 - m^2)R = 0. \tag{10.5.37}$$

Changing from r to $x = kr$ we get

$$x^2 R'' + x R' + (x^2 - m^2)R = 0 \tag{10.5.38}$$

(where primes now denote derivatives with respect to x), which we recognize to be Bessel's equation (10.4.59) with $\nu = m$. Since roots of the indicial equation are $s = \pm m$, we will have the regular solution which goes as x^s times a power series and the singular one, the Neumann function which blows up at the origin. We set the coefficient of the latter to zero since the membrane distortion is finite at the origin. Thus the solution regular at the origin is

$$\psi(r, \phi, t) = J_m(kr)(A e^{im\phi} + B e^{-im\phi}) \cos \nu k t. \tag{10.5.39}$$

Problem 10.5.4. First and second-order Bessel functions. *Show, by going back to Eqn. (10.4.62) that*

$$J_0(x) = \sum_0^\infty \frac{(-1)^n}{(n!)^2} \left(\frac{x}{2}\right)^{2n} \tag{10.5.40}$$

$$J_1(x) = \sum_0^\infty \frac{(-1)^n}{n!(n+1)!} \left(\frac{x}{2}\right)^{2n+1}. \tag{10.5.41}$$

Observe that J_0 does not vanish at the origin and is even while J_1 is odd and hence vanishes at the origin. Look at Fig. 10.2.

We have so far not addressed the boundary condition that $\psi = 0$ at $r = a$. Now the Bessel function $J_m(x)$ is known to have an infinite number of zeros at the points x_{mj}. These are however not evenly spaced as in the trigonometric functions as is clear from Fig. 10.2.

We must then demand that at each m,

$$k = k_{mj} = \frac{x_{mj}}{a} \tag{10.5.42}$$

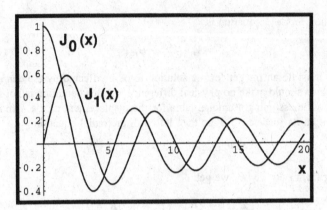

Figure 10.2. The first two Bessel functions J_0 and J_1. Note that the zeros are not evenly placed as in trigonometric functions.

which in turn controls the allowed frequencies

$$\omega = vk = \frac{vx_{mj}}{a}. \tag{10.5.43}$$

For example if $m = 0$, J_0 has its first zero at $x_{01} \simeq 2.4$ so that the lowest frequency of vibration is

$$\omega = (2.4v)/a. \tag{10.5.44}$$

From the linearity of the equation, the general solution is some linear combination of these solutions (Eqn. (10.5.39)).

Problem 10.5.5. *In quantum mechanics one defines a box as a region to which the particle is confined. Consider a square box in two dimensions of sides L with its lower left corner at the origin. The allowed energies for the particle of mass m are given by E, which is defined by the following eigenvalue equation:*

$$-\frac{\hbar^2}{2m}\nabla^2 \psi = E\psi \tag{10.5.45}$$

and the boundary condition that $\psi = 0$ at the walls of the box. Show that the allowed energies are $E = \frac{\hbar^2 \pi^2 (n_x^2 + n_y^2)}{2mL^2}$, where n's are positive integers. Find the lowest energy state in this box. Assuming the solution is rotationally invariant, show that the lowest energy state in a circular box of radius a is $E = \frac{\hbar^2 x_{01}^2}{2ma^2}$, where $x_{01} \simeq 2.4$ is the first zero of $J_0(x)$.

10.5.2. The heat equation in one and two dimensions

Consider a homogeneous body in three dimensions. Let $u(x,y,z,t)$ be the temperature at time t at the point (x,y,z). I will sketch the derivation of the heat equation, which governs the evolution of u. First it is clear that any decline in u with time is due to the net outflow of heat from the vicinity of the point in question, which in turn is given by the divergence of $\mathbf{j}$, the heat flux or current (in calories per unit area per second). In other words, we have the equation of continuity:

$$\frac{\partial u}{\partial t} + \nabla \cdot \mathbf{j} = 0, \tag{10.5.46}$$

where all positive constants of proportionality have been set to unity. Next we ask why heat ever flows: it flows in response to a temperature gradient, from hot to the cold regions. In other words

$$\mathbf{j} = -\nabla u, \tag{10.5.47}$$

where once again all positive constants of proportionality have been set to unity and the minus sign tells us that the current flows from hot to cold, i.e., opposite to the gradient. Combining this with the continuity equation we get the *heat equation*

$$\frac{\partial u}{\partial t} = \nabla^2 u. \tag{10.5.48}$$

We will study this equation in one and two spatial dimensions to learn the basics. In one dimension

$$\frac{\partial u}{\partial t} = \frac{\partial^2 u}{\partial x^2}. \tag{10.5.49}$$

Let us apply this to a rod of length L whose ends are held at some fixed temperature $u_{r/l}$ at the right and left ends. Let us assume that the rod is in equilibrium so that nothing is changing with time. In this case (since u depends only on x),

$$\frac{d^2 u}{dx^2} = 0 \tag{10.5.50}$$

$$u = a + bx \tag{10.5.51}$$

$$= u_l + \frac{u_r - u_l}{L}x. \tag{10.5.52}$$

(You may wish to think about how one solves for the free parameters a and b in terms of $u_{r/l}$.) Observe that the temperature rises linearly from one end to the other.

Consider now the following nonequilibrium problem. At $t = 0$, the rod has some temperature distribution $u(x,0)$. At this instant its ends are jammed into ice trays at $u = 0$. It is clear that asymptotically the whole rod will reach zero temperature. The question is how this limit is approached in time. This is what the heat equation will tell us.

First we look for solutions to Eqn. (10.5.49) of the form:

$$u(x,t) = X(x)T(t). \tag{10.5.53}$$

Following the same routine and using a separation constant $-k^2$ we get two ODEs:

$$\frac{dT}{dt} = -k^2 T \tag{10.5.54}$$

$$X'' + k^2 X = 0. \tag{10.5.55}$$

The first equation has an exponential solution while second equation has the usual solution

$$X = \sin kx \qquad k = n\pi/L \tag{10.5.56}$$

when we insist u vanish at the ends. This gives us the general solution

$$u(x,t) = \sum_1^\infty u_n \sin \frac{n\pi x}{L} e^{-(n\pi/L)^2 t}. \tag{10.5.57}$$

Now we proceed as we did with the string and fit the initial value data $u(x,0)$ to a sum of sines with appropriate coefficients. The subsequent evolution is given by the expansion above with these coefficients.

Let us consider an example. Say a rod of length 100 has its ends at zero degrees and starts with the following temperature distribution:

$$u(x,0) = 150^\circ \qquad 0 < x \le 50 \tag{10.5.58}$$

$$= 0^\circ \qquad 50 < x \le 100. \tag{10.5.59}$$

At $t = 0$ its ends are jammed into ice trays at zero degrees.

Our goal is to find $u(x,t)$. All we need is a_n of Eqn. (10.5.57). Multiplying both sides of it by $\sin \frac{n\pi x}{100}$ at $t = 0$ and integrating, we see (upon recalling that the sine-squared averages to a half over half a period),

$$a_n \cdot \frac{100}{2} = \int_0^{100} u(x,0) \sin \frac{n\pi x}{100} dx \tag{10.5.60}$$

Thus

$$a_n = \frac{1}{50} \int_0^{50} 150 \sin \frac{n\pi x}{100} dx = \frac{300}{n\pi} \left[1 - \cos \frac{n\pi}{2} \right] \tag{10.5.61}$$

so that

$$u(x,t) = \sum_1^\infty \frac{300}{n\pi} \left[1 - \cos \frac{n\pi}{2} \right] \sin \frac{n\pi x}{L} e^{-(n\pi/L)^2 t}. \tag{10.5.62}$$

Rather than show the cooling of the rod with time, I show in Fig. 10.3 what happens when we try to fit the initial condition to a sum of sine waves. Note that the series has a hard time with the discontinuities. As we increase the number of terms in the sum, there is overshoot equal to roughly 9% of the jump or discontinuity in the function. By bringing in more terms, we can reduce the width of these "Gibbs Oscillations" but not eliminate them.

Problem 10.5.6. *Consider a bar of length L. Given $u(x,0)$ rises linearly from 0 at the left end to U at the middle and comes down linearly to zero at the right end, find $u(x,t)$ for all future times. (The ends are jammed into ice trays that hold them fixed at zero degrees.) Note that the sines are not normalized to unity in Eqn. (10.5.57).*

Consider now a variation of the problem where at $t = 0$ the two ends are placed in contact with reservoirs at temperatures u_l and u_r, respectively. Now the limiting temperature at infinite time is

$$u(x,\infty) = u_r + \frac{u_r - u_l}{L}x. \tag{10.5.63}$$

Since the ends are no longer at zero temperature so that it appears as if the old solutions with vanishing temperature at the ends are no longer appropriate. But the following trick allows us to use the old solutions. First define

$$v(x,t) = u(x,t) - u(x,\infty) \tag{10.5.64}$$

and observe that

- $v(x,t)$ obeys the heat equation since is a difference of two functions which do.

- v tend to zero as $t \to \infty$.

- For all positive times, $v(0,t) = v(L,t) = 0$ since the values u is forced to take by the reservoirs placed at the ends coincide with the values that will prevail at $t = \infty$.

What all this means is that we may apply to v the solution we previously used for u. Thus

$$v(x,t) = \sum_1^\infty v_n \sin\frac{n\pi x}{L}e^{-(n\pi/L)^2 t}. \tag{10.5.65}$$

If you want to see it all in terms of u, here it is:

$$u(x,t) = u(x,\infty) + \sum_1^\infty v_n \sin\frac{n\pi x}{L}e^{-(n\pi/L)^2 t}, \tag{10.5.66}$$

where v_n is the expansion coefficient of $v(x,0) = u(x,0) - u(x,\infty)$.

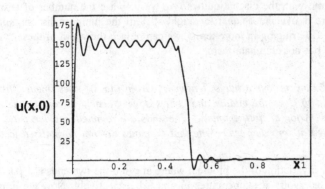

Series with 40 terms

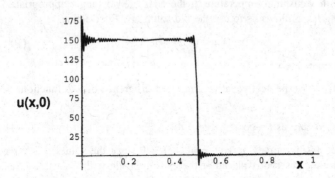

Series with 200 terms

Figure 10.3. The sum of sines is not able to really deal with the discontinuous initial condition. No matter how many terms there are, there will always be an overshoot at the discontinuity points of about 9%. This is called the Gibbs phenomenon.

Problem 10.5.7. *Consider once again the initial temperature given in problem (10.5.6.) (isosceles triangle of height U) but assume that at t = 0 the reservoirs connected at the left and right ends are at u = 0 and u = 2U respectively. Find the temperature for all future times. Remember that the sines are not normalized to unity in Eqn. (10.5.66).*

Consider finally the heat equation in two dimensions but in equilibrium, so that the operative equation becomes Laplace's equation:

$$\frac{\partial^2 u}{\partial x^2} + \frac{\partial^2 u}{\partial y^2} = 0. \tag{10.5.67}$$

Consider a two-dimensional region of width L in the x-direction and semi-infinite in the y-direction, with its lower left-hand corner at the origin. We are given that the edge at $y = 0$ is held at some distribution $u(x, 0)$ and the other two edges parallel to the y-axis (at $x = 0$ and $x = L$) as well as the edge at $y = \infty$ are at zero temperature. We are to determine u in the interior given this boundary condition and the heat equation.

Separation of variables gives us:

$$\underbrace{\frac{1}{X}X''}_{-k^2} + \underbrace{\frac{1}{Y}Y''}_{k^2} = 0 \tag{10.5.68}$$

with the solutions

$$
\begin{aligned}
X(x) &= A\sin kx \qquad k = n\pi/L \quad n = 1, 2.. \tag{10.5.69}\\
Y(y) &= e^{-ky}. \tag{10.5.70}
\end{aligned}
$$

Notice that we reject the exponentially growing solution in y since u must vanish as $y \to \infty$. (Note also that had we chosen hyperbolic functions for the x-direction and trigonometric functions for the y-direction, we would not have been able to meet the boundary conditions.)

The general solution is then

$$u(x, y) = \sum_{1}^{\infty} a_n \sin \frac{n\pi x}{L} e^{-n\pi y/L}, \tag{10.5.71}$$

where a_n are the Fourier coefficients of $u(x, 0)$. Let us consider an example where

$$u(x, 0) = x(L - x) \tag{10.5.72}$$

We already know from Eqns. (10.5.25-10.5.26) that

$$a_n = \frac{8L^2}{n^3\pi^3} \quad [n \text{ odd}] \tag{10.5.73}$$

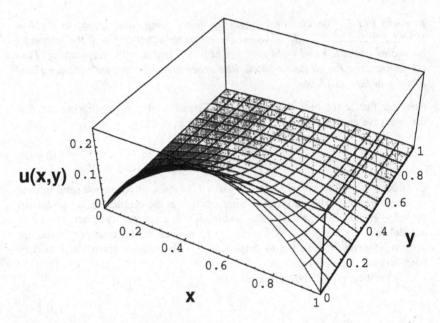

Figure 10.4. The temperature distribution in a square sheet with $L = 1$ and boundary condition $u(x, 0) = x(L - x)$ and $u = 0$ on the other sides.

from which it follows that

$$u(x, y) = \sum_{n=\text{odd}}^{\infty} \frac{8L^2}{n^3 \pi^3} \sin \frac{n \pi x}{L} \, e^{-n \pi y / L} \tag{10.5.74}$$

The solution is depicted in Fig. 10.4.

Problem 10.5.8. *Consider the problem of the semi-infinite strip of width L given above. (i) Given $u(x, 0) = x$, and $u = 0$ on the other three sides, find u at all interior points. (ii) Repeat with $u(x, 0) = \cos(x\pi/L)$. Hint: In integrating trigonometric functions, write them in terms of exponentials, use $\sin x = \text{Im} \, e^{ix}$.*

There are more and more variations possible such as the case where the two-dimensional body is finite in both directions or that the temperature is non-zero on two edges. But we move on to the next topic, leaving the interested to try the following problems.

Problem 10.5.9. *Consider a sheet of width a along the x axis and b along the y-axis, with its lower left corner at the origin. You are given that the edge at $y = b$ is at $u(x, b) = 100$ and that the other edges are at zero degrees. Find the temperature* *in the interior. Here are some suggestions: Use separation of variables, in the y-direction admit a superposition of hyperbolic functions and finally kill the $\cosh y$ part using the conditions at $y = 0$ to reach an expansion of the form*

$$u(x,y) = \sum A_n \sin(n\pi x/a) \sinh(n\pi y/b).$$

Set $y = b$, use the boundary conditions and solve for A_n.

Problem 10.5.10. *So far we have only considered cases where the sheet has nonzero temperature on only one of its edges. What about a plate which has $T \neq 0$ on all four edges? We simply solve four problems in each of which $T \neq 0$ on just one edge and then we* add *these solutions. It is clear that the sum has the right boundary conditions and satisfies the heat equation due to linearity.*

Consider a square plate of edge 10 with its lower left corner at the origin and with $T = 100$ on the edges running along the axes and $T = 0$ on the other two. Show that

$$u(x,y) = \frac{100}{\pi} \sum_{odd} \frac{1}{n} \left[\frac{\sinh(n\pi(10-y)/10)}{\sinh n\pi} \right] \sin[n\pi x/10] + [x \rightarrow y].$$

Consider finally the solution of the *heat equation in polar coordinates*. Let us say we have a circular sheet of radius a with a temperature distribution $u(a, \theta)$ on the circumference. This determines the temperature inside. Laplace's equation for the radial function is (for assumed angular dependence $e^{im\theta}$)

$$r^2 R'' + rR' - m^2 R = 0, \tag{10.5.75}$$

the solution to which is

$$R(r) = Ar^{|m|} + Br^{-|m|} \tag{10.5.76}$$

The details are left to the following exercise.

Problem 10.5.11. *Derive the above equation by trying a series solution. Only one term will survive. Notice that the equation has scale invariance: it is unaffected by $r \rightarrow \alpha r$. This means that if $R(r)$ is a solution, so is $R(\alpha r)$. Pure powers have this property. Analyze the special case $m = 0$ separately. Show that the general solution is $(A + B \ln r)(C + D\theta)$. In other words, besides r^0, $\ln r$ is a possible solution for the radial function $R(r)$ and likewise, the angular function for $m = 0$ is not just e^0 but of the form $A + B\theta$. Why must we choose $B = D = 0$?*

Returning to the Eqn. (10.5.76), we drop the negative powers since they blow up at the origin. The general solution is then

$$u(r, \theta) = \sum_{-\infty}^{\infty} A_m r^{|m|} e^{im\theta} \tag{10.5.77}$$

where

$$A_m = a^{-|m|} \frac{1}{2\pi} \int_0^{2\pi} u(a, \theta') e^{-im\theta'} \frac{d\theta'}{2\pi}, \tag{10.5.78}$$

which we get by setting $r = a$ in Eqn. (10.5.77) and dotting both sides with the exponential function.

If we feed this result into Eqn. (10.5.77) we find

$$u(r, \theta) = \sum_{-\infty}^{\infty} (r/a)^{|m|} \int_0^{2\pi} e^{im[\theta-\theta']} u(a, \theta') d\theta'. \tag{10.5.79}$$

It is straightforward to do the geometric sum over m and to obtain

$$u(r, \theta) = \frac{1}{2\pi} \int_0^{2\pi} \left[\frac{a^2 - r^2}{a^2 + r^2 - 2ra \cos(\theta - \theta')} \right] u(a, \theta') d\theta', \tag{10.5.80}$$

which is a result due to Poisson.

Problem 10.5.12. *Do the sum over the geometric series to obtain the above formula.*

Note that the above formula expresses u in the interior of the circle in terms of its values at the boundary. This sounds like complex variables are at work, and they indeed are! Recall that the real and imaginary parts of an analytic function $f = u + iv$ obey Laplace's equation and that an analytic function is determined inside a closed contour by its values on the boundary. Let this boundary be our circle of radius a. Then

$$f(re^{i\theta}) = \frac{1}{2\pi i} \int_0^{2\pi} \frac{f(ae^{i\theta'})}{ae^{i\theta'} - re^{i\theta}} [ae^{i\theta'} i d\theta']. \tag{10.5.81}$$

At this point you may be tempted to take the real part of both sides to obtain a formula for $u(re^{i\theta})$. Unfortunately the real part in the integral will involve both u and the imaginary part v. We get around this as follows. We note that if we replace $re^{i\theta}$ in the denominator of the right-hand side by another number which lies *outside the circle of radius* a, the integral will give zero by Cauchy's Theorem, since there are no singularities inside the contour of integration. So we can subtract such an integral for free. Let choose for the outside point one at $a^2/(re^{-i\theta})$. By subtracting the corresponding integral (which is zero) from the right side of Eqn. (10.5.81), we will obtain Poisson's formula.

Problem 10.5.13. *Provide the missing steps between Eqn. (10.5.81) and the Poisson formula.*

10.6. Green's Function Method

When we combine the Maxwell equation

$$\nabla \cdot \mathbf{E} = \frac{\rho}{\varepsilon_0} \tag{10.6.1}$$

with

$$\mathbf{E} = -\nabla \phi - \frac{\partial \mathbf{A}}{\partial t} \tag{10.6.2}$$

in the static case we find *Poisson's Equation*

$$-\nabla^2 \phi = \frac{\rho}{\varepsilon_0} \tag{10.6.3}$$

relating the scalar potential to the charge density that produces it. A generalization of this equation is the *Screened Poisson Equation*

$$(-\nabla^2 + m^2)\phi = \frac{\rho}{\varepsilon_0}. \tag{10.6.4}$$

(Note that m is not an integer here.)

Our goal is to find ϕ corresponding to a given ρ. Suppose we can find a *Green's Function* $G(\mathbf{r} - \mathbf{r}')$ such that

$$(-\nabla^2 + m^2)G(\mathbf{r} - \mathbf{r}') = \delta^3(\mathbf{r} - \mathbf{r}') \tag{10.6.5}$$

where

$$\delta^3(\mathbf{r} - \mathbf{r}') = \delta(x - x')\delta(y - y')\delta(z - z') \tag{10.6.6}$$

vanishes unless all $\mathbf{r} = \mathbf{r}'$ in which case it blows up in such a way that

$$\int \delta^3(\mathbf{r} - \mathbf{r}')d^3r' = 1 \tag{10.6.7}$$

assuming the region of integration includes $\mathbf{r}$. *Thus the Green's function $G(\mathbf{r} - \mathbf{r}'))$ is, up to a factor ε_0, the potential at $\mathbf{r}$ due to a point charge at $\mathbf{r}'$.*

Given G we can write down the solution to the problem as follows:

$$\phi(\mathbf{r}) = \int G(\mathbf{r} - \mathbf{r}')\frac{\rho(\mathbf{r}')}{\varepsilon_0}d^3r'. \tag{10.6.8}$$

To verify this act on both sides with $(-\nabla^2 + m^2)$, where ∇ acts on the variable r. On the right-hand side this operator goes into the integral (where r enters as a parameter) and acts on $G(\mathbf{r} - \mathbf{r}')$ to give $\delta^3(\mathbf{r} - \mathbf{r}')$, which pulls out $\frac{\rho(\mathbf{r})}{\varepsilon_0}$ upon doing the r' integration.

So let us find the Green's function obeying

$$(-\nabla^2 + m^2)G(\mathbf{r} - \mathbf{r}') = \delta^3(\mathbf{r} - \mathbf{r}'). \tag{10.6.9}$$

Since everything depends on just $\mathbf{r} - \mathbf{r}'$, let us choose $\mathbf{r}' = 0$. Next we write G and the delta function in terms of their Fourier integrals:[3]

$$G(\mathbf{r}) = \int \frac{d^3k}{(2\pi)^3} G(\mathbf{k})e^{i\mathbf{k}\cdot\mathbf{r}} \tag{10.6.10}$$

$$\delta^3(\mathbf{r}) = \int \frac{d^3k}{(2\pi)^3} e^{i\mathbf{k}\cdot\mathbf{r}}. \tag{10.6.11}$$

---

[3] We are using the obvious generalization of the one-dimensional Fourier transform to three dimensions.

We now feed these expressions into Eqn. (10.6.8) and use the fact that

$$-\nabla^2 e^{i\mathbf{k}\cdot\mathbf{r}} = k^2 e^{i\mathbf{k}\cdot\mathbf{r}} \tag{10.6.12}$$

to arrive at

$$\int \frac{d^3k}{(2\pi)^3}(k^2+m^2)G(\mathbf{k})e^{i\mathbf{k}\cdot\mathbf{r}} = \int \frac{d^3k}{(2\pi)^3}e^{i\mathbf{k}\cdot\mathbf{r}}. \tag{10.6.13}$$

Since the exponentials at different $\mathbf{k}$ are linearly independent, we may equate their coefficients on both sides:

$$(k^2+m^2)G(\mathbf{k}) = 1 \tag{10.6.14}$$

$$G(\mathbf{k}) = \frac{1}{k^2+m^2} \tag{10.6.15}$$

which gives us the transform of G. We just have to invert the transform and go back to $G(\mathbf{r})$ using

$$G(\mathbf{r}) = \int \frac{d^3k}{(2\pi)^3}\frac{1}{k^2+m^2}e^{i\mathbf{k}\cdot\mathbf{r}}. \tag{10.6.16}$$

We are trying to do an integral in k-space. Let us choose the k_z-axis along the direction of $\mathbf{r}$ so that $\mathbf{k}\cdot\mathbf{r} = krz$, where $z \equiv \cos\theta$. Since nothing depends on ϕ we can do that integral knocking off a 2π to get

$$G(r) = \int \frac{k^2 dk dz}{(2\pi)^2}\frac{1}{k^2+m^2}e^{ikrz}, \tag{10.6.17}$$

where we write $G(r)$ instead of $G(\mathbf{r})$ since the answer is seen to depend only on r. Doing the z-integral between -1 and 1, we obtain

$$G(r) = \frac{1}{2\pi^2}\int_0^\infty \frac{k^2 dk}{k^2+m^2}\frac{\sin kr}{kr} \tag{10.6.18}$$

We now use the fact that the integrand is even to extend the limits to $\pm\infty$ (taking care to divide by 2) and evaluate the integral by the residue theorem to obtain:

$$G(r) = \frac{e^{-mr}}{4\pi r}. \tag{10.6.19}$$

Problem 10.6.1. *Provide the missing steps involving contour integrations that link Eqn. (10.6.18) to Eqn. (10.6.19). First write the integral in question as the imaginary part of an integral with e^{ikr} instead of $\sin kr$. Note that for large k, the integrand behaves as e^{ikr}/kr. Argue that on a semicircle in the upper-half-plane of radius $R \to \infty$, the exponential kills the integrand except in an infinitesimal angular range near the real axis which makes an infinitesimal contribution.*

To get the Green's function for Poisson's equation we set $m = 0$ above and obtain

$$G(r) = \frac{1}{4\pi r} \tag{10.6.20}$$

so that the potential corresponding to a given charge density is

$$\phi(\mathbf{r}) = \int \frac{1}{4\pi\varepsilon_0|\mathbf{r}-\mathbf{r}'|}\rho(\mathbf{r}')d^3r'. \tag{10.6.21}$$

Let us test this on a unit point charge sitting at the origin:

$$\rho(\mathbf{r}') = \delta^3(\mathbf{r}'). \tag{10.6.22}$$

Feeding this into the r' integral we find

$$\phi(\mathbf{r}) = \frac{1}{4\pi\varepsilon_0 r}, \tag{10.6.23}$$

which agrees with Coulomb's law and the idea that G is the potential of a point charge.

10.7. Summary

Here are the highlights of Chapter 10.

- Learn to identify the nature of the differential equation. If the dependent variable and its derivatives occur linearly, it is a linear equation and a superposition of solutions with constant coefficients is also a solution. If it is nonlinear, go find another problem.

- If the equation has constant coefficients and is of the form

$$(a_n D^n + a_{n-1} D^{n-1} + \cdots + a_0) x(t) = 0$$

then the solution is

$$x(t) = \sum_{i=1}^{n} A_i e^{i\alpha_i t}$$

where A_i are constants and α_i are the roots of the algebraic equation

$$(a_n \alpha^n + a_{n-1} \alpha^{n-1} + \cdots + a_0) = 0.$$

If $\alpha_1 = \alpha_2$, the solution is of the form $(A + Bt)e^{\alpha t}$.

- If the equation is inhomogeneous, i.e., has nonzero function $f(t)$ on the right-hand side, the solution is a sum $x(t) = x_c(t) + x_p(t)$, where x_c, the complementary function, is the solution with $f = 0$; and the particular solution x_p, which is the response to f. The latter can be found easily if $f(t)$ is a sum of exponentials. Thus if

$$(D - \alpha_1) \ldots (D - \alpha_n) x(t) = e^{\alpha t}$$

we can assume $x_P(t) = Ae^{\alpha t}$, where A is a constant and solve for A:

$$A = \frac{1}{(\alpha - \alpha_1) \ldots (\alpha - \alpha_n)}$$

as long as $\alpha \neq \alpha_i$. If it is,

$$x(t) = \frac{te^{\alpha t}}{(\alpha - \alpha_1) \ldots (\alpha - \alpha_{n-1})} + x_c(t).$$

- The solution to

$$\frac{dy}{dx} + p(x)y = q(x)$$

is

$$y(x) = e^{-P(x)} \left[\int_{x_0}^{x} q(x')e^{P(x')}dx' + y_0 e^{P(x_0)} \right],$$

where $P(x) = \int^x p(x')dx'$.

- If the equation is of the form

$$y'' + p(x)y' + q(x)y = 0$$

and p and q are analytic at the origin, try a solution of the form $y = \sum_0^\infty c_n x^n$. Find a recursion relation between coefficients. If you are lucky, it will be a two-term relation so that given c_0 and c_1, the rest will follow. In many physical problems the requirement that the solution behave well all over the region of interest may require the parameters in the equation to be quantized to some special values that turn the infinite series into a polynomial.

- If

$$x^2 y'' + x p(x)y' + q(x)y = 0$$

and p and q are analytic at the origin, try a generalized series $y = x^s \sum_0^\infty c_n x^n$. Solve the indicial equation for s. If you get two distinct roots, not differing by an integer, you are done—you have your two solutions. If the roots coincide or differ by an integer, go back to some book (like this one) and see what is to be done.

- Let us recall the partial differential equations we have discussed

$$\frac{1}{v^2} \frac{\partial^2 \psi}{\partial t^2} = \frac{\partial^2 \psi}{\partial x^2}$$

$$\psi(x = 0, t) = \psi(x = L; t) = 0,$$

which is the one-dimensional wave equation for a string of length L,

$$\frac{1}{v^2} \frac{\partial^2 \psi}{\partial t^2} = \frac{\partial^2 \psi}{\partial x^2} + \frac{\partial^2 \psi}{\partial y^2} \equiv \nabla^2 \psi$$

$$\psi(x = 0, yt) = \psi(x = L, y, t) = \psi(x, y = 0, t) = \psi(x, y = L, t) = 0,$$

which is the two-dimensional wave equation for a square membrane of sides L,

$$\frac{1}{v^2} \frac{\partial^2 \psi}{\partial t^2} = \frac{1}{r} \frac{\partial}{\partial r} \left(r \frac{\partial \psi}{\partial r} \right) + \frac{1}{r^2} \frac{\partial^2 \psi}{\partial \phi^2},$$

which is the membrane equation, but in polar coordinates,

$$\frac{\partial u}{\partial t} = \nabla^2 u,$$

which is the heat equation in its general form. (The Laplacian will assume different forms in different dimensions in different coordinate systems.)

There is just one set of rules for dealing with them. Assume a solution of the product form. For example in the case of $\psi(x, y, z, t)$ assume

$$\psi(x, y, z, t) = X(x)Y(y)Z(z)T(t).$$

Feed into the equation and decouple it into ordinary differential equations, one for each factor. Solve each one subject to the boundary conditions. Throw out options that blow up in the region of interest (at the origin, at infinity etc.). The most general solution is a linear superposition of such factorized solutions. The coefficients in the linear superposition are determined by boundary conditions.

ANSWERS

Chapter 1

1.6.1. $x/3 - x^3/9, .0332224$ (from series) and $.0332222$ (calculator).

1.6.2. *(i)* $3x^2 \cos(2 + x^3)$, *(ii)* $-2\cos[\cos(2x)]\sin(2x)$, *(iii)* $3\sec^2 x \tan^2 x$, *(iv)* $\tanh x$, *(v)* $1/(1 + x^2)$, *(vi)* $1/(1 - x^2)$, *(vii)* 0, *(viii)* $1/(1 + \cos x)$.

1.6.3. $100e^{.12} \simeq \$112.72$.

1.6.4. $v = \tanh \theta$

1.6.7. Square of side $L/4$.

1.6.9. 1,3

1.6.12. $-\frac{1}{2}$

1.6.13. $x = 1$ is a minimum; $x = -1$ is a maximum.

Chapter 2

2.1.2. $x \ln x - x$

2.2.1. $\sin^{-1} \frac{x_2}{a} - \sin^{-1} \frac{x_1}{a}$

2.2.3. 2

2.2.5. $\ln(3/2)$

2.2.6. 1/3

2.2.7. $1/(\ln 2)$

2.2.9. $I_3 = 1/(2a^2)$, $I_4 = \frac{3}{8} \frac{\sqrt{\pi}}{a^{5/2}}$

2.2.11. $\frac{2ak}{(a^2+k^2)^2}$; $\frac{a^2-k^2}{(a^2+k^2)^2}$

Chapter 3

3.1.2. $f_x = 3x^2 + 2xy^5$, $f_y = 5x^2y^4 + 4y^3$, $f_{xy} = 10xy^4 = f_{yx}$

3.1.5. $d = 4/\sqrt{5}$

3.2.5. $I = MR^2/2$, $(3MR^2)/5$

3.2.8. $V = 128\pi$

Chapter 4

4.2.3. Diverges

4.2.4. *(i)* ratio test is inconclusive, integral test says divergent.

(ii) $r = e^3/27$, convergent by ratio test.

(iii) Divergent by integral test.

(iv) Ratio test inconclusive, integral test says divergent.

(v) $r = e$, divergent.

(vi) Ratio test inconclusive, integral test says divergent.

4.2.5. *(i)* C, *(ii)* C, *(iii)* C *(iv)* C.

4.2.6. C, C, D, D, D.

4.3.1. $R =$ *(i)* $\sqrt{2}$, *(ii)* 1, *(iii)* 1, *(iv)* 1, *(v)* Converges for $|x| > 1$, *(vi)* $|x| > \frac{1}{2}$.

4.3.4. $E = m(1 + \frac{v^2}{2} + \frac{3v^4}{8} + \frac{5v^6}{16} + \ldots)$, $P = vE$.

4.3.6. $T = 2\pi\sqrt{l/g}(1 + \frac{k^2}{4} + \ldots)$, $\delta T/T = 1/16$.

4.3.7. *(i)* $\frac{1}{\sqrt{2}}[1 + x - x^2/2 - x^3/6 + x^4/24]$, *(ii)* $\frac{1}{e}[1 + x + x^2/2 + x^3/6 + x^4/24]$,

(iii) $\ln 2 + x/2 - x^2/8 + x^3/24 - x^4/64$.

Chapter 5

5.2.3. $z = \frac{137}{3517} - \frac{761}{3517}i$

5.2.4. *(i)* $\operatorname{Re} z = 6/25$, $\operatorname{Im} z = -8/25$, $|z| = 2/5$, $z^* = 6/25 + 8/25\,i$, $1/z = 3/2 + 2i$, *(ii)* $\operatorname{Re} z = -7$, $\operatorname{Im} z = 24$, $|z| = 25$, $z^* = -7 - 24i$, $1/z = -(7/625) - (24/625)\,i$, *(iii)* $\operatorname{Re} z = -7/25$, $\operatorname{Im} z = 24/25$, $|z| = 1$, $z^* = -(7/25) - (24/25)i$, $1/z = -(7/25) - (24/25)\,i$,

(iv) $\operatorname{Re} z = \frac{1 - \sqrt{6}}{4}$, $\operatorname{Im} z = \frac{\sqrt{3} + \sqrt{2}}{4}$, $|z| = \frac{\sqrt{3}}{2}$, $1/z = \frac{1 - \sqrt{6}}{3} - i\frac{\sqrt{3} + \sqrt{2}}{3}$,

(v) $\mathrm{Re}\, z = \cos\theta$, $\mathrm{Im}\, z = \sin\theta$, $1/z = z^* = \cos\theta - i\sin\theta$.

5.3.2. *(i)* $z_1 = e^{i\pi/4}$, $z_2 = 2e^{-i\pi/6}$, $z_1 z_2 = 2e^{i\pi/12}$, $z_1/z_2 = .5e^{5\pi i/12}$.

(ii) $z_1 = e^{2i\arctan 4/3}$, $z_2 = .5e^{-i\pi/2}$.

5.3.3. $|z_1 + z_2| = 7.95$, Phase is .98 radians.

5.4.4. $z = 1 - i = \sqrt{2}e^{-i\pi/4}$, $I_0 = 100/\sqrt{2}$, current leads by $\pi/4$, resonance at $\omega = 223.6$ rads.

5.4.5. $Z = \frac{31+33i}{41}$.

5.4.6. $Z = \frac{R}{1+iR(\omega C - 1/(\omega L))}$.

5.4.7. $Q(t) = e^{-6t}[4\cos 8t + 3\sin 8t] - 4\cos 10t$

Chapter 6

6.1.2. $u_x = -v_y \quad u_y = v_x$

6.1.3. $f_x = f_y$

6.1.7. Poles at $(\pm 1 \pm i)/\sqrt{2}$, double pole at $z = \pm i$

6.1.8. $u_r = v_\theta/r$, $v_r = -u_\theta/r$

6.1.12. *(i)* $f = z^3$, *(ii)* $f = e^{iz}$, *(iii)* u is not harmonic.

6.1.13. Hint: Relate the Laplacian to $\frac{\partial^2}{\partial z \partial z^*}$

6.1.14. Hint: Consider f^2.

6.2.3. $\sin x \cosh y$, $\cos x \sinh y$, $\sqrt{\sin^2 x + \sinh^2 y}$, $[x = n\pi, y = 0]$

6.2.4. $z = n\pi$, $(n+1/2)\pi$, $in\pi$, $i(n+1/2)\pi$

6.2.9. $e^{2\pi i n/N}$, $n = 0, 1, ... N - 1$. Roots add to zero.

6.2.10. $3 + 4i$, $12 + 5i$

6.2.11. *(i)* $\pm(1+i)/\sqrt{2}$, *(ii)* $\pm(2+i)$

6.2.12. $\ln 2 + (2m+1)i\pi$

6.2.13. *(i)* $\cos\ln 3 + i\sin\ln 3$ *(ii)* $i\pi/2$ *(iii)* $\pm 5e^{i\pi/4}$ *(iv)* $\pm e^{i\pi/3}$

6.2.14. *(i)* Repeated twice: $\pm i$. *(ii)* Repeated twice: $[e^{i\pi/3}, -1, e^{-i\pi/3}]$

6.2.15. $\sqrt{2}e^{i\pi/4}$. The cube roots are $(2)^{1/6}(e^{i\pi/12}, e^{i(\pi/12)\pm(2\pi i/3)})$

6.2.16. $e^{-\pi[2n+1/2]}$

6.4.1. π/e, $-\pi e$

6.4.5. *(i)* $\pi/(2^{3/2})$, *(ii)* 0,0 *(iii)* $\pi/18$, *(iv)* $-\pi/3$, *(v)* 0, *(vi)* $\pi/4$

6.4.8. $\pi/(27e^3)$

6.5.3. Yes

6.5.4. No

Chapter 7

7.4.3. *(i)* 5/6, 0.

7.5.1. $h = \sqrt{R^2 - x^2 - y^2}$, $\nabla h = -\mathbf{i}\dfrac{x}{\sqrt{R^2 - x^2 - y^2}} - \mathbf{j}\dfrac{y}{\sqrt{R^2 - x^2 - y^2}}$

7.5.4. Cylindrical: $h_\rho = 1$, $h_\phi = \rho$ $h_z = 1$. Spherical: $h_r = 1$, $h_\theta = r$, $h_\phi = r\sin\theta$.

7.5.5. *(ii)* $\sqrt{13}$, *(iii)* $12/\sqrt{13}$, *(iv)* 2.

7.5.7. *(i)* Towards the origin *(ii)* $\Delta T = -3\sqrt{2}/10$

7.5.8. *(i)* $2\sqrt{3}$, *(ii)* $\sqrt{3}e^3$, *(iii)* $\sqrt{3}$, *(iv)* All gradients were radial, this direction is perpendicular to radial.

7.6.2. *(i)* no, *(ii)* yes, *(iii)* yes *(iv)* yes

7.6.3. *(i)* 1, *(ii)* 1, *(iii)* Possibly. *(iv)* $\phi = x^2 y$.

7.6.4. *(i)* 1, *(ii)* $\phi = x^3 y$

7.6.5. *(i)* 1/2, *(ii)* no.

7.6.11. *(i)* 0, *(ii)*, 0, *(iii)* conservative, $\mathbf{F} = \nabla\phi, \phi = (x^2 + y^2)/2$

7.6.13. 2π

7.6.14. The curl has no component in the plane containing the contours for two line integrals.

7.7.1. $-\sin x + 2z$

7.7.3. 1

7.7.4. $\frac{4\pi^4}{3}$

7.7.6. *(i)* $4\pi R^3$. *(ii)* $2\pi R^3$, *(iii)* 3.

Chapter 8

8.1.7. $M + N = \begin{bmatrix} 6 & 8 \\ 10 & 12 \end{bmatrix}$ $M^2 = \begin{bmatrix} 7 & 10 \\ 15 & 22 \end{bmatrix}$

$MN = \begin{bmatrix} 19 & 22 \\ 43 & 50 \end{bmatrix}$ $[M, N] = \begin{bmatrix} -4 & -12 \\ 12 & 4 \end{bmatrix}$.

8.3.4. $(1, 2, -1); (-3, -4, 8)$

8.3.5. $\begin{bmatrix} -\frac{2}{3} & -\frac{4}{3} & 1 \\ -\frac{2}{3} & \frac{11}{3} & -2 \\ 1 & -2 & 1 \end{bmatrix}$.

Chapter 9

9.1.1. yes, real scalars, no, $U_1 + U_2$ is not unitary, yes with integer field

9.1.2. yes, yes,no

9.1.5. No: $|3\rangle = |1\rangle - 2|2\rangle$

9.2.1. (i) $v_I = (2 + i(1 - \sqrt{3}))/\sqrt{2}$, $v_{II} = (\sqrt{3} + 1)i/\sqrt{2}$.
(ii) $v_I = (i - 1)(i + \sqrt{3})(\sqrt{6} - 1)/4$, $v_{II} = 1 + \sqrt{3/2}$.

9.2.2. $\frac{1}{5}(3i + 4j)$ $((104)^2 + (78)^2)^{-1/2}[104i - 78j]$

9.2.4. When $|V\rangle = c|W\rangle$

9.2.6. $\alpha = \frac{9+i}{\sqrt{2}}$, $\beta = -\frac{3+9i}{\sqrt{2}}$, $\gamma = 8$

9.5.2. [eigenvalue](eigenvector components)

[1] $(0, 0, 1)$ $[e^{\pm i\theta}]$ $\frac{1}{\sqrt{2}}(1, \mp i, 0)$

9.5.3. The answers are given in the same format as the matrices in the assigned problem. The three eigenvalues and eigenvectors for each matrix are given one below another.

[1] $(1, 0, 0)$ $[-1]$ $\frac{1}{\sqrt{2}}(-1, 0, 1)$ $[-1]\frac{1}{\sqrt{6}}(1, -2, 1)$
[2] $\frac{1}{\sqrt{30}}(-5, -2, 1)$ $[0]$ $(0, 1, 0)$ $[1]$ $\frac{1}{\sqrt{2}}(-1, 0, 1)$
[4] $\frac{1}{\sqrt{10}}(1, 0, 3)$ $[1]$ $\frac{1}{\sqrt{2}}(1, 0, 1)$ $[2]$ $\frac{1}{\sqrt{3}}(1, 1, 1)$

[0] $\frac{1}{\sqrt{2}}(-1, 0, 1)$ $[2]$ $\frac{1}{\sqrt{2}}(-1, 0, 1)$ $[2]$ $\frac{1}{\sqrt{12}}(-\sqrt{3}, 0, 3)$
$[-\sqrt{2}]$ $\frac{1}{2}(1, -\sqrt{2}, 1)$ $[2(1 - \sqrt{2})]$, $\frac{1}{2}(1, -\sqrt{2}, 1)$ $[3]$ $(0, 1, 0)$
$[\sqrt{2}]$ $\frac{1}{2}(1, \sqrt{2}, 1)$ $[2(1 + \sqrt{2})]$ $\frac{1}{2}(1, \sqrt{2}, 1)$ $[6]$ $\frac{1}{2}(\sqrt{3}, 0, 1)$

9.5.7. $[0]$ $\frac{1}{\sqrt{2}}(-1, 0, 1)$ $[2]$ $\frac{1}{\sqrt{2}}(1, 0, 1)$ $[2]$ $(0, 1, 0)$

9.5.8. $[1]$ $(1, 0, 0)$ $[0, 0]$ $(a^2 + b^2)^{-1/2}(0, a, b)$
$[-1]$ $\frac{1}{\sqrt{3}}(1, 1, 1)$ $[1]$ $\frac{1}{\sqrt{5}}(-1, 0, 2)$ $[2]$ $(0, 0, 1)$
9.5.9. $[2]$ $\frac{1}{\sqrt{2}}(-1, 0, 1)$ $[1]$ $(0, 1, 0)$ $[2]$ $\frac{1}{\sqrt{2}}(-1, 1, 0)$
$[2]$ $\frac{1}{\sqrt{2}}(-1, 1, 0)$ $[6]$ $\frac{1}{\sqrt{5}}(2, 0, 1)$ $[4]$ $\frac{1}{\sqrt{2}}(1, 1, 0)$

9.5.10. $[-1]\frac{1}{\sqrt{6}}(-1, 2, 1)$ $[2]\frac{1}{\sqrt{3}}(-1, -1, 1)$ $[3]\frac{1}{\sqrt{2}}(1, 0, 1)$ for N

9.5.11. (a) $[1, 0, -1]$, (b) $[1, 0, -1]$, (c) $(\frac{1}{2}, \frac{1}{\sqrt{2}}, \frac{1}{2})$,

(d) $P(1) = \frac{1}{4} = P(-1)$, $P(0) = \frac{1}{2}$; $(1, 0, 0)$; no, (e) 2, (f) two,

(g) $|V\rangle = \frac{1}{\sqrt{14}}(1, 2, 3)$ $P(1) = \frac{1}{14}$ $P(0) = \frac{4}{14}$ $P(-1) = \frac{9}{14}$

Average $= -\frac{4}{7} = \langle V|S_z|V\rangle$.

9.6.1. $|x(t)\rangle = \frac{1}{2}\begin{bmatrix} \cos\sqrt{k/mt} + \cos\sqrt{3k/mt} \\ \cos\sqrt{k/mt} - \cos\sqrt{3k/mt} \end{bmatrix}$

9.6.2. Same as in worked example, but with $\sqrt{3k/m} \to \sqrt{5k/m}$.

9.6.3. The eigenfrequencies (in square brackets) are followed by unnormalized eigenvectors:

$[\sqrt{2}]$ $(-1, 0, 1)$; $[\sqrt{2-\sqrt{2}}]$ $(1, \sqrt{2}, 1)$; $[\sqrt{2+\sqrt{2}}]$ $(1, -\sqrt{2}, 1)$.

9.6.4. $(x = 0, y = 1)$ saddle point, $f'' = \begin{bmatrix} 0 & -1 \\ -1 & -1 \end{bmatrix}$

Eigenvalue and eigenvector $\left[\frac{-1\pm\sqrt{5}}{2}\right] = (1, \frac{1\mp\sqrt{5}}{2})$

9.6.7. $(-12/7), 32/7)$ is a maximum.

9.7.3. (a) $f(x) = \frac{h}{2}(1 - \sum_{\text{odd}} \frac{4}{\pi^2 m^2} e^{2mi\pi x/L})$. (b) $\frac{\pi^2}{8}$ Look at $x = L/2$

9.7.4. (a) $\frac{1}{2} + \frac{i}{\pi}\sum_{\text{odd}} \frac{e^{inx}}{n}$.

(b) $\sum_{-\infty}^{\infty}(-1)^n \frac{(1-n\pi i)\sinh 1}{1+n^2\pi^2} e^{i\pi nx}$

9.7.5. $I = \sum_{n \neq 0} \frac{100i e^{200n\pi it}}{60n\pi + i(200n^2\pi^2 - 2500)}$

9.7.8. (i) $f = \frac{e-1}{e} + 2\sum_1^{\infty} \frac{e-(-1)^n}{e(1+n^2\pi^2)}\cos n\pi x$

$f = \sinh 1 \left[1 + 2\sum_1^{\infty} \frac{(-1)^n \cos n\pi x}{1+n^2\pi^2}\right]$.

9.7.12. $\psi(x, t) = \frac{8h}{\pi^2}\sum_1^{\infty} \frac{1}{m^2}\sin\frac{m\pi x}{L}\sin\frac{m\pi}{2}\cos\frac{m\pi vt}{L}$.

9.7.13. $u(x, t) = \frac{32}{\pi^2}\sum_0^{\infty} \frac{(-1)^n}{[2(2n+1)]^2}\sin(2 + 4n)x \cos(2 + 4n)t$.

9.7.14. $\psi(x, t) = \frac{2h}{\pi}\sum_1^{\infty} \frac{1}{n}(1 - \cos\frac{n\pi}{2})\sin\frac{n\pi x}{L}\cos\frac{n\pi vt}{L}$

9.9.2.(a) $I_{ij} = 4m\delta_{ij}$

(b) $\frac{m}{2}\begin{bmatrix} 2 & -1 & -1 \\ -1 & 2 & -1 \\ -1 & -1 & 2 \end{bmatrix}$

$[0]$ $(1, 1, 1)$ Masses lined up along axis through origin.

$[\frac{3m}{2}, \frac{3m}{2}]$ $(-1, 0, 1)$ and $(-1, 1, 0)$ Anything in plane perpendicular to $(1,1,1)$.

Chapter 10

10.2.4. $2\cosh\omega t$

10.2.5. $Ae^t + Be^{it} + Ce^{-it}$

10.2.8. *(i)* $(1+t)e^{-t}$, *(ii)* $A \exp[\frac{1+i}{\sqrt{2}}t] + B \exp[-\frac{1+i}{\sqrt{2}}t] + C \exp[\frac{1-i}{\sqrt{2}}t] + D \exp[\frac{-1+i}{\sqrt{2}}t]$,

(iii) $Ae^{5t} + (Bt + C)e^{-t}$, *(iv)* $(A + Bt)e^{-t} + Ce^{4t} + De^{-4t} + Ee^{4it} + De^{-4it}$.

10.2.10. *(i)* **Critically damped** *(ii)* $e^{-t}[3 - 5t] - 3\cos 2t + 4\sin 2t$ *(iii)* $e^{-t}[4 - 4t] - 3\cos 2t + 4\sin 2t$ *(iv)* $e^{-t}[3 - 4t] - 3\cos 2t + 4\sin 2t$

10.2.11. *(i)* $(2/9)e^{2x} + (7/9)e^{-x} + \frac{xe^{2x}}{3}$, *(ii)* $e^x - \sin x$, *(iii)* $\cos 4x + 2x\sin 4x$,

(iv) $\cosh x + \frac{x\sinh x}{2}$

10.3.3. $y = \frac{1}{2} - \frac{1}{x} + \frac{1}{2x^2}$

10.3.4. $y = x^2 + 1 - e^{-x}$

10.3.5. $y = \frac{1}{x^2}(\cosh x + 2 - \cosh 1)$.

10.3.6. $y = (1/2)(x - 1)(x^2 - 2x + C - 2\ln|x - 1|)$

10.3.7. $y = (1/2)(x - 1)(x^2 + 2C)$

10.3.8. $y = x + C\sqrt{1 + x^2}$

10.3.9. a: $y = (1 + Ce^{x^2/2})^{-1}$ b: $y = (x^2 + Cx)^{-\frac{1}{3}}$

10.4.8. a: $y = (1 + x)(A + B\ln x)$

b: $y = \frac{(Ax+B)}{x(x-1)}$

c: $y = Ax^3 + Bx^{-3}$

d: $y = A(1 - 4x + \frac{8}{3}x^2 - \frac{32}{45}x^3 + \frac{32}{315}x^4 \cdots) + Bx^{1/2}(1 - \frac{4}{3}x + \frac{8}{15}x^2 - \frac{32}{315}x^3 + \frac{32}{2835}x^4 \cdots)$

e: $y = x(A + B\ln x)$

f: $y = A[1 + 2x - \frac{(2x)^2}{2!} + \frac{(2x)^3}{3\cdot3!} - \frac{(2x)^4}{3\cdot5\cdot4!} + \cdots] + Bx^{3/2}[1 - \frac{2x}{5} + \frac{(2x)^2}{5\cdot7\cdot2!} - \frac{(2x)^3}{5\cdot7\cdot9\cdot3!} + \frac{(2x)^4}{5\cdot7\cdot9\cdot11\cdot4!} + \cdots]$

g: $y = A(x^2 + 2x) + B((x^2 + 2x)\ln x + 1 + 5x - \frac{x^3}{6} + \frac{x^4}{72} + \ldots]$

h: $y = \frac{Ax}{1-x} + B[\frac{2x\ln x}{1-x} + 1 + x]$

10.4.9. *(i)*: $y = x^2(x + C)$

(ii): $y = \frac{c}{x^2} + \frac{x^4}{6}$

10.4.10. $L_0 = 1$

$L_1 = 1 - x$

$L_2 = \frac{2 - 4x + x^2}{2}$

$L_3 = \frac{6 - 18x + 9x^2 - x^3}{6}$

10.5.6. $u(x,t) = \frac{8U}{\pi^2} \sum_0^\infty \frac{(-1)^m}{(2m+1)^2} \sin\frac{(2m+1)\pi x}{L} e^{-\frac{(2m+1)^2\pi^2}{L^2}t}$.

10.5.8. *(i)* $u(x,y) = \frac{2L}{\pi} \sum_1^\infty \frac{(-1)^{n+1}}{n} \sin\frac{n\pi x}{L} e^{-\frac{n\pi y}{L}}$

(ii) $u(x,y) = \frac{4}{\pi} \sum_1^\infty \frac{2n}{4n^2-1} \sin\frac{2n\pi x}{L} \exp\left[-\frac{2n\pi y}{L}\right]$

10.5.9. $u(x,y) = \frac{400}{\pi} \sum_{n=\text{odd}} \frac{1}{n} \sin\left(\frac{n\pi x}{a}\right) \frac{\sinh n\pi y/a}{\sinh n\pi b/a}$

INDEX

9 780306 450365